Lecture Notes in Computer Science

Commenced Publication in 1973
Founding and Former Series Editors:
Gerhard Goos, Juris Hartmanis, and Jan van Leeuwen

Ioannis Stavrakakis   Michael Smirnov (Eds.)

Autonomic Communication

Second International IFIP Workshop, WAC 2005
Athens, Greece, October 2-5, 2005
Revised Selected Papers

 Springer

Volume Editors

Ioannis Stavrakakis
National and Kapodistrian University of Athens
Department of Informatics and Telecommunications
Panepistimiopolis, Ilisia, 15784 Athens, Greece
E-mail: ioannis@di.uoa.gr

Michael Smirnov
Fraunhofer Institut FOKUS
Kaiserin-Augusta Allee 31, 10589 Berlin, Germany
E-mail: smirnov@fokus.fraunhofer.de

Library of Congress Control Number: 2006921997

CR Subject Classification (1998): C.2, H.3, H.4

LNCS Sublibrary: SL 5 – Computer Communication Networks and Telecommunications

ISSN   0302-9743
ISBN-10   3-540-32992-7 Springer Berlin Heidelberg New York
ISBN-13   978-3-540-32992-3 Springer Berlin Heidelberg New York

Springer is a part of Springer Science+Business Media

springer.com

© 2006 IFIP International Federation for Information Processing, Hofstrasse 3, 2361 Laxenburg, Austria
Printed in Germany

Typesetting: Camera-ready by author, data conversion by Scientific Publishing Services, Chennai, India
Printed on acid-free paper   SPIN: 11687818   06/3142   5 4 3 2 1 0

Preface

The Second IFIP Workshop on Autonomic Communication (WAC 2005) took place on October 2–5, 2005, in Athens, Greece. The previous (and first) edition of WAC took place in Berlin in 2004 and its next (and third) edition in Paris in 2006. The workshop was organized by the National and Kapodistrian University of Athens and was supported by the EU-funded IST-FET Autonomic Communication Coordination Action (ACCA – IST-6475). Additional support was provided by the EU-funded IST Network of Excellence E-NEXT (IST-506869). Finally, IFIP TC6 provided scientific sponsorship through Working Groups IFIP WG6.6 (Management of Networks and Distributed Systems) and IFIP WG6.3 (Performance of Communication Systems).

The workshop was organized at a time when the – yet to be well defined – field of autonomic communication (AC) is attracting the interest of both the scientific community and the research funding organizations. The latter is manifested, on one hand, by the numerous recent relevant research exploratory forums, workshop panels, preliminary forward-looking position papers, research outlooks and frameworks and, on the other hand, by the commitment of the FET program of the European Commission in Europe to funding long-term research in this area for the next four years. Consequently, the second edition of WAC was highly exploratory and included a nice mix of technical work addressing some already identified problems and well-articulated ideas on the direction this field should take and the fundamental problems whose solution would enable autonomicity.

For a relatively new – and not yet established – workshop series that also focuses on an immature field, it is important that every effort is put into securing and establishing its quality. For this reason, the Technical Program Committee (TPC), the paper evaluation process and the overall program were all carefully set up. The 35-member TPC included predominately highly regarded, established researchers, with a few highly recommended and trusted younger and promising researchers with quality record. The TPC members were asked to review from two to five papers, depending on the thematic area, the amount of work affordable by the reviewer at the time and the desire to identify (through re-assignments) the most appropriate reviewer. The TPC co-chairs did not formally review any paper, but read some of them as needed and took care of the paper selection process. All papers received at least three reviews, and some papers received four reviews. The review scores were summarized in a table, containing for each paper: the scores for each of the questions asked and for each of the reviewer, numerical averages by each reviewer, names of reviewers and major comments by each reviewer. There was no pre-set cut-off threshold or number of papers to admit. Papers were classified in three groups based on the grades and the consistency of the grades and comments: (A) clearly accepted; (B) to be discussed carefully; (C) rejected. There were 13 papers in category A, 15 papers in category B and 7 in category C. Papers in category B were carefully considered, by reading the reviews carefully, reading the paper briefly and discussing extensively the paper and the reviews between the TPC co-chairs; 9 papers from this second class were accepted. All reviews were returned to the authors and the authors of the accepted papers were required to return a response document to the reviewers' comments, indicating how they took the criticisms (if any) into account in the final paper and pointing to and discussing any criticism they disagreed with. The previous step is an unusual one encountered typically in journal editorial processes (responses to the comments of the reviewers) and helped improve the quality of the papers that were finally presented at the workshop. Finally, the authors were given ample time and were requested to revise their paper after the workshop taking into consideration the feedback from the paper presentation at the workshop and any latest enhancements to their work or its presentation.

In addition to the 22 technical paper presentations (organized in 7 sessions) selected by following the aforementioned evaluation process, the program also included 1 keynote presentation, 3 invited presentations and 2 panels.

The keynote talk was delivered by Paul Spirakis (University of Patras - Research Academic Computer Technology Institute, Greece) and addressed algorithmic aspects of sensor networks with emphasis on complexity. The first invited presentation discussed research challenges on opportunistic spectrum access for wireless ad hoc networks and was delivered by Cesar Santivanez (BBN Technologies, USA). The second invited presentation discussed incentive schemes in memory-less P2P systems and was delivered by Costas Courcoubetis (Athens University of Economics and Business, Greece). The third invited presentation focused on coordination and resilience in ad hoc and sensor networks and was delivered by Leandros Tassiulas (University of Thessaly, Greece). Summaries of all the above presentations are included in these proceedings.

The first panel in WAC 2005 focused on the relation between autonomicity and complexity and discussed the extent to which autonomicity reduces management complexity and possibly increases overall (system) complexity. The panel was composed of the following researchers from academia, research organizations and the industry: Paul Spirakis of the University of Patras - Research Academic Computer Technology Institute in Greece (coordinator), Radu Popescu-Zeletin and Mikhail Smirnov of Fraunhofer FOKUS in Germany, David Lewis of Trinity College Dublin in Ireland, Tom Pfeifer of Waterford IT in Ireland, Stefan Schmid of NEC Europe in Germany and Cesar Santivanez of BBN Technologies in USA. An extended report on the deliberations and conclusions of this panel is included in this volume.

The second panel posed several interesting questions on presented ideas in an effort to discuss and define a meaningful and effective autonomic communication roadmap. The panelists were predominately researchers participating in the IST FET Autonomic Communication Coordination Action (ACCA) who have been involved in the last year or two in a European-wide effort to define and promote this research field. Specifically, these panellists were: Mikhail Smirnov of Fraunhofer FOKUS in Germany (Chair), Lidia Yamamoto of the University of Basel in Switzerland, Spyros Denazis of the University of Patras in Greece and Hitachi SAL in France, Simon Dobson of University College Dublin in Ireland, Ioannis Stavrakakis of NKUA in Greece, James Scott of Intel Corporation (UK) Ltd., David Lewis of Trinity College Dublin in Ireland, Jaouhar Ayadi of CSEM in Switzerland, and Serge Fdida of UPMC in France. In addition to the aforementioned ACCA researchers, the following speakers were invited: Fabrizio Sestini of European Commission Future and Emerging Technologies and Nancy Alonistioti of NKUA in Greece, also representing the IST integrated project E2R. An extended report on the deliberations and conclusions of this panel is included in these proceedings.

The help and contributions of several people – that made WAC 2005 possible and successful – are highly appreciated and acknowledged: the TPC members and the reviewers, the authors and presenters of the papers, the invited speakers and the panelists, as well as the officers of the Future and Emerging Technologies (FET) Program European Commission, the researchers of the EU-funded IST-FET Autonomic Communication Coordination Action (ACCA), the EU-funded IST Network of Excellence E-NEXT, the Autonomic Communication Forum (ACF), IFIP TC6 and all the individuals involved from the National and Kapodistrian University of Athens.

December 2005   Ioannis Stavrakakis
                                                                  Michael Smirnov

About This Book

This is the post-workshop proceedings of the Second IFIP TC6 WG6.3 and WG6.6 International Workshop on Autonomic Communication (WAC2 2005); it includes 22 full papers presented at WAC 2005 and revised by the authors based on the workshop discussions, and summaries of the one keynote talk and three invited talks and two panel reports.

Workshop Chairs

General Chair   I. Stavrakakis, National & Kapodistrian University of Athens, Greece

TPC Chair   I. Stavrakakis, National & Kapodistrian University of Athens, Greece
M. Smirnov, Fraunhofer FOKUS, Germany

Technical Program Committee

R. Battiti, U. Trento, Italy
R. Boutaba, U. Waterloo, Canada
L. Chapin, Interisle, USA
L. Chlamtac, Create-Net, Italy
C. Diot, INTEL, UK
S. Denazis, Hitachi, France
S. Dobson, UCD, Ireland
C. Douligeris, U. Pireas, Greece
H. Einsiedler, DTAG, Germany
S. Fdida, UPMC, France
M. Gerla, UCLA, USA
E. Gregori, IIT-CNR, Italy
S. Hadjyefdymiades, U. Athens, Greece
D. Hutchison, U. Lancaster, UK
G. Karlsson, KTH, Sweden
O. Koufopavlou, U. Patras, Greece
G. Leduc, ULG, Belgium
D. Lewis, TCD, Ireland
I. Matta, U. Boston, USA
M. Mulvenna, U. Ulster, UK
B. Plattner, ETHZ, Switzerland
G. Polyzos, AUBE, Greece
G. Pujolle, UPMC, France
S. P. Romano, U. Napoli, Italy

C. Santivanez, BBN, USA
I. Schieferdecker, TUB, Germany
F. Sestini, European Commission
V. Siris, FORTH, Greece
O. Spaniol, U. Aachen, Germany
C. Tschudin, U. Basel, Switzerland
J. Vicente, Intel, USA
L. Wolf, U. Braunschweig, Germany
L. Yamamoto, U. Basel, Switzerland

Local Organizing Committee (National and Kapodistrian University of Athens)

E. Tsoukali
A. Panagakis
C. Vassilakis
L. Tzevelekas

Sponsoring Organizations

The National and Kapodistrian University of Athens, Greece.
The European Commission, FET Program, Autonomic Communication Coordination Action, ACCA (IST-6475).
The European Commission, E-NEXT Network of Excellence (IST-506869).
IFIP TC6 (WG6.3 and WG6.6).

Table of Contents

Autonomic Session 4

Autonomic Session 5

Autonomic Session 6

Autonomic Session 7

Invited Program

Panel Reports

Pocket Switched Networking: Challenges, Feasibility and Implementation Issues

Pan Hui[1], Augustin Chaintreau[2], Richard Gass[2],
James Scott[2], Jon Crowcroft[1], and Christophe Diot[2]

[1] Cambridge University
[2] Intel Research
{pan.hui, jon.crowcroft}@cl.cam.ac.uk
{augustin.chaintreau, richard.gass,
james.w.scott, christophe.diot}@intel.com

Abstract. The Internet is built around the assumption of contemporaneous end-to-end connectivity. This is at odds with what typically happens in mobile networking, where mobile devices move between islands of connectivity, having opportunity to transmit packets through their wireless interface or simply carrying the data toward a connectivity island. We propose *Pocket Switched Networking*, a communication paradigm which reflects the reality faced by the mobile user. Pocket Networking falls under DTN. We describe the challenges that this approach entails and provide evidence that it is feasible with today's technology.

1 Introduction

Mobile networking is finally becoming ubiquitously deployed, due in large part to the convergence of mobile telephony and handheld computing. Current mobile devices typically have one or more wireless interfaces (e.g. Bluetooth, WiFi). The applications which are commonly deployed on such devices (e.g. email, web browsing), however, are rarely able to fully exploit this local wireless connectivity, and instead use it only as a means of acquiring global connectivity via access points.

Therefore, there is currently a large amount of wireless bandwidth capacity that remains unused because the current communication paradigm (i.e. the Internet) has not been designed to take advantage of local and intermittent connectivity. The underlying reason for this failure is that *IP-centric networking* (a term covering everything from the IP network layer through to application-layer protocols such as HTTP) relies on several assumptions which do not hold for mobile users. One such assumption is that the source and recipient of a datagram are *contemporaneously connected*, i.e. that throughout a communication there exists a complete path between the two parties communicating. Another assumption, based on the end-to-end argument, is that it is sensible to determine the precise endpoints of a connection before any application data is transferred, and to have intermediate nodes in the network simply perform best-effort routing.

I. Stavrakakis and M. Smirnov (Eds.): WAC 2005, LNCS 3854, pp. 1–12, 2006.
© IFIP International Federation for Information Processing 2006

We propose a new set of assumptions for mobile networking. We argue that these assumptions lead to a new networking model, which we term *Pocket Switched Networking* (or PSN) since it relies on both occasional transmission opportunities and user mobility to carry data to their destination. These assumptions are as follows. Mobile networking users carry one or more devices having significant storage capacity. Their mobility may be useful as a data-carrying mechanism. Devices have local networking interfaces, with which they can exchange data with neighbors. Devices may have access to one or more global networks (e.g., Internet, GSM), which differ in price, bandwidth, and availability. Both global and local connections may provide *opportunities* to transfer data.

We identify two classes of communications that users demand. Local communication allows wireless devices to use their communication infrastructure to provide communication services in the absence of end-to-end infrastructure. Local services are currently not provided by the Internet. Examples are prevention of natural risks and disasters, security, localization, messaging. Global services extent legacy communication services such as those provided by GSM, GPRS, or the Internet. They make these legacy services available to mobile users. Note that some services can make use of both local and global communication paradigms. Examples are "ad-hoc google" and asynchronous messaging.

In the next section, we position Pocket Switched Networking with regard to related initiatives in mobile networking. We then discuss the challenges that Pocket Networking must solve, and present experiments into the feasibility of Pocket Networking.

2   Related Architectures

Pocket Switched Networking falls under Delay Tolerant Networking (DTN) umbrella. The delay Tolerant Networking research Group [1] defines itself as follows: "The Delay-Tolerant Networking Research Group (DTNRG) is concerned with how to address the architectural and protocol design principles arising from the need to provide interoperable communications with and among extreme and performance-challenged environments where continuous end-to-end connectivity cannot be assumed. Examples of such environments include spacecraft, military/tactical, some forms of disaster response, underwater, and some forms of ad-hoc sensor/actuator networks". The Delay Tolerant Networking (DTN) architecture, routes self-contained messages ("bundles") through networks with long delays, high error links, and intermittently connected, pre-scheduled, or opportunistic link availability. DTN messages contain information about service requirements and setup, though there is little notion of using application-level information to assist in forwarding decisions. However, the DTN RG does not make the assumption that the current DTN architecture is the only one possible.

Therefore, Pocket Switched Networking is a specific application domain of DTN. However, we take a radically different approach than most of the DTN

---

[1] www.dtnrg.org

related work to date. Instead of trying to extend the Internet legacy applications to support intermittently connected communication environment, we choose to design a new communication architecture, orthogonal to the Internet, that can use the Internet (as any other local communication) when available.

We believe that under the PSN assumptions described above, IP-centric networking is not a sensible approach. Reasons are abundant, from the need for the sender to determine the IP address of the recipient before sending data, to the use of closed-loop protocols such as TCP, SMTP and HTTP which employ a sequence of end-to-end exchanges for data transfer. In addition, IP-centric networking often relies on the availability of infrastructure services (e.g. DNS) that are not systematically available to mobile users. We assert that most attempts in this area are designed to extend IP-centric networking to new environments, and rely on the same invalid end-to-end assumptions.

Mobile Ad-Hoc Networks (MANET)[2] attempt to utilize local bandwidth without the presence of an infrastructure provider. However, they are IP-centric and aim to provide Internet style routes. For example, protocols such as AODV [9] depend on contemporaneous connectivity between the endpoints, and do not work if the only connectivity available is asynchronous and depends on mobility of nodes. Both MANET and DTN require a sender to know the recipient address for a given communication. In PSN, such an assumption cannot be made, as the destination may be a particular node, a class of nodes, or any node able to service the request.

Some sensor networks act in an opportunistic fashion. One example is Zebranet, which uses intermittent connections between zebra-mounted nodes to transfer sensor data and collect statistics about zebra populations. This and other similar projects do not target the mobile user domain of PSN, and thus do not address challenges such as trust and usability.

There is an interesting synergy between PSN and pervasive computing [11]. Both are user-centric, and face the challenges of trust, usability, and the need to collapse layered networking models to accomplish their goals. It is our belief that PSN provides a networking abstraction which serves the needs of pervasive computing much more cleanly than IP-centric approaches.

3   Challenges

We have defined PSN as a communication paradigm capable of taking advantage of both local and global connectivity, as well as device mobility to convey messages or queries[3] to an appropriate endpoint, in the absence of contemporaneous end-to-end connectivity and global services. In this section, we identify the challenges that have to be addressed to successfully implement PSN. We outline previous attempts to address these challenges, and highlight the key problems of PSN yet to be solved.

---

[2] www.ietf.org/html.charters/manet-charter.html

[3] These two terms are used interchangeably to mean "transmission data units."

3.1 Usability

Opportunities are surprising, and users often dislike surprises. The success of PSN will depend on our ability to address two concerns. First, we need to provide some level of predictability of the behavior of PSN, although we cannot usually provide deterministic performance or 100% provable reliability. Because of this, the second concern is to provide appropriate feedback to users about the state of the system.

These concerns can only be addressed through the development of applications that perform useful tasks.

3.2 Naming

Naming fulfills two basic functions: it provides a level of indirection, and a way to identify things meaningfully. Names are bound to identifiers, typically by a *name service*. The service takes the name as a key, possibly with some attributes that provide more semantic clues or hints, and returns a more specific "lower level" identifier.

Traditionally, naming is implemented by some distributed set of services. It is questionable whether a *name service* per se is actually necessary in the context of PSN. A naming scheme is needed so that communication between named entities concerning named objects can be carried out [3]. Such names may need to be constructed dynamically or modified by attributes. Name construction may not need to be specified in advance; it can be an emergent property of the node behavior or state. In this sense PSN has features of distributed systems such as LIME [6] and Content Addressable Networks [10].

3.3 Security

PSN operates in an environment where a number of resources are at risk. Adversaries have several targets, including messages, nodes, transfer opportunities, and the models of user mobility embedded in individual nodes. Potential classes attacks include redirection, impersonation, eavesdropping, piercing of anonymity, fabrication, denial of service, and poisoning.

Solutions designed for ad-hoc networks may not be appropriate. Techniques which rely upon on-demand access to a centralized service cannot be used, nor can the assumption be made that all intermediate nodes are trusted. Admission control and in-network authentication, although effective in other contexts, are not sufficient to protect against malicious nodes in PSN, as all nodes are potentially malicious.

The DTN Research Group (DTN-RG) has suggested the use of identity-based encryption (IBE) [1], which has the property that public keys can be generated off-line on the basis of an arbitrary string (often a node identifier) and before the private key is calculated. Naming and addressing increase in importance in such a network. Private keys could be obtained while global connectivity is available, either before or after receipt of an encrypted message.

There are many opportunities for innovative work in the area of securing PSN. Locality information could be used to prevent Sybil and other identity attacks. Nearby (either logically or physically) nodes could create localized incentive or reputation systems. Finally, it will be necessary to develop mechanisms for preserving a user's privacy (both of location and identity) whilst still allowing messages to reach them.

3.4 Forwarding

Forwarding is the key challenge in opportunistic networking, as the utility of PSN is strongly correlated with the number of messages that reach their destination. The problem of forwarding is simple to describe: when nodes have a local or global connection opportunity, messages are forwarded according to some policy, with the intention that they are brought "closer" to their destination.

Local forwarding makes use of intermittent and mobility-based connectivity. This precludes formation of routes; instead, nodes must forward messages according to knowledge of their local environment and of the messages themselves. How best to acquire and interpret this information is a difficult problem. In addition, availability of storage and energy may affect the willingness of a node to forward messages, as discussed in Section 3.6.

When global connectivity is available to a node, messages can be forwarded directly to suitable nodes which are also globally connected, or to available proxies for those recipients (e.g. by encapsulating a message as an email and sending it to a recipient's IMAP server). The latter allows for a recipient's device to retrieve the message during a subsequent period of global connectivity. However, the sending device should not necessarily discard the message after forwarding it, as it may encounter the recipient directly before the recipient has had a global connection opportunity.

Prior work on message forwarding has focused on making Internet services available in a disconnected setting, exploiting nearby resources where possible; the 7DS system [8] is an example of this approach. An initial scheme for true message forwarding was proposed by Davis et al. [4] on the basis of last seen nodes, and variants of the algorithm were later presented by others [5]. There have been algorithms reported for rumor- and gossip-type communication at the application layer [4]. There have also been biologically and physically inspired schemes for communication in specific problem domains (e.g., sensor nets). Zhao, et al. [12], Burns, et al. [2] and DTNRG examined the use of a series of predictable, reliable, but non-contemporaneous links for routing.

For PSN, the challenge is in developing methods to determine which neighboring nodes provide good forwarding opportunities for a given message. To guide this process, meaningful communication and mobility data for the problem domain are required. Real world data of this sort are scarce, and random models are inappropriate as there is no structure for intelligent forwarding algorithms to exploit. Real systems must be built, measured, and learned from in order to make progress on this most important facet of PSN.

---

[4] www.grapewineproject.org

3.5 Mobility

A number of major challenges arise from the mobility of nodes. Intermittent communication links, the associated long messages lifetimes, and the movement of nodes within the network all pose problems for the timely, reliable and efficient delivery of messages. The short-lived nature of links presents a further problem, as currently deployed wireless technologies such as Bluetooth and WiFi were not designed with short-lived connection opportunities between power-limited devices in mind.

Short-lived connection opportunities are an inevitable consequence of high mobility and short radio range. In a realistic environment, Class 2 Bluetooth devices provide usable throughput at distances of about ten meters (see Section 4). Class 1 Bluetooth devices and 802.11 radios are too power hungry for continuous use in battery powered devices, though they may see use in other environments. These ranges support connection opportunities which last on the order of tens of seconds with typical pedestrian and vehicle speeds.

Searching for and connecting to other nodes opportunistically must be made efficient in the absence of a central coordinator. Protocols that minimize transmission delay and maximize the amount of data and sent over short-lived, error-prone links must be developed and evaluated. As mentioned in Section 3.4, traces of real-world mobility must be collected and analyzed, potentially leading to the development of more realistic synthetic models.

3.6 Resource Management

There are two main resource management issues in PSN: network scheduling and energy conservation. With opportunistic forwarding, network interface scheduling becomes much more complex than IP-centric outgoing queues. It includes issues such as balancing time spent discovering neighbors with time spent transmitting data, handling limits on transfer opportunities imposed by mobility, and ensuring fair sharing of available radio spectrum with other devices. These problems are similar to those of congestion control in global end-to-end networks.

Energy concerns are likely to lead to culling of potential transfers, where the energy cost outweighs the expected benefit of the transfer. Other conservation techniques include duty cycling and wake-on-LAN which avoid the continuous powering of network interface receivers. Among the techniques known are using a high-power and low-power radio, adaptation to observed temporal and spatial availability of power, and preferentially forwarding via powered nodes when they are available.

4 Feasibility of Pocket Switched Networking

As described in Section 3.5, transfer opportunities are time-limited, and existing technologies and protocols are not designed for this case. Nonetheless, devices supporting Bluetooth and 802.11 are widely deployed, and their numbers are expected to increase rapidly in the future [5]. In this section, we use measurements

---

[5] www.bluetooth.org

and/or simulations of Bluetooth and 802.11 data transfers. The amount of data that can be transferred between two mobile nodes that encounter one another is dependent upon several factors: the time required for the nodes to discover one another, the time that the nodes are within radio range of one another, and the variation of throughput with range and operating environment. We measured each of these factors for both Bluetooth and 802.11. Results are described below.

4.1   Bluetooth Transfer Opportunities

We performed our measurements using the PC laptops running Windows XP and class II Bluetooth USB devices manufactured by MSI and Belkin. We observed no significant performance differences between the two varieties of device. The measurements presented here were obtained using the MSI devices, and were taken one meter above the ground. All other wireless devices in the machines were removed. The goodput between devices was measured at various distances by opening an RFCOMM connection between the machines and sending 64 kilobyte messages from the initiator to the slave. We performed this measurement in two environments: indoors, in an office corridor in the presence of background 802.11 and Bluetooth interference, and outdoors, in a field far from such interference. Finally, we created a simulation of limited-duration transfer opportunities. In the simulation, two nodes performing inquiry approach one another head on, pass at some relative speed, and eventually move out of range. Once inquiry is successful, one node sends data until out of radio range. The average number of kilobytes transferred (in 50,000 experiments at each point) in two cases is shown in Fig. 1. The solid line indicates the results obtained using the best case values from the experiments. The dashed line indicates the results obtained using the worst case values. At walking speed, around 1Mb of data can be exchanged during a contact opportunity that would last approximately 10s. Although these results are preliminary, they indicate that Bluetooth is usable for opportunistic data transmission.

4.2   802.11 Transfer Opportunities

In [7], experiments are performed with a wireless Host in a car passing by a 802.11 Access Point at various speeds. They show that even at 180kph, 1.5Mb of data can be sent from the fixed point to the mobile host with both TCP and UDP in a single transfer opportunity (i.e within a 10s time interval). At 80kph, they could transfer up to 6Mb of data in one transfer opportunity (i.e. 36s).

We have also performed our own 802.11 experiment Fig. 2. We observe similar results as in [7]. below 20 meters per second (around 70kph), above 10Mb of data can be transferred with TCP. And a couple of Mb can still be transferred at and above 100 kph.

In both Bluetooth and WiFi case, the numbers above, despite realistic, should be considered upper bounds as more protocols can interfere (e.g. VPN, encryption, etc.) to reduce the amount of data transferable. However, new technologies

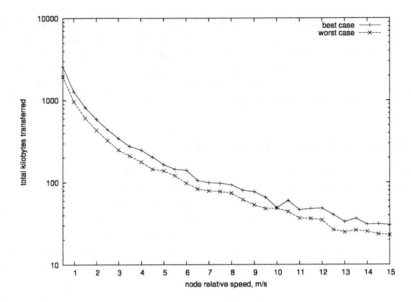

Fig. 1. Expected volume sent during a transfer opportunity with Bluetooth

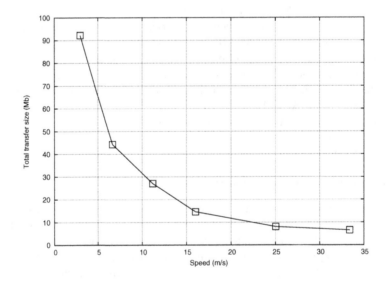

Fig. 2. Expected volume sent during a transfer opportunity with 802.11

can be designed that allow more data to be transferred in short time intervals. The main observation to retain from this section is that both Bluetooth or WiFi can be used in the context of PSN.

5   Enabling Technologies

We now discuss components that need to be provided in order to support the challenges of PSN.

5.1   Community-Based Networking

We expect PSN to enable a new family of applications with a high degree of spatial or logical locality. We refer to these areas of networking as communities, and provide explicit support for community formation and management with PSN. We believe that this notion of community will make propagation of information in PSN easier to achieve and control. Service requests may have more success if forwarded preferentially through a community. Communities can help improving the security of transactions. A node might give preferential treatment to a service request related to a community for which this node has state information. Examples of such communities are participants to a symposium, fans of Elvis Presley, London subway users, etc.

Communities are formed in a distributed fashion. The impetus for community formation may be implicit or explicit. An example of the former would be a common geographical location or a common interest, while communities may form in the latter manner around a given event, such as a conference. To form communities based upon space-time proximity, a TTL field in messages may be appropriate. Other communities with more tightly controlled access may allow open membership, vote to decide upon membership, or appoint a trusted authority to control the membership of the community. Communities might provide some information that can later be used by a node to make a decision (i.e. in a conference, the list of registered participants can be provided prior to the events). Community is also a convenient paradigm to secure PSN: nodes can create shared keys to allow for private communications, and utilize known techniques for key revocation within the community.

5.2   Security

Security is a major issue in PSN and several important security services must be provided to users. In this section, we discuss issues related to secure distributed naming, authentication, trust, reputation systems, and incentive to cooperate.

Identity and Trust. A user can have one or more identities. Some of these identities will be specific to a community. On the other hand, they are not device-specific and a PSN user can use the same identity on various devices. An identity can be (non exhaustive list) an email address, a URL, a names, a picture, or any combination of the above.

Identities will be tied to public/private key pairs generated by the user, and may be shared or moved between nodes. In traditional public key cryptosystems, the bootstrapping of trust[6] is a difficult problem, solved by either a trusted third

---

[6] Trust is the belief that a person is who his identity claims he is.

party or a distributed web of trust, such as that in PGP. PSN is particularly well suited to easy deployment of a web of trust, as users will be carrying small devices and can quickly and conveniently bootstrap trust between one another. In the case of more managed communities, a trusted third party can serve as an authenticator of identity. This notion of identity will serve as a building block for more advanced security services, as other users will attach notions of trust, reputation, content, and capability to an identity.

Reputation System and Incentive to Cooperate. A device implementing PSN will provide a reputation system to discourage malicious behaviors. Reputation systems have a different (thus complementary) purpose than trust control. Trust control is about getting the guarantee a user really is who he claims to be. While reputation is about quantifying how good a citizen a node or user is.

In PSN, we must be able to adjust reputations not only on the basis of performance, but on degree of trust, on the community, on the kind of service request, and other metrics. Application level behavior may draw upon this reputation among other mechanisms. For example, a user might trust anyone to forward his messages, but only accept address book update from identities whose reputation exceeds a certain threshold.

Therefore, reputation systems will be a strong components (not necessarily the only) in creating an incentive to cooperate. Encouraging users to contribute resource (memory, battery, time, etc.) will be a major problem in PSN. The problem is solved in Kazaa by limiting the amount of information that is made available to those that do not want to contribute, but just consume. We can implement the same kind of mechanism in PSN. However, we believe that communities will be a strong element in getting a given user more cooperative.

5.3   Localization

Most PSN applications will rely on locality information such as geographical location, or neighborhood. However, localization does not necessarily mean GPS. There are numerous localization algorithm that could be implemented. Each of them will match specific community needs, and have an impact on the way services can be provided and service requests are forwarded.

5.4   User Interface

In the current networking world, users are often forced to make routing decisions when trying to send data to local recipients, e.g. having to pick one of email, infrared transfer, Bluetooth transfer, USB key, or another method when wishing to transfer files to other local recipients — and the list keeps growing. This is precisely opposed to the goals of transparency and ease of use which are held dear to computer users. In the PSN, we may be able to offer the advantage that both local and wide-area connectivity are made transparent to users.

However, transparency is not necessarily the only goal. It is important that users remain appraised of the status of the delivery of the data they send and request. With end to end communications, this is relatively simple to determine and to display for the users convenience, e.g. using a spinning globe icon in a web browser, which stops spinning when page-load is complete. In PSN, not only are there many more states which one may wish to communicate, but a PSN client may also have little indication of the network status, since some nodes it was communicating with are no longer visible. Allowing PSN users to achieve and maintain an intuitive mental model of the status of their on-going data transfers may be a key issue in providing usable and deployable PSN applications.

Explicit user involvement in certain situations is also necessary, for example in determining a trust relationship if there is no prior community network of trust to draw upon, or in mapping a device's public key to a user that it claims to represent (similar to the way ssh maps keys to hosts). Users should also retain control over all operations since they may involve the spending of scarce resources such as storage, bandwidth or battery.

5.5   Monitoring

A monitoring module will collect traffic information to make it possible to analyze a number of "dispatching" or "forwarding" strategies, applications behavior and so forth. Monitoring is also key to PSN as the information collected by the monitoring modules could be used to optimize forwarding decisions or to take part to decisions on the trustability of some node.

6   Conclusion

We aim to implement PSN for *mobile computing devices*, which obviously includes notebook PCs and PDAs. Recently, this term has also become applicable to mobile phones, which now have significant storage, computing power, local networking (generally in the form of Bluetooth), and support for dynamically-loaded applications. Our implementation of PSN is known as Haggle.

Over the next few years, we aim to address the research challenges described above and to build Haggle-based applications including distributed Usenet-style newsgroups, messaging, file-sharing, and web browsing with automatic use of neighbors' caches. We plan to test these applications by rolling out proto-types to both local users and larger groups such as conference attendees. This will enable us to study aspects of Haggle including usability, scalability, network congestion, and user behavior, which can only be conclusively studied in deployments.

We also expect to release our implementation of the Haggle infrastructure and example applications under an open source license, and to encourage downloads and additional deployments by users as well as other research groups.

Please visit the Haggle project web[7] for access to all project resources.

---

[7] www.cambridge.intel-research.net/haggle/

Acknowledgments

David Blunden and Neeraj Sharma (with their team) helped us collect the 802.11 in-motion data. Marc Liberatore and Brian Levine from University of Massachusetts Amherst contributed to Bluetooth data collection and analysis. Ralph Kling and the SNO team helped us with the iMotes.

This work has been partly supported by the European Union under the E-Next NoE FP6-506869, and under the ACCA CA FP6-IST-6475.

References

1. D. Boneh and M. Franklin. Identity based encryption from the Weil pairing. *SIAM Journal of Computing*, 32(3):586–615, 2003.
2. B. Burns, O. Brock, and B. N. Levine. *MV* routing and capacity building in disruption tolerant networks. Technical Report TR-04-68, University of Massachusetts at Amherst, July 2004.
3. J. Crowcroft, S. Hand, R. Mortier, T. Roscoe, and A. Warfield. Plutarch: An argument for network pluralism. In *ACM SIGCOMM*, Aug. 2003.
4. J. A. Davis, A. Fagg, and B. N. Levine. Wearable computers as packet transfer mechanisms in ad-hoc networks. In *International Symposium on Wearable Computing*, Oct. 2001.
5. M. Grossglauser and M. Vetterli. Locating nodes with EASE: Mobility diffusion of last encounters in ad hoc networks. In *IEEE INFOCOM*, 2003.
6. A. L. Murphy, G. P. Picco, and G.-C. Roman. LIME: A middleware for physical and logical mobility. In *International Conference on Distributed Computing Systems*, pages 524–233, June 2000.
7. J. Ott and D. Kutscher. Drive-thru internet: IEEE 802.11b for automobile users. In *IEEE INFOCOM*, 2004.
8. M. Papadopouli and H. Schulzrinne. Performance of data dissemination among mobile devices. Technical Report 005, Columbia University, 2001.
9. C. E. Perkins and E. M. Royer. Ad hoc on-demand distance vector routing. In *IEEE Workshop on Mobile Compting Systems and Applications*, pages 90–100, feb 1999.
10. S. Ratnasamy, P. Francis, M. Handley, R. Karp, and S. Shenker. A scalable content addressable network. In *ACM SIGCOMM*, 2001.
11. M. Satyanarayanan. Pervasive computing: Vision and challenges. *IEEE Personal Communications*, Aug. 2001.
12. W. Zhao and M. Ammar. Message ferrying: Proactive routing in highly-partitioned wireless ad hoc networks. In *IEEE Workshop on Future Trends in Distributed Computing Systems*, May 2003.

Experiments on the Automatic Evolution of Protocols Using Genetic Programming

Lidia Yamamoto and Christian Tschudin

Computer Science Department, University of Basel,
Bernoullistrasse 16, CH-4056 Basel, Switzerland
Lidia.Yamamoto@unibas.ch, Christian.Tschudin@unibas.ch

Abstract. Truly autonomic networks ultimately require *self-modifying, evolving* protocol software. Otherwise humans must intervene in every situation that has not been anticipated at design time. For this to become feasible autonomic systems must ensure non-disruptive on-line software evolution. We investigate related code steering techniques in two directions: One is the fully automatic selection of protocol service elements where, depending on device characteristics and current operation environment, each communication entity has to select among a potentially wide variety of protocol implementations providing similar services. The other direction relates to the automatic synthesis of new protocol elements which are the result of optimizing existing implementations for a specific context. In both cases we look at genetic programming as a tool to generate new code and software configurations automatically. In this paper we propose a framework for such a resilient protocol evolution and report on first exploratory results on the adaptation and re-adaptation to environmental conditions, and the elimination of superfluous code.

Keywords: protocol synthesis, protocol evolution, genetic programming.

1 Introduction

Managing change in a network and its services is currently a labor intensive task which is not automated. Any new algorithm must be engineered, then programmed, and deployed in the network. Today this process is slow and requires the effort of many people (network managers, engineers, programmers), which is outside the scope of autonomic networks. Networking software must be able to adapt and reconfigure – i.e., to evolve – by itself in the most autonomous way possible.

Ultimately, protocols and algorithms for autonomic networks should evolve during their own execution, with minimum service disruption. Such long term run-time automated code evolution is useful in two main situations: a need to optimize a given network service at run time, that cannot be satisfied by just optimizing service parameters; in response to steady changes in the environment or internal errors that require modifications within already deployed code.

At the same time, autonomic networks should be able to resist disruptions (hence change), including the actions of malicious or erroneous entities which

I. Stavrakakis and M. Smirnov (Eds.): WAC 2005, LNCS 3854, pp. 13–28, 2006.
© IFIP International Federation for Information Processing 2006

try to disturb the network's functional blocks in any possible way. Ideally, these blocks would react by detecting and defeating such attacks, and would then recover and heal themselves to continue providing the required services. In case of failures, alternative service blocks would replace the non-functioning ones in a reactive and non-supervised way.

With these problems in mind (simultaneous pressure to evolve and the requirement to resist changes) we describe in this paper our framework for protocol evolution based on genetic programming. We concentrate on two research directions: the first one is to automatically select combinations of protocol modules adapted to given network conditions; the second is the automatic synthesis of new protocols optimized for a specific context. The contribution of this paper is to show the feasibility of automatic network software selection based on service agnostic target functions. This result is based on the introduction of competition at the level of functional blocks and the use of genetic algorithms to steer the selection process. We report our experimental results using simple case studies, still in a simulated, off-line environment, but with considerations and parameters intended to progressively detach the framework from the off-line simulation out into the real world. We show the feasibility of code trimming, context aware selection of protocol variants and their re-adaption to changing environments using the proposed genetic programming framework.

This paper is structured as follows: Section 2 summarizes the state of the art in program and protocol evolution techniques. Section 3 states our position and describes our framework for protocol evolution. Section 4 reports the experimental results obtained so far. Section 5 concludes the paper with our outlook for this new area.

2   State of the Art and Related Work

Automatic programming or *program synthesis* refers to any method for automated generation of a computer program that is able to solve a given problem expressed in a high-level form. Examples include variations of meta-programming, deductive program synthesis [1], and evolutionary methods such as genetic programming.

Genetic Programming (GP) [2] is a machine learning method to evolve computer programs automatically from random initial code, using genetic operations such as crossover and mutation, and evolution by natural selection ("survival of the fittest") to select the solutions that best satisfy specified criteria. GP is typically employed when the solution to a problem is not known or very difficult to program by hand.

Although GP has been mostly applied to off-line solution of problems, it has also been used to evolve new programs at run-time, in domains such as evolvable hardware [3] and robotics [4, 5]. However, to the best of our knowledge, on-line evolution of networking protocol code has not been tried yet.

In [6] genetic algorithms are applied in a decentralized way to evolve agents that provide network services. Although their work is still implemented via simulations, their design aims at on-line evolution. Their results show that evolution can improve agent performance. However, in their scheme, the code itself does not change. They focus on the evolution of parameters that trigger certain predefined behaviors.

Protocol synthesis [7] aims to generate a valid protocol specification that satisfies a supplied service specification. A survey of synthesis methods is provided in [7]. The methods must guarantee the safety and liveness properties of the synthesized protocols, meaning that these must be guaranteed free from syntactic, logical and semantic design errors. Since these methods must guarantee error-free code, they are still not feasible for on-line evolution.

Examples of machine learning methods applied to protocol synthesis include [8, 9, 10, 11, 12]. In [8, 9] an iterative deepening search approach is used to find protocol specifications that satisfy a given set of security properties.

In [10] genetic search is used to synthesize protocol implementations from scratch. The synthesized protocols are expressed as communicating finite state machines. This research is extended in [11] and shows that relatively complex protocols can be synthesized in this way, and in certain cases these protocols can even outperform a reference protocol designed and validated by human beings. However in most cases the fitness of synthesized protocols is significantly lower than the reference protocol.

In [12] an evolutionary method to synthesize communication protocols is proposed. Similar to [10, 11], it also synthesizes finite state machines. Moreover it includes a method to derive a set of input/output training sequences that assures semantic correctness of the generated protocol. They show that optimum protocols can be generated for the simple case of a connection establishment task.

In most of the existing work, protocol synthesis is regarded as a protocol engineering method to be applied at the design phase. In contrast, we are investigating protocol synthesis as a tool for automated protocol evolution, to be incorporated as part of the tasks that an autonomic network must handle during run-time, on a routine basis.

3   Evolving Communication Protocols

The main premise underlying our work is that software in an autonomic network must be *self-modifying*. If the software was not self-modifying, it would mean that humans had to cater for the software's adaption every time that a case is encountered which was not anticipated at design time. Our aim is to find a framework where software self-modification is carried out in a goal oriented and non-disruptive way. Hence, we seek a mechanism which is agnostic to which function it adapts as long as the mechanism is capable of steering the whole network into optimal configurations.

We envisage different levels at which self-modification of software takes place and different time scales at which such modifications can happen. In order to cope with the constraints of a realistic run-time environment, we aim first at optimizing existing working protocol code, as opposed to full protocol synthesis from scratch. A first step, aimed at a shorter time scale, is the configuration of function blocks, where the challenge consists in selecting the right combinations from ready-made modules. Today, this is mostly controlled by standardization process and interoperability tests. Although several systems able to dynamically reconfigure software have been proposed, for instance [13, 14], most of these systems still rely on humans to program exactly what kind of reconfiguration should be performed under which circumstances. Considerable effort has also been spent on configurable protocol stacks [15, 16] but here again the reconfigurations were not fully autonomic.

In the future we imagine that a network "settles" by itself on different protocol sets without having humans to intervene. For example, depending on the available hardware, different "stack profiles" could be selected for sensors, PCs or core routers. This selection process is also applicable at finer time scales where for example an ad hoc network can switch among different routing algorithms, depending on the current topology. Another example would be the downloading of networking code, as exemplified by instantiating TCP flavors inside a TCP connection [17], where end nodes have to settle on the optimal combination of options.

At a longer time scale, these self-modification scenarios could in principle be extended down to the level of single instructions where the autonomic network would have the power to create new implementation variants, instead of just manipulating coarse grained functional blocks. At first, these new variants would emerge out of existing implementations. Eventually, full protocol synthesis from scratch, at the level of single instructions, could become possible, leading to fully autonomic networks.

3.1   Resilience and Competition

For such an autonomic selection process to work we need a modus operandi that permits adaption (medium time scale) as well as evolution (long term). Adaption relates to the configuration of existing functionality while evolution refers to the modification of old and generation of new functions. We believe that two attributes of such a system are key for its viability: resilience and competition.

The network must start with *inherent resilience*, otherwise there is a risk that (malicious or erroneous) function blocks can be inserted that disrupt the network's operation. In other words: adaption and evolution have to be activities that are running in parallel with the network and which, in the worst case, may temporarily disturb the network but cannot inhibit its operation.

The second attribute is *competition*: the autonomic network operates in a constant optimization mode where it picks those function blocks and code variants which are best suited.

Both attributes are currently implemented by having humans performing the adaption and evolution, and by writing and selecting those software bundles which provide the best value. Often, this human activity is not solely based on detailed analysis but also includes a simple trial–and–error strategy. Our goal is to rely on the later selection process only and to provide an environment where new functionality or function profiles can be evaluated and selected without disrupting the network.

3.2 Software Hardening and Genetic Programming

We have started to explore the feasibility of self-modifying communication software by demonstrating protocol resilience, where protocol implementations can survive the removal of an arbitrary code line [18]. In the current paper we explore genetic programming as a tool for modifying, recombining and erasing protocol modules. Other machine learning methods or heuristics could also be envisaged, for example, as has been demonstrated for the synthesis of security protocols [8, 9]. However, plain genetic programming lends itself for our project because it is agnostic to the functions adapted, and naturally extends to the finer grained code evolution that enables long-term synthesis and evolution.

Another choice we have made relates to the execution environment for the protocol software, which should be amenable to genetic programming. Sequential code, for example, is less suitable than a "chemical soup of rules" execution model [19, 20] because the executability of a linear code sequence depends on almost each of its instructions. For our experiments we are currently using our "Fraglets" chemical model [21], which also permits to express code mobility e.g., for evolving code deployment logic. Section 3.5 gives a quick overview of the Fraglet model and describes its useful properties which make it our model of choice for protocol synthesis and evolution.

3.3 A Framework for Automated Code Steering

Ideally, a software environment for an autonomic network should feature continuous adaption and evolution: Alternative code variants should co-exist in parallel with the currently best selection of protocol implementations. In terms of code steering, there would be a mechanism in place for on-line evaluation and selection of the alternatives. This on-line evolution has to be a continuously ongoing process that is decentralized and asynchronous, working on each node and at many levels inside the graph of functional modules.

Figure 1 shows a conceptual model of how resilience and competition work together to enable the automatic evolution of protocol implementations and configurations. Applications (or any client protocol) delegate service provisioning to a resilient protocol implementation, and from time to time or in parallel give a chance to test candidates. Based on their performance, new service implementation variants can increase their chance to be selected a next time. Service variations do include different ways of combining sub-services. Because the evaluation and selection mechanism takes into account the overall performance of a service implementation, it will give preference to the service with the most optimal internal composition and configuration of sub-services.

Our current implementation of the model of Fig. 1 is still limited to off-line evolution, i.e. to the case of synchronous evaluation and selection, so there are

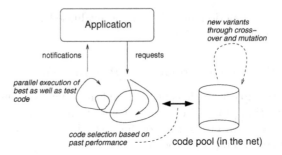

Fig. 1. Conceptual framework for automatic protocol evolution

no concurrent services yet. However we plan to progressively detach it from the off-line sphere in favor of the long-term goal of on-line evolution.

3.4   Genetic Programming Set-Up for Protocol Evolution

We apply Genetic Programming to evolve communication protocols or protocol structures, which are regarded as individuals in a GP population. A major difference between our system and classical genetic programming is that our GP run starts with a population of working or partially working solutions, which may or may not be adapted to the task in question. Another difference is that our GP run is a continuous optimization process: the system must continuously adapt and readapt. This is in contrast with classical off-line GP where the system runs until a termination condition is satisfied; it then outputs the solution and stops.

The genotype is the metaphor for the protocol implementation code, and is manipulated from one generation to the next through well-known genetic operators such as crossover, mutation and cloning. The crossover operator in our set-up is a simplified implementation of the genetic concept of homologous recombination. Homologous recombination states that the exchange of genetic material can only occur between functionally compatible DNA segments, and is only triggered when the two DNA strands are completely aligned. This form of recombination preserves gene functionality, promotes genetic stability, and increases the probability of producing viable offspring. We implement this concept by dividing the protocol genotype into modules that make up the "genes" of the individual, and by allowing crossover to occur only at gene (module) boundaries and between functionally equivalent modules.

Homologous recombination is a step towards program transformations that formally maintain program properties. If the system starts with a population of programs that contain only functionally correct modules, then homologous recombination among these programs can only produce new program variations that implement similar functionality in different ways (some might be better adapted to given situations than others), but which are still functionally correct.

The fitness measure is the performance of the protocol as perceived by the applications. They reward correct behavior and punish incorrect one when detected. For instance, the score of an individual is incremented when it performs the correct operation (e.g. successfully delivering a packet), and it is decremented when an error is detected (e.g. an acknowledgment is issued for a data item that has never been actually received). Resource consumption, in terms of memory occupied by the genotype, is proportionally penalized. Fitness evaluation also helps keeping the system controllable, as humans can steer it through applications able to translate user input into fitness functions.

We now describe the GP algorithm. For each generation, a tournament selection is held, as follows:

1. Insert each individual of the population into its execution context (i.e. connect it to its application and network environment), and run each of them for the same fixed amount of time or execution cycles.
2. Extract the fitness scores for each individual in the population.
3. Select the n_b best fit individuals and add them to the population of the new generation.
4. From the set of n_c fittest individuals, with $n_c > n_b$, select $n_p \leq n_c/2$ pairs of individuals at random.
5. Perform crossover for each pair, producing $2 \cdot n_p$ new output code streams, which are then added to the pool of new generation individuals.
6. If mutation is enabled, select a small number n_m of individuals at random within the set of n_c fittest, and perform a mutation on each of them. Add the resulting individuals to the population of the next generation.

Traditional genetic programming models perform an off-line genetic search in which production of offspring is synchronous and fitness evaluation is centralized. Our current experiments are still limited to an off-line set-up, since we first need to demonstrate the basic viability of an automatic selection process.

3.5   Fraglets

The Fraglet paradigm [21] has been proposed as part of our search for feasible ways to achieve automated synthesis of protocol implementations. It is an instance of Gamma systems [19, 20], a chemical model where "molecules" interact with each other or undergo some internal transformation. A fraglet is a string of symbols $[s_1 : s_2 : \ldots : s_n]$ representing data and/or protocol logic. It is a fragment of a distributed computation, that may be carried in packets or stored inside a network node. The fraglet processing engine continuously executes tag matching operations on the fraglets in the store, in order to determine the actions that should be applied to them. The fraglet instruction set contains two types of actions: transformation of a single fraglet, and "chemical reaction" between two fraglets. The instruction set is described in [21, 18], along with examples of processing and protocol functions. Table 1 summarizes the reaction and transformation rules used in the examples of Section 4.

The fraglets model has many relevant properties that must be highlighted in connection with automated protocol synthesis and evolution. First of all, any string of symbols is a valid fraglet, therefore fraglets can be split at arbitrary

Table 1. Fraglet reaction and transformation rules

Reaction	Input	Output	Semantics
match	$match : s : tail_1$], $s : tail_2$]	$tail_1 : tail_2$]	concatenates two fraglets with matching tags
matchp	$matchp : s : tail_1$], $s : tail_2$]	$tail_1 : tail_2$] $matchp : s : tail_1$]	persistent match (preserves matchp rule)
Transf.			
dup	$dup : t : u : tail$]	$t : u : u : tail$]	duplicates a symbol
exch	$exch : t : u : v : tail$]	$t : v : u : tail$]	swaps two symbols
split	$split : t : \ldots : * : tail$]	$t : \ldots$], $tail$]	breaks fraglet at * position
send	$_A$[$send : B : tail$]	$_B$[$tail$] (unreliably)	sends fraglet from A to B
wait	$wait : tail$]	$tail$] (after interval)	waits a predefined interval
nul	$nul : tail$]	[ ]	fraglet is removed

places and merged with other fraglets to produce different code. A second property is the ability to express code and data in a uniform way. Code is manipulated just like any other form of data, and it is easy to express rules that generate and delete code from the running pool. A third aspect is the ability to express code mobility in a natural way: any fraglet can be regarded as either a set of packet header tags that can be processed by a header processing engine, or as a program fragment that is executed at a given node. This facilitates the dynamic deployment of new code logic.

A fourth property of the fraglet environment stems from its roots in Gamma systems: it enables programs to be expressed in a highly parallel way that is very close to their specification, without artificial sequentiality constraints. This is relevant for automated program synthesis and evolution, in two ways: first, this parallelism can be used to produce resilient programs as shown in [18], which tolerate the loss of parts of their code stream, due to fallback alternatives running in parallel. This can be used to diminish the impact of malfunctioning code. Secondly, the fact that programs are relatively compact and close to their specification could open up potential avenues for deterministic synthesis techniques based on specification.

4   Experiments

We have performed a few experiments using the fraglet environment to verify whether software configurations can adapt to their environment, by the mere application of generic and service agnostic GP methods. We start with a description of the protocols involved in the experiment (Section 4.1), and then describe the results for three experiments: testing the capacity to eliminate superfluous code (Section 4.2), adaptation to the environment (Section 4.3), and re-adaptation (Section 4.4).

4.1   Protocol Implementations

A simple case is considered where a reliable delivery service must be provided over different channel characteristics. The task is to transmit all packets from the client application, with acknowledgment of correct delivery. Two types of underlying transmission channels are considered:

- *Perfectly reliable channel:* In this case, the protocol does not need to retransmit packets. A simple implementation of this in fraglets is the confirmed delivery protocol (CDP) presented in [21]. It simply transmits a given payload from node A to node B and returns an acknowledgment from B to A.
- *Unreliable channel:* In this case, the protocol must retransmit lost packets. A reliable delivery protocol (RDP) has been implemented for this purpose. It takes an input payload from the application, sends it to the destination, stores a copy locally, and sets a waiting timer. When the timer expires, and the corresponding local copy of the information is still stored, the packet is retransmitted. When an acknowledgment is received, the local copy is destroyed; this cancels any pending retransmissions scheduled for the item. For simplification, no losses from sink to source are modeled.

Each protocol is encoded as a fraglet genotype made up of constituent modules or genes. The genotype is the concatenation of all the modules (and their constituent fraglets) that implement the protocol. Each module starts with an *"m"* marker followed by the module name.

Fig. 2 shows the fraglet code for CDP, both sender and receiver sides. When presented with an application payload of the form $A[data : payload]$, the first *matchp* rule in the *send* module will be activated, and the resulting reaction will produce a rule $A[send : B : deliver : payload]$, which will send the fraglet $[deliver : payload]$ to B, where the *deliver* tag will cause *payload* to be delivered to the application. The application will respond by injecting a $B[ack]$ fraglet, which will react with the *matchp* rule of the *receive* module, causing the *ack* to be delivered to the source application on node A. Note that the *deliver* tag can be implemented as a predefined rule that takes the tail symbol string out of the fraglet environment (towards an external application), or can be caught by a $[matchp : deliver : ...]$ rule as part of a fraglet application.

m send   **m receive**

$A[matchp : data : send : B : deliver]$ $B[matchp : ack : send : A : deliver : ack]$

Fig. 2. CDP implementation in fraglets

The RDP implementation is shown in Fig. 3. It has exactly the same interface with the application as CDP, so that both protocols can be interchanged in a transparent way. A $[data : payload]$ fraglet injected by the application activates the *send* module, producing two fraglets: $[retransmit : payload]$ and $[mack : payload]$. The first one triggers a retransmission loop (*retransmit* module). The second one triggers a series of reactions which produce a new rule able to treat an incoming *ack* and cancel any corresponding retransmission.

Several variants of CDP and RDP have been implemented to make up a reasonably sized initial population for the GP run. Figures 2 and 3 show examples

m send
[$matchp : data : dup : data3$]
[$matchp : data3 : exch : data2 : mack$]
[$matchp : data2 : exch : data1 : *$]
[$matchp : data1 : split : retransmit$]
[$matchp : mack : exch : mack5 : nul$]
[$matchp : mack5 : exch : mack4 : *$]
[$matchp : mack4 : dup : mack3$]
[$matchp : mack3 : exch : mack2 : wait$]
[$matchp : mack2 : exch : mack1 : split$]
[$matchp : mack1 : match : ack : split :$
  $deliver : ack : * : match$]

m retransmit
[$matchp : retransmit : dup : t91$]
[$matchp : t91 : exch : t92 : t94$]
[$matchp : t92 : exch : t93 : *$]
[$matchp : t93 : split : transmit$]
[$matchp : t94 : dup : t95$]
[$matchp : t95 : exch : t96 : retransmit$]
[$matchp : t96 : dup : t97$]
[$matchp : t97 : exch : t98 : *$]
[$matchp : t98 : split : wait : match$]
[$matchp : transmit : send : B$]

Fig. 3. RDP implementation in fraglets (sender side)

of correct implementations. Other correct variants are also present in the experiments, as well as variants that introduce arbitrary delays, consume more memory, contain useless code segments, pollute the code pool with byproduct debris of reactions, and so on.

Crossover by homologous recombination is implemented by swapping modules of the same name in different protocol implementations. Since the interface of each module is the same regardless of its internal implementation, modules are compatible and crossover produces viable individuals. Mutation is applied with a low probability, changing a symbol at random in the fraglet pool.

4.2 Stripping Protocol Implementations

In this first baseline experiment we test whether the system is able to strip exceeding code, by eliminating garbage that is arbitrarily added to the programs. We take the CDP implementation and add several modules, some of which are empty, and some which perform random but non-disruptive actions consuming CPU cycles.

We generate 10 such "polluted" individuals, and perform repeated GP runs of 50 generations each, and $n_b = 4$, $n_c = 8$, $n_p = 3$, $n_m = 0$. A typical result from these runs is that roughly 75% of the garbage modules are eliminated. In a sample run, a relatively clean individual (with a single garbage module remaining) emerges around the second generation, and progressively propagates to the rest of the population. By the 7th generation, all individuals have a single garbage module. In this example the system does not improve beyond that, because all the individuals have the same garbage module, therefore homologous crossover is not able to eliminate it.

4.3 Adaptation

The goal of this experiment is to verify whether a mixed population of protocols is able to adapt to a given environment. Our mixed population is composed of eight CDP and eight RDP variants. These are alternative implementations of the same functionality. Some of them are perfect with no known bugs, others are deliberately made inefficient to different degrees, for instance, by not retransmitting packets correctly, or retransmitting too much, or spending a lot of time on bogus tasks.

We insert this population into two GP runs. In the first run, the population faces a reliable channel with no packet loss. In the second run a rather lossy channel (25% packet loss) is introduced. For each run we choose $n_b = 6$, $n_c = 14$, $n_p = 4$, $n_m = 2$. This results in a population size of $N = n_b + 2 \cdot n_p + n_m = 16$ individuals per generation, which is the same size as the original (hand-made) population.

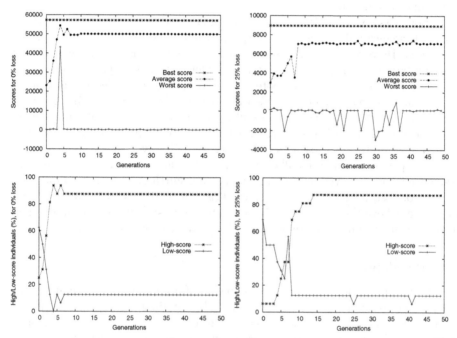

Fig. 4. Absolute scores and percentage of high/low scores for different packet loss rates

Figure 4 shows the adaptation of the initially mixed population to these two loss environments. The upper part shows the fitness scores for the different link loss rates, and the lower part shows the percentage of high and low-score individuals. A high-score individual is an individual that has achieved a score equivalent to at least 80% of the best score from its generation. A low-score one scores less than 40% of the best of its generation.

For the non-lossy channel (Fig. 4 left), the population starts with a low average score, but after a few generations most of the individuals have a score close to the best, and the percentage of individuals with very low score is small. In this case, the best individual is also the optimum (hand-designed), and the GP selection process succeeds to keep it in the population through the successive generations. After four or five generations the retransmission code is eliminated, and the surviving individuals are all instances of CDP.

In the lossy channel the retransmission code spreads very quickly through the entire population: all the individuals contain it after the first couple of generations. In Fig. 4 (top right) we can notice that the best score achieved by RDP is much lower than its equivalent in CDP. This is because the retransmit logic and associated timers consume execution cycles. Since all the individuals are allowed to consume the same amount of cycles, the simple code achieves much higher score. The adaptation to the environment can be observed in Fig. 4 (bottom right): after roughly 15 generations, more than 80% of the population is made up of high-score individuals. At the same time, the number of low-score individuals is reduced to a minimum.

In both lossy and non-lossy cases, mutations are mostly responsible for these low-performance individuals. The purpose of mutations is to introduce genetic variability. However, it is well known that most mutations are harmful. In our case, mutations are kept in the system in order to test its capacity to produce new code, and its resilience to potentially disrupting code. The production of new useful code has not been verified in such short runs though. On the other hand, the fact that the system can still adapt in spite of harmful mutations is an indication that resilience at the population level is possible even with the high rate of mutation chosen ($n_m/N = 12.5\%$). However this system is obviously not perfect. There are still clients affected by low-performance individuals: resilience is not achieved at the individual level. Furthermore, as it adapts, the population also loses genetic variability (this will be discussed in the next section). We believe this sort of drawback can be diminished if resilient individuals incorporating redundancy are used in place of the current non-resilient ones.

4.4   Re-adaptation

In this experiment we investigate the capacity of a population to readapt to an environment different from the one where it has originally evolved. We inject a population evolved in a 25% loss environment into a no-loss and vice-versa, and repeat the GP run with the same parameters as described in Section 4.3.

Figure 5 shows the obtained scores. These results clearly show that the population is not able to readapt. The lost retransmission modules cannot be recreated in such a short time by genetic operators only. The homologous crossover used only recombines existing modules, and mutations of individual symbols is simply a too slow and randomized process. The search space for the solution is far too vast, even though GP has shown to remarkably focus the search when compared to pure random search. For example, in the RDP example of Fig. 3, there are about 20 different symbols that may be placed at about 100 positions, leading to a search space of size 20^{100}. This is still too vast for short-term on-line GP. A similar problem may also occur in nature, when genetic variability is lost in small populations adapted to a fairly stable environment.

Nevertheless, if we inject a single optimally adapted individual in the population, it instantly redeploys and the entire population readapts. This can be observed in Fig. 6. After about 15 generations, more than 80% of its individuals achieve scores comparable to those of the best individuals of Section 4.3.

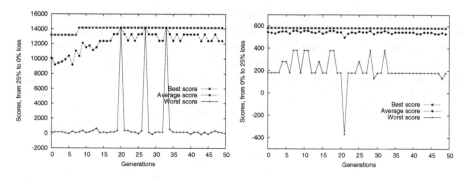

Fig. 5. Scores for two different re-adaptation situations: Left: from 25% loss to 0% loss. Right: from 0% loss to 25% loss.

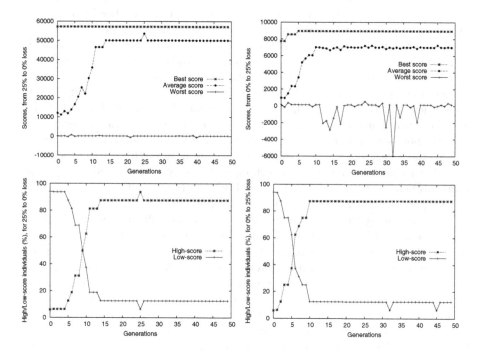

Fig. 6. Inserting a single adapted individual. Scores (top) and percentage of high/low scores (bottom) for different adaptation situations: Left: from 25% loss to 0% loss. Right: from 0% loss to 25% loss.

4.5   Discussion

We can extract several lessons from these early experiments. We first discuss the aspects related to genetic operators and other GP parameters. We then discuss future issues of resilience and on-line evolution.

We have modeled homologous recombination which is generally overlooked in GP. By restricting crossover to functionally compatible genes only, we have a high probability of producing viable individuals. In a few earlier experiments we had tried crossover at arbitrary points, and the result was poor score evolution combined with the well-known code bloat phenomenon in GP [2, 22], in which code tends to grow across generations, leading to large, inefficient programs in the long run. A widespread theory to explain the phenomenon says that GP code accumulates *introns* [2], i.e. portions of code that serve no functional purpose. These introns would then act as a protection against destructive crossover, as the probability of crossover points falling inside an intron (and therefore not breaking existing useful functionality) increases with the percentage of introns in the individual. Experimental results [22] show that code growth occurs even when crossover within introns is not allowed. However these results are valid only for tree-based GP, which is not our case. Anyway, independently of the actual causes of code bloat in a general sense, in our experiments the phenomenon disappeared as soon as we introduced homologous recombination.

However, homologous recombination in a limited population of simple individuals with few genes, as shown in the experiments, leads to low genetic variability, and after a few generations most of the variability is lost.

Mutation is usually regarded as the main source of genetic variability in GP populations [2]. However, the benefits of mutation can only be observed at the very long run, since most mutations are lethal. In our short-run experiments, we have not been able to observe really productive mutations. We have to interpret these very preliminary results with caution; nevertheless, they seem to indicate that new, more intelligent techniques for evolving populations of genetic protocols need to be devised to make on-line evolution a reality.

The parameters of a GP run clearly have an impact on the evolutionary process. Adjusting these parameters is a well-known difficult problem in GP. Some researchers have inserted GP parameters into the genotypes evolved such that the best combination of parameters can also emerge from the evolutionary algorithm itself. This is a path we intend to explore in our future work.

In our current experimental set-up, fitness evaluation still has a centralized component. This prevents the emergence of cheat programs, e.g. programs that lie about transmitted or acknowledged packets. Fitness evaluation is a non-trivial issue in a real distributed on-line environment. Perhaps redundancy and reputation mechanisms could be combined to to provide a safe and reliable way to evaluate the behavior of protocols at run-time.

The next immediate step towards on-line evolution that we are starting to investigate is how to combine our previous resilience work [18] with genetic programming in order to add resilience at the level of individuals, as opposed to the level of entire populations as described in the experiments above. Each protocol is modeled as tuples of redundant genetic code. This should in principle improve resilience, and help preserving genetic variability in small populations.

5   Conclusions and Outlook

In this paper we propose an intrinsic approach to the automated evolution of network software. The goal is to enable automatic code deployment, self-configuration of functional modules and even automatic synthesis of protocol implementations in an autonomic network. We argue that the automated selection of protocols becomes feasible if the networking code is *resilient* such that we can have *competing* protocol variants running in parallel.

Using a concept known as homologous recombination, we have carried out exploratory adaption experiments using genetic programming. They show that a networking system can automatically and gradually evolve depending on the environment it is confronted with, provided that a minimum variability of code instances is kept. This observation relates both to identifying an optimal protocol implementation for a given context, as well as to finding the most efficient combination of several software modules.

A more complex task, that has yet to be demonstrated, is an on-line version where software evolution is a continuous activity. Our experiments have provided first insights on the obstacles that have to be overcome: For instance, fitness evaluation in a decentralized and competitive environment is a non-trivial issue. Another fundamental issue is to devise new and potentially correctness preserving genetic operators beyond homologous crossover which are able to evolve genuine new code for unforeseen situations.

References

1. Manna, Z., Waldinger, R.: Fundamentals of Deductive Program Synthesis. IEEE Transactions on Software Engineering **18** (1992) 674 – 704
2. Banzhaf, W., Nordin, P., Keller, R.E., Francone, F.D.: Genetic Programming, An Introduction. ISBN 155860510X. Morgan Kaufmann Publishers, Inc. (1998)
3. Sipper, M., Sanchez, E., Mange, D., Tomassini, M., Perez-Uribe, A., Stauffer, A.: A Phylogenetic, Ontogenetic, and Epigenetic View of Bio-Inspired Hardware Systems. IEEE Transactions on Evolutionary Computation **1** (1997)
4. Andersson, B., Svensson, P., Nordin, P., Nordahl, M.: On-line Evolution of Control for a Four-Legged Robot Using Genetic Programming. In: Real-World Applications of Evolutionary Computing – EvoWorkshops 2000. Springer-Verlag LNCS 1803, Edinburgh, Scotland (2000) 319–326
5. Steels, L.: Emergent functionality in robotic agents through on-line evolution. In: Proceedings of AlifeIV, Cambridge, MIT Press. (1994)
6. Nakano, T., Suda, T.: Adaptive and Evolvable Network Services. In: Proc. Genetic and Evolutionary Computation Conference (GECCO-2004). Springer-Verlag LNCS 3102 (2004) 151–162
7. Probert, R.L., Saleh, K.: Synthesis of Communication Protocols: Survey and Assessment. IEEE Transactions on Computers **40** (1991) 468 – 476
8. Perrig, A., Song, D.: A First Step towards the Automatic Generation of Security Protocols. In: Proc. Network and Distributed System Security (NDSS 2000). (2000)
9. D. Song, A.P., Phan, D.: AGVI – Automatic Generation, Verification, and Implementation of Security Protocols. In: Proc. 13th Conference on Computer Aided Verification (CAV 2001). (2001)

10. Sharples, N., Wakeman, I.: Protocol construction using genetic search techniques. In: Real-World Applications of Evolutionary Computing – EvoWorkshops 2000. Springer-Verlag LNCS 1803, Edinburgh, Scotland (2000)
11. Sharples, N.: Evolutionary Approaches to Adaptive Protocol Design. PhD dissertation, University of Sussex, UK (2001)
12. Araújo, S.G., Pedroza, A.C.P., Mesquita, A.C.: Evolutionary Synthesis of Communication Protocols. 10th International Conference on Telecommunications (ICT 2003) **2** (2003) 986–993
13. Whisnant, K., Kalbarczyk, Z.T., Iyer, R.K.: A system model for dynamically reconfigurable software. IBM Systems Journal **42** (2003) 45–59
14. Ruf, L., Keller, R., Plattner, B.: A Scalable High-performance Router Platform Supporting Dynamic Service Extensibility On Network And Host Processors. In: Proceedings of the 2004 ACS/IEEE International Conference on Pervasive Services (ICPS'2004), Beirut, Lebanon (2004) 19–23
15. Hutchinson, N.C., Peterson, L.L.: The x-Kernel: An architecture for implementing network protocols. IEEE Transactions on Software Engineering **17** (1991) 64–76
16. Plagemann, T.: A Framework for Dynamic Protocol Configuration. PhD dissertation, Swiss Federal Institute of Technology Zurich, Zurich, Switzerland (1994)
17. Patel, P., Wetherall, D., Lepreau, J., Whitaker, A.: TCP Meets Mobile Code. In: Proc. of the Ninth Workshop on Hot Topics in Operating Systems. IEEE Computer Society. (2003)
18. Tschudin, C., Yamamoto, L.: A Metabolic Approach to Protocol Resilience. In: Proc. 1st Workshop on Autonomic Communication (WAC 2004). Springer-Verlag LNCS 3457, Berlin, Germany (2004) 190 – 205
19. Banâtre, J.P., Métayer, D.L.: Gamma and the Chemical Reaction Model. Internal Publication PI-984, INRIA, France (1996)
20. Banâtre, J.P., Radenac, Y., Fradet, P.: Chemical Specification of Autonomic Systems. In: Proc 13th International Conference on Intelligent and Adaptive Systems and Software Engineering (IASSE'04). (2004) 72–79
21. Tschudin, C.: Fraglets - a Metabolistic Execution Model for Communication Protocols. In: Proc. 2nd Annual Symposium on Autonomous Intelligent Networks and Systems (AINS), Menlo Park, USA (2003)
22. Luke, S.: Issues in Scaling Genetic Programming: Breeding Strategies, Tree Generation, and Code Bloat. PhD dissertation, University of Maryland, USA (2000)

Service Evolution in a Nomadic Wireless Environment

Iacopo Carreras, Francesco De Pellegrini,
Daniele Miorandi, and Hagen Woesner

CREATE-NET,
via Solteri 38,
Trento – 38100, Italy
name.surname@create-net.it

Abstract. In this paper, we present and analyze a framework for self-evolving autonomic services in a wireless nomadic environment. We present a disconnected network architecture, where users mobility is exploited to achieve a scalable behaviour, and communication is based on localized peer-to-peer interactions among neighboring nodes. Service management is achieved by introducing autonomic services, whose operations are based on a distributed evolution process. The latter relies on the concept of *mating*, i.e., the exchange of information (e.g., code, parameters, data) among service users, which collaborate to enhance their *fitness*, defined as the ability of the actual service to fullfill the environmental features. The core of the evolution process is given by the service mating policy, which defines the way the running services should be modified when mating with other users. We introduce a general framework for analyzing service mating policies and exploit results from martingales theory to study their convergence properties. In particular, we introduce two optimal policies, clone-and-mutate and combine-and-mutate, and analyze their convergence times through extensive numerical simulations, addressing the impact of various parameters (number of nodes, users speed, mobility pattern).

1 Introduction

The emerging of a novel pervasive computing environment imposes big challenges to the current ICT technologies, calling for novel paradigms of communication, computing and service provisioning. The Internet, as we know it, is going to explode, due to the tremendous amount of information exchanged among a massively large number of devices. This problem does not concern only the communication aspect, i.e., the ability of the network to carry information, but also the way in which the network is managed and administrated. New paradigms are needed to face the needs and features of this novel networking environment.

In particular, autonomic communication systems are expected to represent one of the major technological breakthrough in the next decades, enabling the

I. Stavrakakis and M. Smirnov (Eds.): WAC 2005, LNCS 3854, pp. 29–40, 2006.

introduction of novel services and leading to a deep change in people/technology interactions [1]. The term "autonomic" comes from the computing field [2], where it is used to define systems which are inherently self-configuring, self-optimizing, self-healing and self-protecting. As the term itself suggests, an autonomic system should show the same properties of the human nervous system, which controls in an "unconscious" way routine tasks such as blood pressure, hormone levels, heart and breathing rate. Moving to the communication field, a system able to exhibit such features would clearly allow to solve the scalability and management issues related to the deployment of large-scale networks. In particular, we are interested in (i) defining a network architecture able to scale well with the number of nodes (ii) design services that are able to exploit the peculiar features of such networking support in order to evolve, leading, on the whole, to a complete autonomic communication system.

In this work, we first review, along the lines of some previous work by the same authors, a nomadic wireless network infrastructure model [3], which achieves network scalability by exploiting the users mobility [4] and a particular form of information filtering, that spatially limits the diffusion of data in the network [5]. Then, we address the problem of obtaining autonomic services over such a backbone-less networking infrastructure. We will see how such a task can be accomplished borrowing notions and drawing inspiration from the biological concept of evolution [6]. The basic idea is to have a service able to adapt to the surrounding environment, where the environment can include the space/time location of the user, the state of the entire system, the user's requirements etc. The problem becomes then how to achieve service evolution over our infrastructure-less network. The solution we propose is based on a one-to-one mapping from biology to networking, in which a *population* (i.e., the instances of the service running on different users) evolves through *mating* (i.e., the exchange of code/parameters between neighboring nodes made possible by the adoption of a wireless interface), the mating process being driven by *fitness* (i.e., the ability of the service to fit the actual environmental features). In this way, a distributed version of the "survival of the fittest" paradigm can be applied to achieve adaptation by evolution. The result is a service which is able to evolve, without the need of any human intervention nor of a central controller, to adapt to the actual features/needs/requirements. The aim of this paper is to study this evolution process, first introducing a rather general model for the evolution and convergence of the service fitness level, and then addressing, through numerical simulations, the performance of this evolution process, covering a wide range of system parameters. In this way, we are able to obtain some insight into both the performance obtainable by self-evolving services, as well as into some desirable features of the actual evolutionary algorithm to be employed.

The remainder of the paper is organized as follows. Section 2 defines the networking setting we will focus on and introduces concepts and features of the self-evolving services we will consider. Section 3 presents a mathematical model for the fitness evolution process, analyzes three simple evolutionary mating algorithms and presents numerical results on the convergence process. Section 4 concludes the paper with a brief summary of the results obtained and points out directions for future investigations.

2   Autonomic Networks and Services

2.1   A Nomadic Wireless Networking Infrastructure

We consider a scenario where the nodes of a wireless network are attached or otherwise assigned to a human user. They are the central part of the electronic "halo" that will surround the human being in the future, making life easier, more enjoyable and more secure. It is reasonable to assume that the majority of nodes in this halo will fulfill only primitive tasks like sensing the environment and will not be able to perform complex computing as it may be necessary for the rendering of 3D-graphics or database operations. A small number of complex multi-purpose devices will then be controlling the primitive embedded systems, gather and process their information and provide an interface to the user. This separation in the functionality of the nodes follows the technological constraints imposed by the size (and corresponding energy/storage capacity) of the respective devices. In fact, many of those tiny devices will be in part or totally passive, relying on the RFID principle. Consequently only the more complex nodes (named user-nodes in [3]) will be able to exchange information among each other. The nodes of this network are mobile because users are mobile. In addition, the information that is being processed will be of local significance most of the time, which calls for a novel communication paradigm based on localized interactions among peer nodes rather than the conventional end-to-end approach. The access to a fixed Internet backbone is optional and will not be present all the time. If needed to provide a certain service, the user node is responsible for setting up the backbone connection. In addition, the kind of information that is going to be exchanged as well as the addressing (node ID vs. geographic address) will depend on the service running on the user node.

All these considerations lead to a network model where we do not assume packet relaying between the nodes. All communication will take place in a single-hop broadcast. In result we see a fully distributed network of mobile wireless nodes without any backbone connectivity. Routing (or the rules of information exchange) will be part of the service. We are not anymore considering packet forwarding, but instead *information* flow between nodes of the network. Services are self-contained, i.e. they function out of themselves and get information through interaction with the environment. Two basic problems arise: The first one is that of a network-wide information exchange using only single-hop broadcast and the second problem is what we call the management problem. While the authors showed at least the solvability of the first problem in previous publications (using information filters inside example services) this paper is intended to lay the ground to solve the second problem.

Management of a network usually involves different aspects like performance, configuration, accounting, fault, and security management. In traditional networks these tasks are performed in a centralized network management system (NMS). With the network model described above this approach is no longer feasible, and new concepts are needed.

2.2   Self-evolving Autonomic Services

If we target the same level of efficiency of current centralized network management systems, the scenario outlined above represents an harsh environment for two key reasons. First, the nomadic network paradigm is designed to deploy services on large-scale networks [3]. Clearly, centralized management is out of question due to scalability problems with respect to the number of nodes and active services. Second, a large fraction of the information conveyed in such a network will have local scope, as in the case of the estimation of certain physical quantities (temperature, pressure) or activity detection on the battlefield, so that a distributed solution sounds more promising. Furthermore, since the nomadic wireless networking paradigm we propose is based on disconnected operations, this clearly clashes with the needs of a centralized controller. We believe that the adoption of evolving services, i.e., services that are able to configure, regenerate and optimize their behavior, is the natural solution to such challenges. In this way, no centralized management mechanism is needed, and we only need to design supporting mechanisms for evolving and adapting services. Basically, the gain is that the network management is embedded in services and thus, since in the autonomic scenario services are users-situated, this calls for distributed operations and self-adaptation capabilities. Notice that, together with the end-to-end communication paradigm, we drop also the client-server semantics which is at the basis of the current Internet, replacing it with a novel paradigm, where services are carried on users' devices and evolve exploiting local peer-to-peer communications. In the rest of this work we will provide insight into some viable mechanisms to enable self-adaptation of autonomic services over the nomadic wireless infrastructure. Following some preliminary work by the same authors [6], our proposed approach is based on some biologically-inspired techniques. In particular, we focus on evolutionary paradigms for services: services should evolve showing the ability to drift towards better performance, resembling what several biological entities do. The key mechanism is the exchange, through a mating procedure, of data/code/parameters with other users. In other words, we exploit nodes mobility to enable node cooperation, shaped as exchange of code and/or parameters, so that the overall effect is a *distributed evolution process*. The success of this evolution process is quantified through a standard metric. i.e. the *fitness*, which represents also the driving parameter of the mating process. In the next section we shall illustrate that it is possible to provide an abstraction of the concept of service evolution and to obtain, analytically and numerically, results on the performance of the distributed evolution process.

3   On the Fitness Evolution

3.1   The Fitness Convergence Process

According to the framework outlined in the previous section, we are interested in modelling the evolution process of autonomic services over a nomadic wireless network architecture. It is worth remarking that, from the user's point of view, the process of service evolution is completely transparent, since what it experiences, in reality, is the evolution of the degree of satisfaction to the actual service provided, i.e., what we call *fitness*. In order to keep the analysis and simulation scenarios simple, we will employ a simple yet general model for the service and its associated fitness level. This sort of black-box approach to services comes from the observation that the evolution of fitness as an outcome of the evolution of the service code is something which should be evaluated on a case-by-case basis, whereas we are interested in getting insight into a more general framework. This is expected to provide useful information for the design of service mating policies, i.e., the algorithms that will actually drive the service evolution. In the following, we will provide, as examples, three possible service mating policies, derive the associated fitness mating functions (i.e., the functions that defines the fitness evolution process) and study their convergence properties. While this is not meant to be omnicomprehensive, it enables us to individuate stable and optimal service mating policies and to gain insight into the various factors influencing the design of an effective service mating algorithm. In particular, we are interested in understanding how some factors (i.e., the number of nodes, the nodes speed, the mobility model) affect the evolution process. In terms of fitness, we expect that services with a higher degree of fitness will have a higher chance to survive, so that, in the mating process, their *genes* (e.g., routines, code parameters etc.) are likely to be inherited by the offsprings.

   We denote the fitness level of user i at time t as $I_i(t)$, and assume that $0 \leq I_i(t) \leq 1 \; \forall i = 1, \ldots, N$, where N is the total number of users in the network. We can group the fitness level of all users into an N-dimensional vector $\underline{\hat{I}}(t) = [I_1(t), \ldots, I_N(t)]$. Assuming that the users requirements and the environmental features are slowly changing over both time and space dimensions, $\underline{\hat{I}}(t)$ will change only at the mating instants. If the mobile speed is finite, the mating instants form a sequence $\{t_k\}_{k \in \mathbb{N}}$. In general, $\underline{\hat{I}}(t)$ will then be a random jump process defined on a suitable probability space $\{\Omega, \mathcal{F}, \mathbb{P}\}$. We denote by $\mathbb{E}[\cdot]$ the expectation taken with respect to the measure induced by $\mathbb{P}$. By standard arguments, we can transform $\underline{\hat{I}}(t)$ into a right-continuous left-limited (*càdlàg*) process, that will be denoted by $\underline{I}(t)$. In order to study the system evolution, we can then limit our scope to the embedded process $\underline{I}(t_k)$, where t_k denotes the k-th mating time. Please note that the mating times will be defined as a subset of the meeting times (i.e., the time instants two or more nodes get into mutual communication range), depending on the actual service mating policy employed (see below for the definition of the three cases considered). Also, note that the mobility models employed play a crucial role, in that they determine the intensity (and distribution) of the sequence of meeting times, and thus the convergence rate of the service evolution process.

We consider the following two random processes:

$$X(t) = \frac{1}{N} \sum_{i=1}^{N} I_i(t); \tag{1}$$

$$Y(t) = \min_{i=1,\ldots,N} I_i(t). \tag{2}$$

An easy sample-path argument leads to the following:

Lemma 1. *$X(t)$ converges to 1 $\mathbb{P}$-almost surely if and only if $Y(t)$ converges to 1 $\mathbb{P}$-almost surely.*

We then define the following:

Definition 1. *A service mating policy is called stable if it leads to convergence of $X(t)$ [$Y(t)$] with unitary probability.*

Definition 2. *A service mating policy is called optimal if it leads to convergence of $X(t)$ [$Y(t)$] to 1 with unitary probability.*

Please note that the condition of optimality is, in general, not sufficient for a mating policy to be efficient. How it will be discussed in the next section, we actually want a service that is able to converge fast to the optimal operating point, which actually regards the features of the transient behavior of the process $X(\cdot)$, while optimality here refers to a steady-state characteristic. Next, we want to estabilish some general sufficient conditions ensuring convergence of $X(t)$ [$Y(t)$]. We recall from [7] the followings:

Definition 3. *A process Z_n is said to be a submartingale (with respect to its natural filtration) if, $\forall n$, $\mathbb{E}[Z_{n+1}|Z_0,\ldots,Z_n] \geq Z_n$.*

Theorem 1. *Let Z_n be a submartingale such that $\sup_n(|Z_n|) < +\infty$. Then there exist a random variable Z_∞ such that $Z_n \to Z_\infty$ with unitary probability.*

Clearly, if would be desirable to have a service mating policy which leads to $X(\cdot)$ be a submartingale, so that we have convergence of the evolution process. Further, the optimal convergence should be to a random variable with unitary mass at 1, so that the process of evolution will reach the optimal system operating point. We assume that the service of user i can be represented as a binary vector $\underline{v}_i = [v_i(1),\ldots,v_i(T)]$, $v_i(l) \in \{0,1\}$, $l=1,\ldots,T$.[1] The fitness is then taken to be

$$I_i = \frac{\sum_{l=1}^{T} v_i(l)}{T}.$$

---

[1] The representation of the service as a binary string is fully general, in that it applies to any ICT service. This abstraction, while enabling a general tractation, results in a simplistic approach with respect to "real" services. This complies with the main focus of the paper, which is to gain a deep understanding of this distributed evolutionary process; the application of the proposed framework to actual services is not straightforward, and is left for future work.

We introduce the following service mating policies:

Definition 4 (Clonation mating policy). *Let us consider two nodes with respective fitness level I_1 and I_2 that get into mutual communication range. Let us assume, without any loss of generality, $I_1 > I_2$ (if $I_1 = I_2$ no mating takes place). Then user 2 downloads (clones) user 1's service. User 1 keeps its service unchanged.*

Definition 5 (Clone-and-mutate mating policy). *Let us consider two nodes with respective fitness level I_1 and I_2 that get into mutual communication range. Let us assume, without any loss of generality, $I_1 \geq I_2$ (if $I_1 = I_2 = 1$ no mating takes place). If $I_1 = I_2$, we assume node 2 to perform the mating. Then user 2 downloads user 1's service. Mutation is then performed on the new vector $\underline{v}_2$, by changing each digit independently with a given probability p (called the mutation probability). User 1 keeps its service unchanged.*

Definition 6 (Combine-and-mutate mating policy). *Let us consider two nodes with respective fitness level I_1 and I_2 that get into mutual communication range. Let us assume, without any loss of generality, $I_1 \geq I_2$ (if $I_1 = I_2 = 1$ no mating takes place). If $I_1 = I_2$, we assume node 2 to perform the mating. User 2 downloads user 1's service, i.e., the vector $\underline{v}_1$. A number k is uniformly taken in the set $\{1, \ldots, T\}$. Then, a new vector $\underline{v}'_2 = [v_1(1), \ldots, v_1(k), v_2(k+1), \ldots, v_2(T)]$ is formed. Mutation is performed on this vector, by changing each digit independently with probability p (called the mutation probability). User 1 keeps its service unchanged.*

To illustrate the fitness evolution process associated with such policies, let us consider the situation when two nodes, running the same service (but with different parameters), presenting fitness levels I_1 and I_2, respectively, meet at time t_{k+1}. We assume, without any loss of generality, that $I_1(t_k) \geq I_2(t_k)$ and $I_2(t_k) < 1$. Both I_1 and I_2 are taken to be in the interval $[0, 1]$. In the case of $I_1 = I_2$, we assume without any loss of generality node 2 to perform the mating.

In general, we have $(I_1(t_{k+1}), I_2(t_{k+1})) = \phi[I_1(t_k), I_2(t_k)]$, where $\phi[\cdot]$ is what we call the *fitness mating function*, that maps $[0,1] \times [0,1]$ into itself. The mating function is, in general, taken to be a stochastic function, defined on $\{\Omega, \mathcal{F}, \mathbb{P}\}$. For the three examples considered above, the fitness mating function takes the following form:

$$\phi[x, y] = (x, x), \text{ clonation mating policy,} \tag{3}$$

$$\phi[x, y] = (x, x + \xi), \text{ clone-and-mutate mating policy,} \tag{4}$$

$$\phi[x, y] = (x, \psi \cdot x + (1 - \psi) \cdot y + \xi)), \text{ combine-and-mutate mating policy,} \tag{5}$$

where ψ accounts for the combination operator and ξ is a random variable accounting for the mutation operator. From the structure outlined in the definition, it is clear that $\mathbb{E}[\xi] = 0$ and $\mathbb{E}[\psi] = \frac{1}{2}$.

We are interested is in studying the convergence properties of the aforementioned policies. We assume that the initial fitness values are indipendently taken from a continuous distribution $F_0(\cdot)$. We get the following results, whose proofs can be found in [8]:

Proposition 1. *The clonation/clone-and-mutate/combine-and-mutate mating policies are stable.*

Proposition 2. *The clonation mating policy is not optimal.*

Proposition 3. *The clone-and-mutate mating policy is optimal.*

Proposition 4. *The combine-and-mutate mating policy is optimal.*

3.2   Simulation Scenario

In order to study the effectiveness of the distributed evolution process, we run a wide set of simulations using a freely available software tools [9]. We simulated, for the whole range of simulation parameters, the two optimal service mating policies outlined above, and took the mutation parameter p equal to 0.1. We denote by N the total number of nodes, and assume that they are constrained to move in a square of $2000 \times 2000 \ m^2$. Each node is assumed to have a transmission range of 50 m, and IEEE 802.11b-compliant PHY and MAC protocols are used [10]. The nodes are initially dropped according to a uniform distribution on the square, and then start moving according to either a Brownian Motion (BM) or a Random Waypoint Mobility (RWM) at constant speed v taken in the set $\{2, 5, 10, 15\}$ m/s. For the RWM model, the pause time has been set to zero; also since we are interested in the transient behavior of the fitness level, the speed decay phenomenon does not play a significant role in our scenario [11]. Please note that our simulation of RWM is *not* a perfect simulation [12]. Indeed, we do not start the simulation with nodes distributed according to the steady-state distribution, but, rather, with a uniform one. This does indeed represent a pessimistic assumption, in that, as it may be easily understood, a uniform distribution of nodes over the area of interest is the distribution yielding the lowest probability of having nodes connected to one another (or, alternatively, the highest node isolation probability). Nonetheless, we believe that such an assumption leads to meaningful results in terms of comparison of the performance obtainable with BM and RWM. For the BM model, a billiard-like reflection was used when the mobile reached the edge of the domain. The initial fitness values are drawn from a set of i.i.d. random variables having uniform distribution in the interval $[0, 1]$. What we are interested in measuring is the *convergence time*, that will be defined in two ways. First, setting a threshold ξ (in the simulation, we will use $\xi = 0.95$), we want to measure the time it takes for the average fitness level to exceed ξ. Formally, $T_{conv}^{avg} = \min \left(t : \frac{\sum\limits_{i=1}^{N} I_i(t)}{N} \geq \xi \right)$. Then, we are interested in the time it takes for the all the fitness values to exceed ξ, i.e., $T_{conv}^{min} = \min \left(t : \min\limits_{i=1,\dots,N} (I_i(t)) \geq \xi \right)$. Clearly, $T_{conv}^{min} \geq T_{conv}^{avg}$. Further, the smaller such convergence times, the more efficient the evolution process and the ability of the service to adapt to rapidly changing environmental conditions. Indeed, while the framework outlined in the previous section was able to answer our questions regarding the stability (i.e., the steady-state) of the distributed evolution process, we did not get any quantitative result on the convergence time, that is what in reality impacts the user's perception of the service quality.

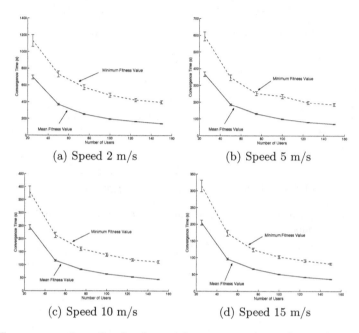

(a) Speed 2 m/s   (b) Speed 5 m/s

(c) Speed 10 m/s   (d) Speed 15 m/s

Fig. 1. Convergence times for the clone-and-mutate mating policy under the random waypoint mobility model

The first issue we want to address is which of the two optimal mating policies described in the previous section is able to achieve the faster convergence. The results, in terms of 95% confidence interval for the convergence times, for the RWM mobility model are reported in Fig. 1 and Fig. 2 for the clone-and-mutate and combine-and-mutate policies, respectively. It can be seen that the clone-and-mutate has, in general, quite lower convergence times, showing thus higher ability to adapt to changing environmental conditions. On the other hand, the combine-and-mutate is able to achieve interesting performance figures when dealing with high-density high-mobility scenarios, the most interesting cases for the pervasive environments we are targeting. The combine-and-mutate policy shows then to represent an interesting choice, and we are currently investigating whether more complex extensions of such a scheme can be actually used to speed up the service evolution process.

We also tested extensively the case of BM mobility model, which resulted in worse performance with respect to the RWM case. In Fig. 3 and Fig. 4 we

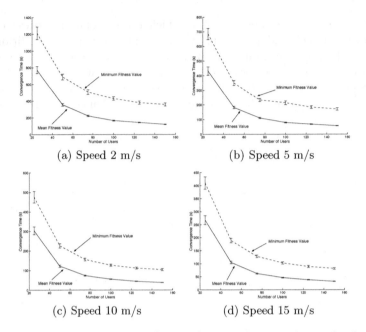

Fig. 2. Convergence times for the combine-and-mutate mating policy under the random waypoint mobility model

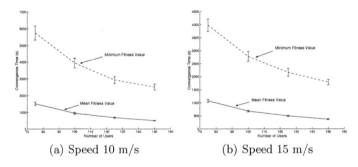

Fig. 3. Convergence times for the clone-and-mutate mating policy under the Brownian motion mobility model

reported the convergence times for the clone-and-mutate and combine-and-mutate policies. The results are reported, in terms of 95% confidence interval, only for the cases of $N = 75, 100, 125, 150$ users moving at a speed $v = 10, 15$ m/s, because of the extremely long convergence times under such a mobility model. As it may be easily seen comparing these results with the ones in Fig. 1 and Fig. 2, the RWM model is able to achieve much better performance (almost one order of magnitude), in terms of convergence time. This phenomenon is worth some comments. In general, it reflects the fact that the inter-meeting times in the RWM are smaller than in the BM (see [13] for an extensive and

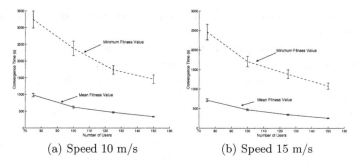

(a) Speed 10 m/s   (b) Speed 15 m/s

Fig. 4. Convergence times for the combine-and-mutate mating policy under the Brownian motion mobility model

in-depth discussion of such phenomena). However, the difference in the inter-meeting times statistics is not sufficient to completely understand the difference between the two. Indeed, an extensive analysis of the trace files shows that the convergence (in particular for the minimum value) is driven by a few nodes which keep isolated for a very long time. This comes from the well-known tendency of the BM model to "move around", without getting far from the initial location. Hence, if a node is very far from all the others in the initial distribution, it will take him a very long time before getting in contact with the rest of the population. Further, nodes will tend to remain in closed clusters; the evolution process remains localized inside the cluster and hence becomes much lower, since such a "local" evolution takes place over a smaller population. On the other hand, in RWM, nodes tend to pass through the center, and to meet more regularly with each other. In particular, a careful analysis of the trace files show that the time it takes for a node to get in touch with *all* other nodes is much lower in RWM than in the BM model. We might thus conclude that regularity in the traffic pattern helps in speeding up the convergence process.

4   Conclusions

In this work, we have proposed a framework for self-evolving autonomic services in a nomadic wireless environment. Freely drawing inspiration from the living world, we outlined a one-to-one mapping from biology to services, and apply the concept of evolution by adaptation to obtain truly autonomic services, able to self-optimize and to self-adapt to changing environmental conditions. The core of this evolution process is represented by the service mating policy. We outlined a general framework for studying service mating policies, and exploited results from martingales theory to assess the convergence properties of the resulting evolution process. We considered three policies and showed that two of them, the clone-and-mutate and the combine-and-mutate ones, are actually optimal, in that they are able to reach with unitary probability the optimal operating point. These two optimal policies have been widely compared through extensive numerical simulations, leading to interesting conclusions in terms of performance impact of parameters such as the number of nodes, the nodes speed and the mobility pattern.

In order to improve the present work, one direction of interest is to build a model for the transient behaviour of the evolution process, in order to get estimates (or, at least, bounds) for the convergence time of the fitness evolution process. Further, an open issue remains the characterization of the mating times sequence for different mobility models, in order to analyze the impact of the mobility pattern on the evolution convergence rate. Finally, we are currently investigating more complex extensions of the combine-and-mutate service mating policy, exploiting results in the area of GAs to find solutions able to speed up the convergence process in highly-dense highly-mobile networks.

References

1. M. Smirnov, "Autonomic communication: research agenda for a new communication paradigm," 2004, white Paper. [Online]. Available: http://www.autonomic-communication.org/publications/doc/WP_v02.pdf
2. J. O. Kephart and D. M. Chess, "The vision of autonomic computing," *IEEE Comp. Mag.*, vol. 36, no. 1, pp. 41–50, Jan. 2003.
3. I. Carreras, I. Chlamtac, H. Woesner, and H. Zhang, "Nomadic sensor networks," in *Proc. of EWSN*, Istanbul, Turkey, 2005.
4. M. Grossglauser and D. Tse, "Mobility increases the capacity of ad hoc wireless networks," *IEEE/ACM Trans. on Netw.*, vol. 10, no. 4, pp. 477–486, Aug. 2002.
5. I. Carreras, F. De Pellegrini, D. Miorandi and I. Chlamtac, "Information filtering in a 2–tier wireless sensor network," in *Proc. of Sensorfusion*, Budapest, HU, 2005.
6. I. Carreras, I. Chlamtac, H. Woesner, and C. Kiraly, "BIONETS: BIO-inspired NExt generaTion networkS," in *Proc. of WAC*, Berlin, DE, 2004.
7. D. Williams, *Probability with Martingales*. Cambridge, UK: Cambridge University Press, 1992.
8. I. Carreras, F. De Pellegrini, D. Miorandi and H. Woesner, "Service evolution in a nomadic wireless environment," CREATE-NET, Tech. Rep., 2005. [Online]. Available: http://www.create-net.org/~dmiorandi
9. AA.VV., "Omnet++." [Online]. Available: www.omnetpp.org
10. *Supplement to 802.11-1999,Wireless LAN MAC and PHY specifications: Higher Speed Physical Layer (PHY) extension in the 2.4 GHz band*, IEEE Std., Sep 1999.
11. J. Yoon, M. Liu, and B. Noble, "Random waypoint considered harmful," in *Proc. of IEEE INFOCOM*, San Francisco, CA, 2003.
12. J.-Y. Le Boudec and M. Vojnović, "Perfect simulation and stationarity of a class of mobility models," in *Proc. of IEEE INFOCOM*, Miami, FL, 2005.
13. R. Groenevelt, "Stochastic models in mobile ad hoc networks," Ph.D. dissertation, INRIA, 2005. [Online]. Available: http://www-sop.inria.fr/maestro/personnel/Robin.Groenevelt/Publications/Thesis.pdf

User Cooperation and Search in Intelligent Networks*

Erol Gelenbe

Dennis Gabor Chair,
Department of Electrical and Electronic Engineering,
Imperial College London SW7 2BT
e.gelenbe@imperial.ac.uk

Abstract. We present a vision of an Intelligent Network in which users dynamically indicate their requests for services, and formulate needs in terms of Quality of Service (QoS), duration, and pricing. Users can also monitor on-line the extent to which their requests are being satisfied. In turn, services will dynamically try to satisfy the users as best as they can, and inform the user of the level at which the requests are being satisfied, and at what cost. The network will provide guidelines and constraints to users and services, to avoid that they impede each others' progress. This intelligent and sensible dialogue between users, services and the network can proceed constantly based on mutual observation, network and user self-observation, and on-line adaptive and distributed feedback control which proceeds at the same speed as changes in traffic flows and the events occurring in the network. We survey some of the technical problems that arise in such networks, illustrate the networked system we propose via an experimental test-bed based on the Cognitive Packet Network (CPN), and discuss the key issue of search for users and services.

Keywords: Network Intelligence, Autonomic Networks, Users and Services, User Goals and Quality of Service, Cognitive Packet Networks.

1   Introduction

Sheer *technological capabilities and intelligence*, on their own, are of limited value if they do not lead to enhanced and cost-effective capabilities that are of value to human – or even beyond humans – to living users. Fixed and then mobile telephony and the Internet have been enablers for major new developments that improve human existence. However advances in telecommunications have also had some undesirable and unexpected outcomes during the past century. A case in point is television broadcasting. It was initially thought that television broadcasting would become a wonderful medium for education. Unfortunately in many instances it has lowered public expectations with regard to the quality of entertainment by limiting the range of programs and content that

---

* Research supported by UK EPSRC under Grant GR/S52360/01 and by the EU FP6 Marie Curie Programme under project MIRG-CT-2004-506602.

I. Stavrakakis and M. Smirnov (Eds.): WAC 2005, LNCS 3854, pp. 41–56, 2006.

are available. Interestingly enough, with few exceptions this effect has appeared both in socio-economic environments where television has been driven by purely commercial considerations, and in other environments where television broadcasting has been driven by monopolistic considerations. This is a good example of a tremendous success in technology which has not been exploited in the most broadly intelligent manner. Since the massively "one-to-very-many" broadcast nature of television has not given communities of users the possibility to dispose of high quality content, one can hope that other models of communications, such as the peer-to-peer concept, can offer a greater degree of user choice and also offer some very high quality content for an enhanced cultural and humanistic environment. Thus we envision Intelligent Networks to which users can ubiquitously and harmoniously connect to offer or receive services. We imagine an unlimited peer-to-peer world in which services, including television broadcasts, voice or video telephony, messaging, libraries and documentation, live theater and entertainment, and services which are based on content, data and information, are available at an affordable cost. In these networks the technical principles that support both the "users" and the "services" will be very similar if they are framed within an autonomic self-managing and self-regulating system. This network will be accessible via open but secure interfaces that are compatible with a wide set of communication standards, including the IP protocol.

We imagine an Intelligent Network (IN) in which users and services play a symmetric role: users of some services can be services of other users, and services can be users of some other services. Users and services can express their requests dynamically to the network in terms of the services that they seek, together with Quality-of-Service (QoS) criteria that they need, their estimate of the quantity or duration of the requested service and the price that they are willing to pay. The users could also have the capability to monitor on-line to what extent their requests are being satisfied. In turn the services and the network would dynamically try to satisfy the user as best as they could, and inform the user of the level at which their requests are being satisfied, and at what cost. The network would also provide guidelines to users to avoid that the latter impede each others' progress. Similarly, network entities and services would also conduct a dialogue, so that they can collectively and autonomously provide a stable, evolving and cost effective network infrastructure. We will sometimes find it useful to distinguish between users and services, merely to indicate the relationship that exists between a specific user requesting a specific service. But we wish to stress that at a certain level of abstraction, these two entities are indeed equivalent. The IN should offer the facilities for a sensible dialogue between all users, including services, and it will adapt to users' needs based on mutual observation, network and user self-observation, and on-line distributed feedback control which acts in response to the events that are being controlled.

1.1 Research Issues

The vision we have described raises many interesting research questions, some of which are discussed below. An obvious research question relates to the network architecture that can physically support the vision we have presented, in particular with respect to system software. We would expect that the IN would be some form of self-managing and programmable overlay network [14, 15] which is discussed in the next Section. The underlying hardware architecture would have to rely on network components, wired or wireless technologies and line speeds that are available at any given point in time, while routers should be programmable and they should not be limited to some pre-defined protocol such as IP. More research is also needed to better identify services and their characteristics, and the technologies which are necessary to support applications in a wide range of heterogeneous networks, with possible feeders coming in from the wide-spread networked sensors of the future. Research is also needed to design network modules whose role will be to recognise and match user needs to the networking context.

Although much excellent work has been done about QoS provisioning, QoS based routing mechanisms, and service differentiation [7, 8, 12], there is still much more that needs to be done in defining broad QoS metrics that are relevant to the end user, and seeing how these translate into mechanisms and policies that exploit the available variability including traffic engineering, and routing and searching methods [10, 11], so that both the users' needs and the networks' objectives can best be met. Understanding the interplay between cooperative or conflicting interests among different users and networks, including issues such as resource utilisation, provisioning, pricing and QoS [8, 9] has received considerable attention. A recent paper provides valuable insight and ideas on some of these issues [18]. Our impression is that a systematic approach to realistic modelling of the dynamic interplay between these different issues can still be of great value to a better understanding of network control. Furthermore we believe that the game theoretic ideas that have been developed should also lead to more experimental research, testing and evaluation in realistic environments, or in large-scale network test-beds.

Although there have been many studies that characterise network traffic, and considerable work has akready been done on various aspects of network observation such as network tomography, we suggest that further research is needed on approaches to network measurement whose primary objective is the real-time control of network performance, of traffic engineering, or of user QoS [17]. Resource provisioning in networks can of course be handled in an *a priori* manner. However when resources are tight, or when the networked environment is imperfectly known, or when the users are accessing very diverse services and resources, or when users, services and network resources are mobile, then the network state can only be observed by real-time measurements of the parameters that are of direct interest. Thus research which combines QoS considerations, with network control and measurement, appears to be of interest [16, 19].

Another important area of research is the design of algorithms that can help users or the network itself to discover and find "things" such as nodes, services, resources, etc. in very large, or even infinite, networked systems. Search problems have long been examined in artificial intelligence and robotics, as well as in the context of combinatorial mathematics [6, 13]. It appears to us that this topic is of increasing importance, because users will operate increasingly in ad hoc networks, sensor networks, or more broadly in large and unknown networks where they will have to discover the best connections, services and modes of operation. For instance, a user may have to connect his/her terminal device automatically to some network in a city that he/she has never visited before and about which he/she only has very sketchy information.

1.2   An Architecture for the Intelligent Network

A sketch of the Intelligent Network (IN) is shown in Figure 1. The IN will based on a standard communication interface derived from the Internet Protocol (IP). Users U (shown with small purple rectangles as U1, U2, etc.) are generally mobile and can be recognised via their ID and password. Users have a credit with the network and with certain network services, as represented by a credit allocation or via a "pay as you go" scheme (e.g. with a credit card), or they can access certain free services or services that may be paid for by the service provider (e.g. advertisements). Users can have a user terminal which may be as simple as a Personal Digital Assistant or mobile phone, or as complex as intelligent network routers (INRs) shown as blue octagons in Figure 1. Users are connected to the IN via INRs or directly to a network cloud (shown as clouds of different colours). Services S (shown as S1, S2, ..) are very similar to users in that they have an ID

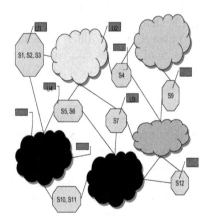

Fig. 1. Architecture of the Intelligent Network

and they may have a credit allocation; they can also receive credit when their services are used by users, just as users may be reimbursed by services or by other users. Services can also be mobile. However:

- Users will in general be light-weight (a mobile phone, a PDA, or just a user ID and password),
- While services will be much more complex and may often be resident on one or more INRs, or they may own one or more INRs for their needs.

When some other user or service asks something of a user, the chances are that there will be an automatic answer saying "sorry no; I am just a simple user". On the other hand, services will often be equipped with authentication schemes to recognise the party who is making a request, billing schemes that allow for payment to be collected, schemes allowing a service to be used simultaneously by many users, and so on, depending on the complexity of the service being considered.

INRs are machines or clusters which can be identified by the community of users and services. Network clouds on the other hand are collections of routers internally interconnected by wire or wireless and which are only identified as far as the users and services are concerned via the ports of INRs which are linked to a cloud; in other words, users and services do not actually know who and what is inside a network cloud. However INRs, and hence users and services, can observe the QoS related to traversing a network cloud; this may include billing of the transport service by the cloud. Also, clouds may refuse traffic, or control and shape the traffic that wishes to access them, depending on the clouds own perception of the traffic.

The IN architecture we have described can be viewed as an overlay network composed of INRs with advanced search, QoS (including pricing and billing), that links different communities of users and services. The networked environment of the future will include numerous INs, and there may be specific INs whose role is to find the best IN for a given user. Some of the se INs may be quite small (e.g. a network for a single extended family), while others would be very large (e.g. a network that provides sources of multimedia entertainment, or educational content). In the three following sub-sections we will discuss three important enabling capabilities of the system: finding services and users, routing through the network, and self-observation and network monitoring to obtain the best QoS and performance.

2 Finding Services and Users

We expect that the IN will have different free or paying directory services that will be used to locate users and services. When appropriate, these directories may provide a "street address and telephone number" for a service that is being sought out; however, since in many cases the services will have a major virtual component, they will especially provide a way to access them virtually, either via an IP address, or more probably via one or more INR addresses or one or more network paths.

The directory services will offer "how to get there" information similar to a street map service, providing a network path in terms of a series of INRs or of network clouds, from the point where the request is made, to the INR where the service can be found. Directory services may have a billing option which is activated by services to reward the directory for being up-to-date, or services or users can subscribe to them, or they may be paid for via advertisement information, and so on. These directories will be updated pro-actively by the services or by the directories themselves, or on demand when the need occurs. Updates would also occur when INR or network cloud landmarks change. Directories can be "smart" in the sense that they offer information about faster or less congested paths to services that are requested, or paths to less expensive services, or paths that are better in some broader sense. An approach for achieving this based on the Cognitive Packet Network (CPN) [10, 20] protocol is described in Section 3.

As a way to understand how a connection can be established between some user U and a service S in the IN, let us go through some of the steps that may be involved, using smart packets (SPs), based on ideas from [17, 20] and some extensions of these ideas.

- U first searches for a directory; assuming he finds one, U formulates his request in the form of (SX, QY, PZ) meaning that he wants a service SX at QoS value QY for a price of PZ. The directory either is unable to answer the request, or it provides one or more paths $\pi(U, SX, QY, PZ)$ which best approximate this request for several possible locations of the service.
- Assuming that the directory does provide the information, U sends out (typically via the INR) a sequence of smart packets SPs which have the desired QoS information, with several following each of the possible designated paths. The first SP for each of the paths will follow it to destination, with the purpose of verifying that the information provided by the directory is correct. Subsequent SPs on each route will be used to search for paths: they will invoke an optimisation algorithm at all or some of the INRs they traverse so as to seek out the best path with respect to the user's QoS and pricing requirements.
- INRs collect measurements and store them in mail boxes (MB). These can concern both short term measurements which proceed at a fast pace comparable to the traffic rates, and long term historical data. INRs will measure packet loss rates on outgoing links and on complete paths, delays to various destinations, possibly security levels along paths (when security is part of a QoS requirement), available power levels at certain mobile nodes, etc.. This constant monitoring can be carried out using the SPs and other user related traffic, or using specific sensing packets generated by the INRs.
- The network monitoring function can also be structured as a special set of users and services whose role is to monitor the network and provide advice to the users and to the directories.
- Each SP also collects measurements from the INRs it visits which are relevant to its users QoS and cost needs, about the path from the INRs which it visits.
- When a SP reaches a service SX, an acknowledgement ACK packet is sent back along the reverse path back to U; the ACK carries the relevant QoS information, as well as path information which was measured by the vSP and by the ACK, back to the INRs and to the user U. The ACK may thus be carrying back a new path which was unknown to the directory.
- For a variety of reasons, both SPs and ACKs may get lost. SPs or ACKs which travel through the network over a number of hops (ERs or total number including routers within the clouds) exceeding a predetrmined fixed

number, will be destroyed by the routers to avoid congesting the IN with "lost" packets.

— Note that the SPs and ACKs may be emitted by the directory itself, rather than by U. This would be an additional service offered by certain directories. One could also imagine that both users and directories have this capability so as to verify that the request is being satisfied.

Some of these features are illustrated in the CPN (Cognitive Packet Network) system [17, 19, 20] test-bed that we have implemented at Imperial College.

2.1 Individual Versus Collective QoS Goals

The usual question that any normally constituted telecommunications engineer will ask with respect to the vision that we have sketched is what will happen when individual goals of users and services conflict with the collective goals of the system. We are allowing for users to set up the best paths they can find, from a selfish perspective, with services, and for services to actually do the same, in parallel with the behaviour of users. This has the potential for:

— Overloading the infrastructure, because services have an interest in maximising their positive response to user's needs, and they may even overdo it in terms of solliciting users; because of the possibility of billing, portions of the infrastructure itself may have an interest in getting overloaded.
— Creating traffic congestion and oscillations between hot spots, as users and services switch constantly to a seemingly better way to channel their traffic.
— Opening the door to malicious traffic whose sole purpose may be to deny service to legitimate users through the focused creation of overload in the services or the infrastructure (e.g. denial of service attacks).

The first of these points, which does not relate to malicious behaviour, can be handled through overall self regulation of the INRs, the users and services:

— When a new part of the infrastructure joins the IN, for instance a INR, it will be allocated an identity within the IN. We could have a virtual regulating agency (VRA) which sets up a dialogue with the INR to provide it with its identity, and which ascertains its type and nature from its technical characteristics. The VRA then enables the INRs operating systm with a set of parameters which in effect limit the number of resident processes and the amount of packet traffic that this particular INR can accept.
— Services and users which join the IN, also need to be identified by the VRA. Just as a shop rents a certain space in a building and on a particular street, the VRA can provide the service with a "footprint", depending on the rent it is willing to pay, and on the VRA's knowledge of currently available resources. This footprint can then determine the fraction and amount of processing power and bandwidth that it is allowed inside the IN and at any given INR.
— Note that the overall quality and seriousness of the VRA will make a particular IN more or less desirable to users and services.

The second point is related to dynamic behaviour. Each INR, in its role as a service support centre enabled by the VRA, will run the dynamic flow and workload control algorithms for each service and user that it hosts. However it will also run a monitoring algorithm which has IN-wide implications.

- For some user U assume that $RU(S)$ is the rank ordered set of best instantaneous choices for some decision (e.g. what is the best way to go to service S with minimum delay).
- At the same time, let $RN(U, S)$ be the rank ordered set of best instantaneous choices for the network (e.g. what is the best way to go to where service S is "sitting" so that overall traffic in the IN is balanced).
- The decision taken by the INR will be some weighted combination of these two rank orders. The weights can depend on the priority of the user, of the price it is willing to pay, and so on.
- Choices which are impossible or unacceptable to either of the two criteria (user or network) will simply be excluded. If there are no mutually possible choices, then the request will be rejected. When there are ties between choices, any one of the tied choices can be selected at random.

As an example, suppose that the ranking indicating the user's preference, in desecending order, among six possible choices is $\{1, 2, 3, 4, 5, 6\}$, while the network's preference ranking could be $\{5, 4, 2, 3, 1, 6\}$. If we use rank order as the decision criterion and weigh the INR and the user equally, then the decision will be to choose 2 whose total rank order is 5. If the network's role is viewed as being twice as important, we can divide the network's rank for some choice by 2 and add the resulting number to the rank that the user has assigned to that choice, which results in a tie between the three top choices $\{1, 2, 5\}$. If the network's role is three times more important, then we get a tie for the top choice between $\{1, 5\}$, and so on.

2.2   The Eternal Problem of Scalability

It is often said that the main impediment to the broad use of QoS mechanisms in the Internet is the issue of scalability. Indeed, if each Internet router were enabled to deal with the QoS needs of each connection, it would have to identify and track the packets of each individual connection that is transiting through it. The routing mechanism we propose for all requests through the IN is based on dynamic source routing[1]. In other words, the burden of determining the path to be used rests with the INR that hosts the service or user. In our proposed scheme, routers have two roles:

- The INR generates SPs for its own use that monitor the IN as a whole, and the user or service process resident at a INR generates the SPs and ACKs which are related to its connections to monitor their individual traffic.

---

[1] Note that MPLS is a form of distributed virtual source routing where label switching at each node maps virtual addresses into physical link addresses.

- As a result of the information that it receives from SPs and ACKs, of the information similarly received by users and services that are resident at the INR, and of the compromise between global (IN) and local (user and service) considerations, the INR generates source routes for its resident users and services.
- Each INR also provides QoS information to SPs and ACKs that are not locally generated but which are transiting through it, such as "what is the loss rate on this line", or "what time is it here now", or "what is the local level of security".

Thus we propose to avoid the scalability issue by making each INR responsible only for local users and services, much as a local telephone exchange handles its local users. Source routing removes the burden of routing decisions from all but the local INR, reducing overhead, and removing the need of "per flow" information handling except at INRs where the flows are resident. However, it comes at the price of being less rapidly responsive to changes that may occur in the network. This last point can be compensated by constant monitoring of the flow that is undertaken with the help of SPs and ACKs. Our scheme also requires that INRs be aware of the overall IN topology in terms of other INRs (but there is no need to know what is inside the "clouds"), although this can be mitigated if one accepts the possibility of staged source routing, i.e. with the source taking decisions up to a given intermediate INR, which then takes decisions as far as some other INR, and so on. Note that this scheme is more general than the one we will describe in the next section which consists of an experimental system that discovers destination nodes, and paths to destination nodes which optimise user specified QoS metrics.

3   Searching and Routing with CPN

Distance Vector and Link-State algorithms are the usual methods for finding the shortest paths to IP addresses in the Internet's Routing Information Protocol (RIP) [7], where a table in each router stores information for each destination in a sub-network with a preferred outgoing link from the node, and an estimate of the time and hop count to the destination. These metrics are updated at regular intervals, or when the network topology changes, via router update messages. This allows each router to update its database with the fastest route being communicated from neighboring routers. However, many factors including non-negligible delay, infrequent link state update due to overhead concerns, and the link state update policy can impact global network state information.

CPN [20] is a distributed algorithm that provides QoS driven routing based on searching for the best path to a given destination. CPN searches for the destination *and* searches for the best path leading to it. It uses *Smart or Cognitive Packets (SP)* that discover routes using a reinforcement learning (RL) algorithm based on a QoS "goal" such as packet delay, loss, hop count, jitter, etc.. The "goal" may be defined by the user, or by the network itself. SPs find routes and collect measurements, but do not carry payload. The RL algorithm uses the observed outcome of a previous decision to "reward" or "punish" the mechanism that lead to the previous choice, so that its future decisions are more likely to meet the QoS goal. When a SP arrives to its destination, an ACK packet is generated; the ACK stores the "reverse route" and the measurement data collected by the SP. It will travel along the "reverse route" which is computed by taking the corresponding SP's route, examining it from right (destination) to left (source), and removing any sequences of nodes which begin and end in the same node. For instance, the path $< a, b, c, d, a, f, g, h, c, l, m >$ will result in the reverse route $< m, l, c, b, a >$. Note that the reverse route is not necessarily the shortest reverse path, nor the one resulting in the best QoS. Finally, *Dumb Packets (DP)* carry payload and use dynamic source routing. The route brought back by an ACK is used as a source route by subsequent DPs of the same QoS class having the same destination, until a newer AND/OR better route is brought back by another ACK. A *Mailbox (MB)* in each node is used to store QoS information. Each MB is organized as a Least-Recently-Used (LRU) stack, with entries listed by QoS class and destination, which are updated when an ACK is received. The QoS information is then used to calculate the reward in the SP routing algorithm. We use recurrent random neural networks(RNN) [5] with reinforcement learning (RNNRL) in order to implement the SP routing algorithm. Each output link of a node is represented by a neuron in the RNN. The arrival of *Smart Packets(SPs)* triggers the execution of RNN and the output link corresponding to the most excited neuron is chosen as the routing decision. The weights of the RNN are updated so that decisions are reinforced or weakened depending on how they have been observed to contribute to the success of the QoS goal. The RNN is an analytically tractable spiked random neural network model whose mathematical structure is akin to that of queuing networks. It has "product form" just like many useful queuing network models, although it is based on non-linear mathematics.

The experimental results concerning search in CPN that we presently from a recent paper [21] use the test-bed consisting of 17 nodes shown in Figure 2. Each pair of INRs is connected by point-to-point 10Mbps Ethernet links. All tests were performed using a flow of UDP packets entering the network at constant bit rate (CBR) with $1024B$ packets. Each measurement point is based on $10,000$ packets that were sent from the source to the destination, and we inserted random background traffic into each link in the network with the possibility of varying its rate. The CPN routing algorithm is used throughout the experiments using three

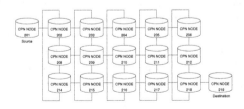

Fig. 2. The test-bed topology used in the experiments

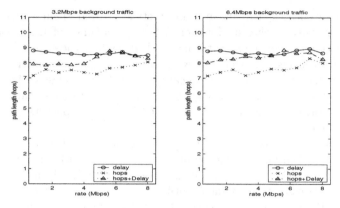

Fig. 3. Path length comparison

different QoS goals: (a) delay [*Algorithm-D*], (b) hop count [*Algorithm-H*] and
(c) the combination of hop count and forward delay [*Algorithm-HD*]. Measurements concern average hop count, the forward delay and packet loss rate under
different background traffic conditions. From Figure 2, we see that the shortest path length from the source node (#201) to the destination node (#219) is
7, and there are only five distinct shortest paths. For example, one of them is
route $\langle 201 \to 202 \to 214 \to 215 \to 216 \to 217 \to 218 \to 219\rangle$. Figure 3 reports
the average number of hops traversed from source to destination when different
algorithms are used. When hop count is used as the QoS goal, we see that the
average number of hops under different background traffic conditions is close to
the minimum of 7. It is interesting to observe that whenn the forward delay is
used as the QoS goal, the average number of hops actually used is no longer the
minimum number. We see that the average path length is close to 7 when the
connection's traffic rate is low or medium, and close to 9 hen it is high. With
respect to path length the experiments confirm our expectations: *Algorithm-H*
is the best, and *Algorithm-HD* is better than *Algorithm-D*. These results show
that one need not use fixed non-adaptive algorithms for routing; it is possible to
find shortest paths adaptively without fixed prior knowledge. The comparison of
an adaptive algorithm using the number of hops, or the path delay, or a mixture
of the two as the way to select paths, also provides some interesting insight.
Note however that in our case we are not always using the *same* shortest path,
since the adaptive routing algorithm will be able to vary the paths it is using
even when it is instructed to use find the shortest path. As a result, even the
adaptive shortest path algorithm should be able to improve observed QoS over a
fixed shortest path, since it will distribute traffic over a larger number of paths.

4   Evaluating the Search Time

As mentioned earlier, searching for objects, geographic locations, data, and so
on, is becoming of fundamental importance in all areas of networking. The sheer
number of different services, nodes and networks, and the mobility and variability

of all of these entities, will make it impractical to find them via some fixed addressing or routing scheme. However, other approximate characteristics such as their location or movement patterns can help us find the objects we seek. If one deals with routing in a packet network, one can augment an internet address with some information about its physical location. In a wireless ad hoc network and in a wireless sensor network, the physical location of nodes is an important element of information when one tries to convey data or packets to or from a geographic area.

For instance, if you consider two physical locations $o = (0,0)$ and $d = (N, M)$ in the integer valued (x, y) plane, in an unconstrained routing scheme (i.e. one that allows the packet to go to any neighbouring point which it can reach) the packet progresses from point o with the aim of reaching d, and at some intermediate time it reaches a location (x, y). Its next step will be to move to one of its reachable and available neighbours, and it will prefer a neighbour which is in the direction of the destination d. An exact probabilistic or combinatorial analysis of the time it would take the packet to reach d from o appears difficult. Thus in this section we develop a model that represents the search process so as to estimate the time it would take to find a destination, or more generally an object, in a search space. The model will be based on a continuous space and time diffusion process.

For the migration process of a packet from o to d that begins at $t = 0$, what matters is the distance of the packet to its destination at some time $t > 0$. This distance will be represented by the real valued stochastic process $Y = \{Y(t) : t \geq 0\}$, where $Y(0) = D$, $D = ||d - o||$. Quantities of interest include:

- $T_1 = inf\{t : Y(t) = 0\}$, the time it takes the packet to reach its destination,
- $\Pi(\delta) = P[Y(t) > \delta, 0 \leq t < T_1]$ for $\delta > D$, the probability that the packet has gotten too far away from its destination, and
- the probability $\pi(\epsilon) = P[Y(t) < \epsilon), 0 \leq t < T_1]$ that the packet is within an $\epsilon - neighbourhood$ of its destination.

To simplify the analysis, we replace the transient process Y by an ergodic process Z which will allow us to compute all these quantities of interest.

Consider the process $Z = \{Z(t) : t \geq 0\}$ which is identical in sample path to Y until time T_1. After T_1 the process Z will reside at the point $z = 0$ for a random time H_1, after which Z jumps to point D and then stochastically repeats its previous behaviour. Let H_i, $i = 1, 2, ..$, be independent and identically distributed positive random variables, and let $T_{i+1} = inf\{t : T_{i+1} > T_i, Z(T_{i+1}) = 0\}$ for $i = 1, 2, ...$ Then Z has the renewal property:

$$P[Z(t) > z] = P[Z(t + T_i + H_i) > z] \tag{1}$$

for any $t \geq 0$, $z \geq 0$, and the instants $\{T_i + H_i\}$ are renewal instants of the process for $i \geq 1$. The random behaviour of the search packet can be represented by a Brownian motion in one dimension, where the distance of the search packet to its destination is the dimension being considered[1, 2]. However, in order to take into account the holding time at the boundary $z = 0$ described above, the process Z

will be represented as a Brownian motion [4] which is modified to have holding times at the boundary $z = 0$ and jumps to interior points as suggested in [3]. Thus in addition to the usual diffusion equation, the process we consider will have a discrete (i.e. not continuous) component as described below. The diffusion process representing the distance of the search packet, or of the searcher, from its destination at time t is defined as follows:

- We assume that on the average each move of the packet gives preference to a direction which reduces the distance to the destination. This is a reasonable assumption because, if the medium is isotropic, it is natural for the packet to select a direction of motion which brings it closer to its destination whenever it can. However, either for lack of knowledge or because other options are impossible, a move may sometimes increase its distance to the destination. Thus the drift parameter b of the diffusion is negative. In the sequel we will also consider the case where $b = 0$, which corresponds to "ignorance on the average".
- Its second moment parameter is some finite quantity $c \geq 0$. Note that $c = 0$ represents the case where the time duration of each step from neighbour to neighbour is constant, while a large value of c would imply a more erratic search process.
- Suppose that at some time $t = T_i$ the destination is reached, i.e. $Z(T_i) = 0$; then after a random time H_i we assume that the search process starts again, so that at $t = T_i + H_i$ the process Z jumps back to the starting point of the search and $Z(T_i + H_i^+) = D$, and the search process is re-initialised. This process repeats itself indefinitely.
- Without loss of generality with respect to the computation of $E[T]$, we will assume that $P[H_i > v] = e^{-v}$, for $v \geq 0$, so that $E[H_i] = 1$.

Result 1. Let $E[T] = E[T_i]$ for any i, and let

$$P = \lim_{t \to \infty} P[Z(t) = 0]. \tag{2}$$

Then:

$$E[T] = \frac{1}{P} - 1. \tag{3}$$

Sketch of Proof. This follows from the fact that the process $\{X(t),\ t \geq 0\}$ defined by $X(t) = 1[Z(t)]$ is a two state (0 and 1) semi-Markov process, where $1[y]$ is the characteristic function $1[y] = 1$ if $y > 0$, and $1[y] = 0$ otherwise.

We will skip the details of the representaton of the process Z which is based on the equations that the probability density function $f_{z,t}dz = P[z < Z(t) \leq z + dz]$, $z > 0$, $t \geq 0$ must satisfy. We will just summarise the main analytical results that we have obtained:

Result 2. The average search time is given by the expression:

$$E[T] = \frac{D}{-b} \tag{4}$$

This result assures us about the feasibility of a search: it tells us that as long as $b < 0$ then the search time will be finite on the average. Although it has a very simple and intuitive form (note that we have assumed that b is negative), it does tell us that the average time it will take for the search to be successful is simply the distance D to the destination, divided by the average distance traversed per unit time. Thus a detailed knowledge of the way the movement occurs towards the destination (e.g. the second moment of the distance traversed per unit time) is not needed. However, we will see that if packets are subject to a time-out so that they self-destroy if they have been travelling for too long a time, then average time to reach the destination will also depend on the second moment of the diffusion process.

In many cases, some mechanism will be incorporated into a search packet so that it is destroyed if it has meandered for too long a time, or too far away, and has not found its destination. We incorporate this property in the diffusion model, so that at any distance z from the origin, $r(z)dt$, with $r(z) \geq 0$, is the probability that the search packet is destroyed in the interval $[t, t + dt[$ when it is at distance z from its destination. We can now use a similar artifact as previously to compute $E[L]$ which is the new value of the average time it takes the packet to find its destination. The artifact now is:

- As before, after the search packet reaches its destination, wait for an exponentially distributed random time of average value one and then generate a new search packet so as to re-start the search process, and
- Generate a new search packet immediately after it is destroyed by the time-out.

Result 3. Assuming an exponentially distributed time-out of average value λ^{-1}, and $b < 0$, the average time for a search packet to reach its destination is given by

$$E[L] = \frac{1}{P} - 1 \tag{5}$$

$$= \frac{-2D}{b - \sqrt{b^2 + 2\lambda c}}.$$

Notice from (6) that, contrary to (3), the average time that a packet reaches destination now depends on the variance parameter c of the diffusion process. Furthermore, when $\lambda = 0$, i.e. when the time-out is in effect removed, we revert as expected to *Result 2* given in (3). Notice also that if $\lambda > 0$ and $b \to 0$, then:

$$E[L] = D\sqrt{\frac{2}{\lambda c}}, \tag{6}$$

which says that even though each step of the search does not, on the average, get the search packet closer to the destination, the fact that we use the time-out mechanism does allow us to get to the destination in a time which is finite on the average due to the repeated usage of the time-out. Furthermore the expression in (6) can also be used as an approximation when $2\lambda c >> |b|$ and $b \leq 0$.

5   Conclusions

We present an architecture for autonomic networks which offers a universal peer-to-peer communication environment for users and services, composed of Intelligent Network Routers capable of supporting the user and service needs. The network allows users to sense and adapt network paths, and identify user to service connections, dynamically as a function of network state and of user and service quality of service needs. The architecture uses smart packets for the search for services, and for on-line dynamic sensing and control. These ideas are extrapolated from an experimental test-bed for QoS driven network routing, the CPN system, which is based on similar concepts with completely decentralised control. We then study the search process itself in order to estimate the time it would take to find another user or a service in the network which was initially at distance D from the object conduction the search. We assume that the search is conducted with a search packet which moves through the network. We model the search via the distance to the destination, some time t after the search begins. Closed form analytical results are derived for the average search time as a function of the initial distance from the point from which the search is being initiated, to the point where the object being looked for is to be found. We consider the case where time-outs are used to destroy packets that have been in the network for too long without reaching their destination, and are then replaced with fresh packets, as well as the case where time-outs are not used.

References

1. A. Einstein. Investigations on the Theory of Brownian Motion. Dutton, New York, 1926, reprinted by Dover, New York, 1956.
2. W. Feller. An Introduction to Probability Theory and its Applications, Vols. I and II. Wiley, New York, 1966.
3. E. Gelenbe. On approximate computer system models. *Journal ACM*, 22 (2), pp. 261-269, April 1975.
4. J. Medhi. Stochastic Models in Queueing Theory. pp. 373 ff. Academic Press, New York, 1991.
5. E. Gelenbe. Learning in the recurrent random neural network. *Neural Computation*, 5 (1), pp. 154–164, 1993.
6. S. Alpern. The rendezvous search problem. *SIAM J. Control & Optimization* 33, 673-683, 1995.
7. D. Williams and G. Apostolopoulos. QoS Routing Mechanisms and OSPF Extensions. RFC 2676, Aug. 1999.
8. H. Yaiche, R.R. Mazumdar and C. Rosenberg. A game theoretic framework for bandwidth allocation and pricing in broadband networks. *IEEE/ACM Transactions on Networks* 8 (5), pp. 667-678, 2000.
9. N. Semret, R. R.-F. Liao, A. T. Campbell and A. A. Lazar Pricing, provisioning and peering: dynamic markets for differentiated Internet services and implications for network interconnections. *IEEE J. Sel. Areas Comms.* 18 (12), pp. 2499-2513, 2000.
10. E. Gelenbe, R. Lent and Z. Xu. Measurement and performance of a cognitive packet network. *Computer Networks*, 37, pp. 691-791, 2001.

11. E. Gelenbe, R. Lent, and Z. Xu. Cognitive Packet Networks: QoS and performance. *Proc. IEEE MASCOTS Conference*, ISBN 0-7695-0728-X, pp. 3-12, Fort Worth, TX, Oct. 2002.
12. N. Christin and J. Liebherr. A QoS architecture for quantitative service differentiation. *IEEE Comms. Mag.* 46 (6), pp. 38–45, 2003.
13. S. Alpern and V. Baston. A common notion of clockwise helps in planar rendezvous. Rendezvous on a Planar Lattice. *CDAM Research Report Series 2004-7*, Centre for Discrete & Applicable Mathematics, London School of Economics, 2004.
14. Autonomic Computing Initiative, http://www.ibm.com/autonomic/
15. A. Galis, S. Denazis, C. Brou and C. Klein (eds). Programmable networks for IP service deployment. Artech House Books, ISBN 1-58053-745-6, 2004.
16. A. Asgari, R. Egan, P. Trimintzios and G. Pavlou. Scalable monitoring support for resource management and service assurance. *IEEE Network* 18 (6), pp. 6–18, 2004.
17. E. Gelenbe, M. Gellman, R. Lent, P. Liu, Pu Su. Autonomous smart routing for network QoS. *Proc. First International Conference on Autonomic Computing*, (IEEE Computer Society), ISBN 0-7695-2114-2, pp. 232-239, May 17-18, 2004, New York.
18. S.K. Das, H. Lin and M. Chaterjee. An econometric model for resource management in competitive wireless data networks. *IEEE Network* 18 (6), pp. 20–26, 2004.
19. E. Gelenbe, R. Lent, A. Nunez. Self-aware networks and QoS. *Proceedings of the IEEE*, 92 (9), pp. 1478-1489, 2004.
20. E. Gelenbe. Cognitive Packet Network. *U.S. Patent No. 6,804,201 B1*, Oct. 12, 2004.
21. E. Gelenbe and P. Liu. Qos and routing in the cognitive packet network. To appear in *Proceedings of the IEEE International Symposium on a World of Wireless, Mobile and Multimedia Networks*, Jun 2005.

Autonomic Wireless Network Management

Kai Zimmermann, Sebastian Felis, Stefan Schmid,
Lars Eggert, and Marcus Brunner*

NEC Europe Ltd., Network Laboratories,
Kurfürstenanlage 36, 69115 Heidelberg, Germany
{zimmermann, felis, schmid, eggert, brunner}@netlab.nec.de
http://www.netlab.nec.de/

Abstract. This paper presents a decentralized approach for the autonomic management of a group of collaborating base stations to provide efficient and effective wireless network access in highly dynamic environments. It provides a
management platform that supports many different management functions
based on common mechanisms for information exchange, transactional semantics and security. A central feature of the system is the inclusion of monitored
feedback information into the autonomic management process, which can enhance the operation of the management system and the quality of its decisions.
An integrated monitoring component provides this feedback information by
monitoring the coverage area and analyzing the measurements in real time. A
preliminary evaluation of the prototype implementation shows that the autonomic management system scales well. Performance is mostly proportional to
the diameter of the network topology and does not heavily depend on the number of base stations present. Further experiments with the wireless monitoring
sub-system demonstrate that it is feasible to automatically detected network
problems caused by radio interference or active attacks.

1 Introduction

Installations and configurations of large wireless networks that consist of multiple,
distributed base stations are challenging, time-consuming and error-prone tasks, even
for experts. Once deployed, such wireless networks require continuous management
to provide a uniform service environment, recover from faults or maximize overall
performance. This is particularly difficult, because wireless environments are typically very dynamic. First, the number, location and traffic patterns of mobile systems
in a wireless network change constantly. Second, wireless networks often use unlicensed, shared frequency spectrum, such as IEEE 802.11b/g base stations that operate
in the unlicensed 2.4 GHz band. Multiple radio applications share this unlicensed
spectrum in an uncoordinated fashion, which causes additional interference on top of
outside interference caused by other electronic equipment.

---

* The authors are partly funded by *Ambient Networks*, a research project supported by the
European Commission under its *Sixth Framework Program*. The views and conclusions
contained herein are those of the authors and should not be interpreted as necessarily representing the official policies or endorsements, either expressed or implied, of the *Ambient
Networks* project or the European Commission.

I. Stavrakakis and M. Smirnov (Eds.): WAC 2005, LNCS 3854, pp. 57–70, 2006.
© IFIP International Federation for Information Processing 2006

To enable effective and efficient networking under these demanding characteristics, continuous network management that proactively and reactively adapts to environmental dynamics is a necessity. Manual network management techniques across a set of diverse, distributed base stations are consequently not an option, requiring fully automated management functionality.

Few wireless network technologies include adequate management mechanisms. Even when such functions exist, they are typically limited to managing physical or link-specific characteristics only and do not cover management of higher-layer inter-networking functions. For example, although existing IEEE 802.11a/b/g WLAN base stations can automatically select an available radio channel, transmission power and link speed, they cannot autonomously configure higher-layer settings such as routed IP connectivity. They cannot even intelligently configure link-specific characteristics that require coordination between neighboring base stations beyond what they can immediately observe themselves.

This paper describes an autonomic approach to the management of wireless base stations. The advantage of an autonomic solution is that new base stations that join an existing wireless network integrate themselves seamlessly. The rest of the system adapts to their presence dynamically. This enables a wireless network to automatically configure itself in accordance to high-level policies that specify what is desired, not how it is accomplished. These policies represent the purpose of the network, its overall goals or business-level objectives.

A key principle of autonomic self-configuration is decentralization. Each component is able to operate in a stand-alone fashion. When several components detect each others presence, they then start to coordinate their management actions to increase the efficiency and effectiveness of their shared access network.

Existing approaches to management of wireless networks consisting of multiple base stations are typically centralized. A central master system periodically uses available information to compute a global configuration for the whole network. It pushes this configuration out to the individual base stations in a piecemeal fashion or they pull their respective configurations from the master. However, such a centralized approach has several disadvantages. First, it creates a central point of failure. Failure of the master can make the whole system unusable. Second, a central master limits scalability due to processing and communication overheads, especially in environments that require frequent configuration changes. Third, it complicates the system, because this approach introduces additional infrastructure, *i.e.*, the central master.

This paper presents a decentralized approach for autonomic management of a group of collaborating base stations. The individual base stations aggregate and share network information. They implement a distributed algorithm that uses the shared information to compute a local configuration at each base station such that the overall network-wide configuration is consistent. An important feature of the proposed system is the wireless monitoring component, which provides the necessary feedback for the autonomic logic to take appropriate management decisions.

Section 2 of this paper presents related work. Section 3 defines the underlying autonomic principles that guide the design of the wireless management system. Section 4 describes the basic functionality and operation of the proposed autonomic management system. Section 5 presents the evaluation results of the prototype systems and Section 6 summarizes and concludes this paper.

2 Related Work

Management of wireless networks is possible through centralized, distributed or hybrid solutions. Whereas centralized systems use a single master device to configure the base stations, decentralized, distributed solutions avoid such a single point of failure and collaboratively implement a fully distributed management solution. With any of the three approaches, the challenge is that all wireless base network stations must arrive at a consistent, system-wide configuration. This section describes existing approaches for all three paradigms and briefly discusses more recent developments that also follow an autonomic approach.

Several companies provide centralized management solutions for groups of base stations [1][2]. The majority of these systems implement link-layer "wireless switches" that connect base stations that act as wireless bridges to a switched wired network. The link-layer switch implements the management component. This centralized, link-layer approach offers traffic and channel management, policy, bandwidth and access control. However, such centralized link-layer solutions also have drawbacks. Link-layer broadcast domains cannot arbitrarily grow due to the scalability issues associated with broadcast traffic. Additionally, the topology of the wired network may not allow direct connection of the management system to the base stations. Centralized network-layer solutions address this shortcoming.

Decentralized management solutions are popular to configure mobile *ad hoc* networks (MANET) [3]. These management systems typically focus on the challenging task of enabling peer-to-peer communication in highly dynamic, mobile environments [4]. Because of their nature – *i.e.*, every base station decides based on its local scope [5] and no central management station exists – they are closely related to the autonomic approach presented in this paper. Ongoing research efforts [6][7] attempt to design self-configuring solutions for MANETs. However, in contrast to those approaches, the autonomic solution presented in this paper focuses on configuring a stationary wireless access network for mobile clients, with the primary goal of improving efficiency and performance.

With respect to decentralized management of infrastructure-based wireless networks, further work [8][9] focuses on the auto-configuration of base stations, with the goal to achieve the best coverage in a given geographical area. Early results suggest using transition rules that are similar to cellular automata to change the local configuration of a base station when receiving the current states of its neighbors. Although these proposed algorithms can support some of the specific applications that the autonomic approach also implements, such as regulating transmission power, they are not a platform for arbitrary management functions. In contrast, the focus of the autonomic management approach is to develop a management platform that can support many types of management functions.

Hybrid approaches to wireless network management, such as the Integrated Access Point of Trapeze Networks [10], push some functionality from a central system into the base stations, which are therefore slightly more complex than the simple wireless bridges of centralized approaches. Although hybrid systems improve scalability, they do not completely address the drawbacks of centralized systems; *e.g.*, they still have central points of failure.

3 Underlying Autonomic Principles

The high-level principles that guide the design and implementation of the proposed autonomic wireless network management systems are *automatic*, *aware* and *adaptive* operation [11].

Automatic operation: an autonomic system must be able to bootstrap itself when it starts and configure its basic functions according to the status and context of its environment, without involving a user or system administrator. This process consequently requires an autonomic system to anticipate the resources needed to perform its tasks and to acquire and use these resources without involvement of a human. For example, a new base station must integrate and configure itself into an existing wireless network without the involvement of a human administrator. Therefore, the base station must automatically configure its frequency, signal strength, network addresses and routing.

Aware operation: To allow an autonomic system to configure and reconfigure itself under dynamic conditions, it is important for the system to be *self-aware*. The system needs detailed knowledge of its components, resources and capabilities, its current context and status, as well as its relation to other systems that are part of its environment, in order to make the correct management decisions. As a result, a key requirement of an autonomic system is a monitoring mechanism that provides the necessary feedback to its control logic. Continuous monitoring is necessary to identify if the system meets its objectives. The feedback information will be logged and forms the basis for adaptation, self-optimization and re-configuration. In addition, monitoring is also important to identify anomalies or erroneous operation in the system, as it provides the basis for safety and security. Finally, for economic reasons, autonomic systems also need to monitor their suppliers and their consumers to ensure that they are providing/obtaining the agreed level of service.

Adaptive operation: The awareness of an autonomic system allows it to adapt according to the continuously changing context of its environment and to the current requirements of its users. Because of this, autonomic system management never finishes; the autonomic system continuously adapts by monitoring its components and fine-tuning its operation.

4 Autonomic Management System

This section describes an autonomic management system for wireless networks that builds on the autonomic principles defined in the previous section. It defines the basic system elements, specifies the target management functions, describes the wireless monitoring system and finally introduces the developed self-configuration and self-management approach. A more detailed description of the management system is available in [12].

4.1 Basic Components and Assumptions

The autonomic management system is completely decentralized across all the base stations of the wireless network. A base station in the decentralized management

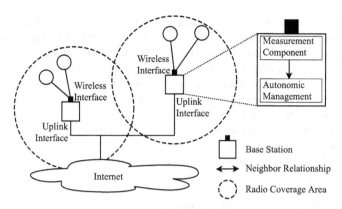

Fig. 1. System Overview

system has to fulfill several requirements. Each base station is a full-fledged IP router for its delegated IP subnet, able to operate stand-alone. It needs at least two network interfaces; one to provide wireless services to its clients and a second interface (wired or wireless) for uplink connectivity. Additional interfaces, when present, can act as probe/measurement interfaces, provide multi-homed uplink connectivity or offer additional client connectivity on different channels or link protocols. Figure 1 illustrates the basic architecture of the system.

Base stations automatically distribute the available address space among themselves, configure subnets for client connectivity on their wireless interfaces and configure the addressing of their wired uplinks. IP auto-configuration occurs through an integrated mechanism that was developed as part of an earlier research effort [13].

4.2 Management Functions

The primary task of the wireless management system is to coordinate radio properties, such as frequency use and transmission power, among a group of neighboring base stations and to implement system-wide functions, such as load balancing. By exchanging utilization information, neighboring base stations can distribute client load by increasing or decreasing transmission power or link speeds. A fully loaded base station, for example, can push clients at the edge of its coverage area off to other base stations by lowering its transmission power. An integrated wireless measurement subsystem (see Section 3.3) uses monitored feedback to enable the management system to adapt to changes in its environment.

A second task of the management system is self-protection of the wireless network. Self-protection also relies on the integrated wireless measurement component. It performs the necessary traffic analyses to detect potential security threats and informs the management system, which in turn can take the appropriate actions. The current system is able to detect and counter act against attacks resulting from rogue base stations and MAC address spoofing. The management system blacklists those malicious nodes and disseminates their presence throughout the system, warning the overall wireless network.

A third system function provides a means to obtain a global view of the system, *i.e.*, retrieve local information from all participating base stations of the system, for logging, administrative and monitoring purposes. The decentralized management system can support this functionality without the need of an explicit logging function. Instead, a *virtual neighbor* can disseminate its ID throughout the system and insert it into each base station's neighbor list. The virtual neighbor will then receive the local information disseminated by each base station as if it was simultaneously in radio range of every individual base station. The virtual neighbor can aggregate and export this system-wide information for a variety of uses.

It is important to note that the current system is a *platform* for autonomic management that can support many other management functions. The platform offers common functionality, such as information exchange, transactional semantics or security functions that can provide many different management capabilities.

4.3 Wireless Monitoring for Self-awareness

A basic capability of a wireless measurement system is capturing and analyzing network traffic, *e.g.*, to identify interference or security attacks, and providing the results to the management system. This feedback forms the basis for the autonomic behavior – self-configuration and self-management – of the system.

Monitoring of a deployed wireless network can pinpoint several causes of problems. For example, inter-channel and cross-channel radio interference can significantly decrease the effective data rate of the network. Traffic measurements can also help to secure and protect wireless networks. For example, analysis of the measurements can detect intrusion attempts and verify that friendly networks are sufficiently protected, *i.e.*, access is authenticated and/or data is encrypted.

One particular challenge for automated configuration of wireless access networks is base stations with overlapping coverage areas that are unable to detect this occurrence because none is visible to the other. Such base stations should become neighbors and coordinate their configurations, but fail to detect each other's presence. Consequently, their configurations will not be coordinated, leading to an inconsistent overall network configuration.

The gathering and analysis of feedback information can address the overlap problem. For example, if clients periodically notify their base station of other clients and base stations within their radio range, the management system can update the neighbor relation when a client enters an overlap area, eliminating or at least significantly reducing the overlap problem. Figure 2 illustrates this feedback process. Moreover, the direct feedback from the monitoring system enables detection of interference or spotty coverage, can identify rogue base stations or aid location tracking.

The autonomic management system presented in this paper uses monitors that provide the results of their continuous measurement efforts as feedback to the autonomic control process (as illustrated in Figure 1). Because dedicated measurements nodes are typically limited to a one or a few wireless interfaces, monitoring the complete spectrum (*i.e.*, on all channels) is difficult. To maximize monitoring effectiveness, the proposed system periodically switches a single interface to scan several frequency bands. When it identifies potential problems, the system focuses its monitoring efforts on the detected occurrence to track the problem at hand.

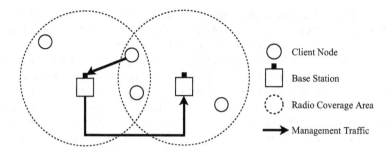

Fig. 2. Integration of monitors into the management system

Apart from the interception of wireless traffic, the measurement system collects additional information from every node that is detected, including the node's wireless mode (infrastructure or *ad hoc*), frame and byte counts, associated base station information such as the Service Set Identifier (SSID), physical-layer information such as signal strength and statistical information about higher-layer protocol use. In addition to many other measurement systems, this system also collects sequence number information for each captured frame. Analysis of patterns in the sequence numbers of captured frames is an effective technique to determine the presence of various anomalies and attacks [14][15].

For intrusion detection, the measurement system maintains also a database of known MAC addresses that it correlates with the IEEE's "organizationally unique identifier" list [16]. This can help to identify spoofed MAC addresses. If the rate of occurrence of new MAC addresses is past a configurable threshold, the autonomic management system is informed to take appropriate action.

4.4 Self-configuration and Self-management

Although each base station is able to operate in a stand-alone fashion, autonomic management of a group of base stations that provide connectivity to a geographic region requires collaboration. This collaboration occurs through periodic information exchange across the uplink interfaces, which allows each individual base station to adapt its local configuration consistently with its peers.

When a base station starts up, it first performs a probing phase – after a brief randomized de-synchronization delay – before configuring itself to provide service to wireless clients. During the first part of this probing phase, it auto-configures its network components for communication, routing and addressing, *i.e.,* it obtains a subnet delegation for its wireless network and configures its uplink interface, routing table and DHCP server for the wireless network appropriately [13].

After the base station has successfully bootstrapped its communication infrastructure, it performs a channel scan to detect other base stations in its immediate neighborhood and determines their identifiers. In other words, the base station identifies its current context as well as its relation to other systems of the embedded environment. Finally, it contacts these neighbor stations over its uplink interface and, after successful authentication and authorization [12], integrates itself into the network-wide information exchange.

Once configured, the base station starts to participate in global and local information exchanges with its peers, which provides the basis for the collaborating base stations to manage the overall wireless networks by themselves.

The system uses different kinds of exchange mechanisms for different kinds of information. Information that is globally important, such as encryption parameters or attack status, is disseminated throughout the network using an epidemic communication mechanism [17]. Information that is of local significance only, such as radio frequencies, transmit power or link utilization, is only disseminated locally among the affected neighboring base stations. This differentiation by information type is crucial for large autonomic systems in order to improve the scalability properties of the system.

The information that each station maintains falls into three different categories. *Private* information, such as logs, is never disseminated. A base station disseminates *local* information, such as its current channel, transmit power or utilization, to its neighbors, *i.e.*, other base stations within wireless range. This allows a group of neighbors to adapt their configurations in response to local events. A base station periodically disseminates updates about its local state to its neighbors every few seconds and likewise receives their updates.

A third kind of information requires *global* dissemination to all cooperating base stations. System-wide parameters, such as wireless protocol, security parameters or attack status are examples of such global information. The system disseminates global information using epidemic communication. Instead of broadcasting such updates to global state, they are piggy-backed onto the periodic information exchanges between neighbors. This technique prevents broadcast storms when global state updates are frequent.

Disseminating a global configuration change throughout the network in a consistent manner requires transactional semantics. This is a well-known challenge in distributed networks and a wide variety of approaches exist [18]. The current system implements a very simple method of guaranteeing global consistency – election of a central locking service. Future revisions will replace this method with a more scalable variant.

5   Evaluation

This section presents a preliminary evaluation of a prototype implementation of the autonomic management system. Due to space limitations, the evaluation focuses on the most important aspects of the autonomic systems, namely the scalability properties of the epidemic management state exchange and the feasibility of wireless network monitoring.

Scalability is a crucial aspect of autonomic management approaches for wireless networks. Because wireless access networks are expected to grow to very large numbers of base stations in the near future, autonomic management becomes particularly challenging and, at the same time, vital to the operation of the system. With respect to network monitoring, the issue of automatic problem detection – without human support – is a major challenge. The remainder of this section focuses on the evaluation of those aspects.

5.1 Autonomic Management Scalability

This section presents preliminary evaluation results of the scalability characteristics of an autonomic management system. The current prototype is a *Perl* daemon that operates on Linux systems with one or more IEEE 802.11a/b/g WLAN interfaces. The management daemon automatically configures and manages a collection of such machines.

For this scalability evaluation, the use of physical devices is impractical. Therefore, the prototype offers a simulation mode, where multiple copies of the same code execute on a single PC inside a simulated topology. During the simulation, each base station runs as a single process. The measurements in this section use this simulation mode to investigate groups of up to 100 base stations.

Also, note that in this simulation mode, the base stations themselves probe and monitor of the wireless network instead of dedicated wireless measurement nodes or interfaces. During the probing, every base station periodically scans, detects and contacts its neighbors to initiate an epidemic information exchange. This operation takes approximately 2 seconds and is repeated every 1800 seconds.

5.1.1 Initial Configuration Convergence

This section evaluates the convergence time of the autonomic measurement system for groups of base stations that all start up within a few seconds of one another, *i.e.*, the time of the initial *self-configuration*, such as after a power failure. The experiments measure convergence times of 500 repetitions and calculate mean performance and standard deviations. Each experiment uses a randomly generated, connected base station topology, *i.e.*, the aggregate coverage area of the base station group is not geographically partitioned. The number of base stations is a parameter of the experiment and grows up to 100 in increments of 10, with two additional group sizes of 5 and 15 to investigate behavior for small groups.

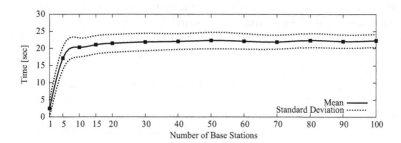

Fig. 3. Initial convergence times of groups of base stations

Figure 3 shows the performance. For smaller groups of 1-20 base stations, the mean initial self-organization time quickly increases from 17 to approximately 20 seconds. For larger groups of 20-100 base stations, the mean initial self-organization time remains between 20 and 25 seconds. As a result, the network convergence time does not grow significantly in relation to the number of base stations present. Although the simulations only demonstrate this effect for small groups of up to 100 base stations, this trend is expected to continue for larger groups. Future simulations will verify this hypothesis.

The results shown here highlight the strength of the decentralized approach in term of scalability with a growing number of attending base stations. However, for larger wireless access networks the distribution time for new *global* configuration settings will grow. The following section will analyze this scenario.

5.1.2 Epidemic Message Spread Time

This section investigates the dissemination times of changes to *global* state for a set of base stations that have already converged. A new *global* configuration setting is inserted at a single, random base station and disseminates throughout the entire network through the epidemic information exchange. The experiments measure the convergence times of 500 repetitions and calculate mean performance and standard deviations. As in the experiments above, the number of base stations is a parameter of the experiment and varies from 1 to 100 in increments of 10, with two additional group sizes of 5 and 15 to investigate behavior for small groups.

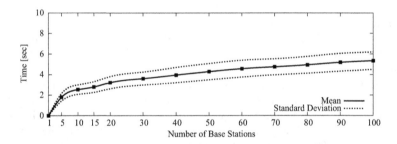

Fig. 4. Dissemination times of changes to global state

Figure 4 shows a growing dissemination time up to 5 seconds. The number appears large, but is a result of the information forward delay of each base station. Each base station informs its neighbors about changes in periodic intervals. The current prototype uses a default of 1 second. That means that a base station forwards a change to its *global* configuration set after at most 1 second.

Additional experiments (omitted here for space reasons) show that the dissemination times of changes to *global* state do not grow significantly with the number of base stations, but instead grow proportional to the topology diameter. This indicates that topology structure has more impact on performance than the number of base stations present.

5.1.3 Management Traffic

The epidemic management approach requires that every node forwards local state changes to its immediate neighbors in order to disseminate the information globally. Figure 5 shows the number of management information exchanges during the initial configuration for different topology sizes. The results indicate that the amount of management traffic grows linearly with the topology size.

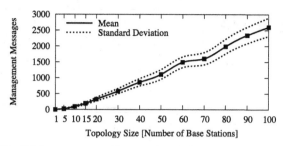

Fig. 5. Management traffic to perform initial self-organization

5.2 Wireless Measurement Nodes

This section evaluates the basic operation of the wireless measurement sub-system described in Section 3. Based on two real-world traffic traces collected inside the NEC Network Laboratories, this evaluation validates the feasibility of using wireless measurements for the autonomic management of wireless networks. Once the measurement nodes detect a problem, *e.g.*, interference or address spoofing, it informs the autonomic management system to take the appropriate actions according to the high-level policies of the network.

5.2.1 Intrusion Detection Using Frame Sequence Numbers

A common technique to attack a wireless network is through MAC address spoofing. If MAC addresses are used to control access to a network, a malicious client could simply probe the MAC address of a trusted client, and then uses this in order to send/receive traffic. The technique evaluated here allows detection of clients that attempt to spoof MAC addresses through analysis of the traffic measurements.

In the 802.11 protocols, a unique sequence number identifies each individual frame sent by a single node to allow detection of duplicates. Sequence numbers are 12-bit counters that monotonically increase from 0 to 4095 and wrap around at overflow. When a network interface starts or is reset, the sequence number counter starts at zero. The minimum time for sequence numbers to wrap around is under one second, but it can be indefinitely longer depending on packet size, send rate and link speed. According to the 802.11 standard, the sequence number counter should be readable but not writable by software. Because of their predictable and hard to spoof order, frame sequence numbers can act as fingerprints that uniquely identify frames sent by a single node over a period of time. Even when a station spoofs its MAC address, the sequence numbers of frames sent with a spoofed MAC address will still continue that stations sequence number pattern. This makes frame sequence numbers much stronger identifiers for specific stations than MAC addresses and thus allows detection MAC address spoofing.

Figure 6 shows a sequence number plot with traffic from two nodes with MAC addresses *A* and *B*. The sequence number curve for MAC address *A*'s at the top of the graph, the plot for MAC address *B* is mainly at the bottom. This trace illustrates how an otherwise well-behaving node with MAC address *A* periodically spoofs traffic to make it appear as if it came from node *B*.

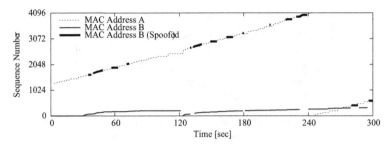

Fig. 6. Plot of sequence numbers over time for two nodes with MAC addresses *A* and *B*

Note that although the two sequence number progressions are clearly distinct, a number of packets are visible that appear to originate at MAC address *B* but have sequence numbers that fit with node *A*'s sequence number pattern. (Figure 6 illustrates these as thicker line segments overlaying node *A*'s curve). Without an analysis of frame sequence numbers, these spoofed packets are difficult to detect; even more difficult is to determine which station originates the spoofed packets.

5.2.2 Detection of Connectivity Problems

Sequence number analysis can also detect connectivity problems in wireless networks. As mentioned above, the measurement system deduces link-layer retransmissions by observing repeated transmissions of *retry* frames with identical sequence numbers. Frequent retransmissions may indicate connectivity issues.

Figure 7 shows the data frame and retransmission rate of the UDP sender (dashed line). It starts transmitting ten seconds into the measurement. The average data rate of the stream is around 75 frames/second until second 24, when the data rate suddenly drops to about half for the next eight seconds before resuming at the original rate. This drop in the data rate goes along with a corresponding increase in the retransmission rate from second 24-32 (solid line).

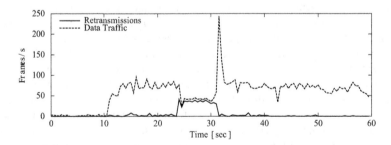

Fig. 7. Data frame and retransmission rates of a constant-bitrate UDP sender

6   Conclusion

This paper presents a decentralized, autonomic configuration and management system for base stations in wireless networks. The proposed system is a generic *platform* for autonomic management that offers generic mechanisms that support many different management functions. This common functionality includes mechanisms for information exchange, transactional semantics or security functions, which are required to realize many different management capabilities.

A novel feature of the autonomic management system is an integrated wireless monitoring component. This component determines common causes of problems through real-time analysis of live network measurements. The monitored feedback provides the system with the necessary awareness of its status and defines context for autonomic control. The feedback also provides a basis for individual base stations to automatically bootstrap and manage themselves.

A preliminary evaluation of the autonomic management systems focuses on the scalability analysis of the epidemic management state exchange and the feasibility of wireless network monitoring for automatic problem detection. The results of the epidemic state exchange show that the time to disseminate global state does not grow significantly with the number of base stations. Instead, it grows proportional to the topology diameter. As a result, the scalability property of the autonomic management system depends primarily on the topology structure and only to a lesser degree on the number of base stations present. For example, the results show that the prototype management system is able disseminate a global state change in a network of 100 base stations in less than 6 seconds, assuming typical connected topologies. The analysis of the measurement component illustrates the effectiveness of the proposed metrics and the measurement system through a series of real-world experiments. The results show that automatic detection of configuration or communication problems is feasible and can aid the autonomic control and management of wireless networks.

A more complete system implementation is currently ongoing. It will investigate the performance of additional system functions such as improved channel allocation, load balancing, rogue detection or location tracking and quantify the quality improvement obtainable by the inclusion of external information. It will also extend the scalability analysis to larger groups of base stations.

Although the current autonomic management system specifically targets WLAN networks, the general idea of decentralized, autonomic management certainly applies to other wireless and wired networks. The proposed system provides a decentralized management middleware built on generic methods for information dissemination that adapt to other network technologies and support many different management functions.

References

1. Airespace Corporation: Putting the Air Space to Work. White Paper (2003)
2. Aruba Wireless Networks: Getting a Grip on Wireless LANs. White Paper (2003)
3. Toh, C-K.: Ad Hoc Mobile Wireless Networks, Protocols and Systems. Prentice Hall Inc., New Jersey, USA (2002)
4. Ji, L., Agre, J., Iwao, T., Fujino, N.: On Providing Secure and Portable Wireless Data Networking Services: Architecture and Data Forwarding Mechanisms. Proc. International Conference on Mobile Computing and Ubiquitous Networking (ICMU'04), Japan (2004)
5. Advanced Cybernetics Group and Meshdynamics: Challenges for 802.15 WPAN Mesh. White Paper (2004)

6. Zhang, H., Arora, A: GS3: scalable self-configuration and self-healing in wireless networks. Proc. 21st Annual Symposium on Principles of Distributed computing, Monterey, California, USA (2002) 58–67

7. Krishnamachari, B., Wicker, S.B., Bejar, R., Fernandez, C.: On the Complexity of Distributed Self-Configuration in Wireless Networks. Telecommunication Systems, Vol. 22 (1-4) (2003) 33-59

8. Mullany, F.J., Ho, L.T.W, Samuel, L.G., Claussen, H.: Self-Deployment, Self-Configuration: Critical Future Paradigms for Wireless Access Networks. Proc. of 1st International Workshop on Autonomic Communications (WAC 2005), Berlin, Germany (2004)

9. Ho, L.T.W, Samuel, L.G., Pitts, J.M.: Applying Emergent Self-Organizing Behaviour for the Coordination of 4G Networks Using Complexity Metrics. Bell Labs Technical Journal, Vol. 8, No. 1 (2003) 5-26

10. Trapeze Networks: Defining An Integrated Access Point. White Paper (2004)

11. Kephart, J., Chess, D.: The Vision of Autonomic Computing. IEEE Computer Magazine (2003)

12. Zimmerman, K.: An Autonomic Approach for Self-Organising Access Points. Diploma Thesis, University of Ulm, Germany (2004)

13. Silva Tobella, J.J., Stiemerling, M., Brunner, M.: Towards Self-Configuration of IPv6 Networks. Proc. Poster Session of IEEE/IFIP Network Operations and Management Symposium (NOMS'04), Seoul, Korea (2004)

14. Wright, J.: Detecting Wireless LAN MAC Address Spoofing. White Paper (2003)

15. Wright, J.: Layer 2 Analysis of WLAN Discovery Applications for Intrusion Detection. White Paper (2002)

16. IEEE: IEEE Organizationally Unique Identifier (OUI) List. December (2004)

17. Demers, A., et al.: Epidemic algorithms for replicated database maintenance. Proc. 6th ACM Sympos. on Principles of Distributed Computing, Vancouver, Canada (1987) 1-12

18. Tanenbaum, A., van Steen, M.: Distributed Systems, Principles and Paradigms. Prentice Hall Inc., NJ, USA (2002)

Context-Driven Self-configuration of Mobile Ad Hoc Networks

Apostolos Malatras and George Pavlou

Centre for Communications Systems Research,
Department of Electronic Engineering, University of Surrey, UK
{a.malatras, g.pavlou}@surrey.ac.uk

Abstract. We present the design and implementation of a working prototype
system that enables self-configuration in mobile ad hoc networks (MANETs)
by exploiting context awareness and cross-layer design principles. The driving
force behind the proposed system is to allow for self-configuration of MANETs
by enabling them to be adaptive to varying conditions. Emphasis is placed on
describing the requirements and specifications of the supporting platform's
functionality. We propose the distributed management of the MANET through
a proactively constructed body of nodes in order to cope with the inherently
dynamic nature of MANETs. We present our work on deploying the designed
system on our experimental MANET testbed and provide results of its
performance based on extended testing.

1 Introduction

The concept of mobile ad hoc networks (MANETs) has brought a new paradigm in
communication networks and acts as an enabler for pervasive computing and
communication environments. In ad hoc networks, the mobile nodes (MNs) are free
to move randomly and organize themselves arbitrarily; thus, the network's wireless
topology may change rapidly and unpredictably. Conventional wireless networks
require some form of fixed network infrastructure (i.e. the core network) and
centralized administration for their operation. In contrast, since MANETs are self-
creating, individual MNs are responsible for dynamically discovering other nodes
they can communicate with. This way of dynamically creating a network often
requires the ability to rapidly create, deploy and manage services and protocols in
response to user demands and surrounding conditions in an equally dynamic manner.

We assert that this highly dynamic environment can benefit from the emerging
context-driven autonomic communications paradigm. There has been no proper
previous research on deploying autonomic communication solutions in MANETs, but
such aspect is important due to their inherent nature. As such, autonomic
communication principles can assist in the self-management of MANETs and enable
network self-configuration and optimization by utilizing context information. The
latter can be used to establish the need for automatic changes (self-configuration) in
accordance to high-level pre-existing rules. Context-information can be used to
trigger cross-layer changes (network and application configurations) according to
predefined rules, leading to autonomic decision-making.

I. Stavrakakis and M. Smirnov (Eds.): WAC 2005, LNCS 3854, pp. 71–85, 2006.

This paper provides conceptual and practical design, implementation and deployment issues regarding a middleware platform used for the self-configuration of MANETs. The structure of the paper is as follows. After this brief introduction, Section 2 reviews basic autonomic communication and computing principles, including pointers to related work. Section 3 gives an overview of the proposed system's design and architecture providing justification for our choices. Details on the implementation of the platform and its deployment on our experimental MANET testbed is the subject of Section 4, where the results of our practical experimentation are also presented. Finally, Section 5 concludes the paper and discusses future research directions.

2 Autonomic Communications Principles and Related Work

Autonomic computing emerged as an initiative by IBM and has generated a very active research stream bringing together interdisciplinary domains. Autonomic computing refers to the self-managed operation of computing systems and networks, without the need for administrators but with high-level objectives dictating the system's functionality. The IBM autonomic computing blueprint [1] defines four distinct concepts behind autonomy, namely self-configuration, self-optimization, self-healing and self-protection [2]. The building block of all autonomic solutions is an autonomic element. This refers to the collection of one or more managed elements that are handled by an autonomic manager. The latter monitors the state of the elements, analyzes it and acting upon high-level objectives (typically defined as policies) imposes the execution of configuration changes on the managed elements. This process is repetitive [2], [3].

Most autonomic computing platforms are targeted to systems with sufficient resources that are relatively stable [1], [4], and [5]. The application of autonomic principles on MANETs has not been adequately researched. In [10] we presented our initial approach and results on self-configuring and optimizing MANETs. In [6] a policy-based network management system for MANETs is proposed but the hierarchical approach adopted assumes the existence of several "thick" nodes in the network, which may not always be the case.

Programmability is a very important aspect of autonomic systems, especially in ad hoc networks given the multitude of potential solutions for routing, quality of service support and other application services. Programmability can be achieved through a variety of means. Active control packets may carry code to be evaluated in routers and this approach has been used for active routing in ad hoc networks [7]. Mobile agents may be used in full mobility scenarios, carrying code and state to manipulate different MNs, or in a constrained mobility mode [8] as a more flexible means for the management by delegation approach [9]; in the latter, code is uploaded and executed in MNs through "elastic management agents", augmenting the node functionality. Programmability is also possible through the provision of suitable management interfaces that allow code to be uploaded to MNs and activated in a controllable fashion. In our recent work [11] we proposed a programmable middleware capable of dynamically deploying services and protocols in ad hoc networks.

3   System Design and Architecture

We propose the deployment of a lightweight, context-aware middleware platform on every MN of a MANET and a distributed management approach based on the existence of an adaptive set of nodes called Management Body (MB). The middleware platform is responsible for monitoring the individual MN context individually and the context of the MANET as a whole. Context information is handled locally at each MN and aggregated information is passed to the management body of the MANET. The latter reaches management decisions based on this aggregated context and in accordance with predefined rules. The corresponding configuration changes are autonomously deployed on the MNs through software plugins that carry the desired functionality.

3.1   A Hybrid Approach to MANET Management

There exist two diverse approaches regarding the management approach to be deployed in a MANET. In the hierarchical approach the MANET is grouped into clusters, each electing a local leader or cluster head (CH). The CHs act in cooperation and elect a global leader or network head (NH) that is responsible for deciding on key management issues. This approach bears similarities to the one undertaken by routing protocols such as OSPF and scales well, limiting the MN interactions within a cluster or among CHs. Moreover, it allows operation in a controlled distributed fashion, where decisions are taken not only by the NH but through cooperation and "voting" among the CHs. A diametrically different approach is a fully distributed one, in which all the nodes are deemed as equal and determine collectively any management decisions to be taken. This approach requires more complex cooperation protocols and may not scale for large networks with many MNs. On the other hand the hierarchical approach suffers from the existence of single points of failure, i.e. the CHs. In case a CH leaves the MANET or moves to a different location (and thus changes cluster), the clustering process will have to be re-initiated, an option not suitable for dynamically formed MANETs.

We chose to use a hybrid approach for our management scheme. Our approach resembles the hierarchical approach by dividing the MANET in clusters; a collaborative Management Body of MNs replaces the CH. The MB has collectively the functionality of the CH but does not suffer from single node movements as these are mitigated from interactions with the other MNs forming the MB. In a similar fashion, a collaborative body comprising selected nodes from the management body replaces the network head (Figure 1). The management decisions are taken collaboratively by the MNs assigned to the management body. Our scheme is inspired from the formation of virtual backbones in MANET routing protocols and service provisioning. The idea of using a virtual backbone to serve as a management entity in a MANET is not new. There have been several approaches in the literature that have considered similar schemes [12], [13], [14].

We chose this hybrid approach due to the fact that neither of the existing approaches suits the MANET features completely. The hierarchical approach does not perform well when node mobility is involved and is thus applicable to longer-term, relatively stable MANETs. In contrast, the fully distributed approach is very demanding as far as message exchanges are concerned and can be applied to small MANETs with few nodes. The combined approach we chose has the following benefits. The management decisions are devised by a number of nodes in the MANET and not by a single one. This distributes the load across the MANET, which is necessary for both resource conservation and reliability & robustness reasons (i.e. avoiding single points of failure). The hierarchical features of this scheme allow for the deployment of a uniform management approach over the MANET as desired. The MBs are constructed so as to be relatively stable, while there is support for nodes leaving the MB. The MB is reconstructed only if a significant amount of MNs that comprise it leave. This ensures the avoidance of dangerous situations, with any node potentially triggering the MB formation process unnecessarily. We realise the overhead imposed on the MANET from the cooperative management architecture but we consider this a fair trade-off given the robustness achieved.

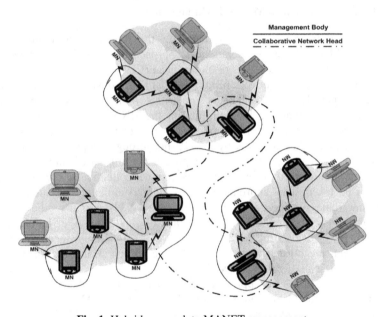

Fig. 1. Hybrid approach to MANET management

The virtual backbone used in MANETs is usually constructed as the Minimum Connected Dominating Set (MCDS) of the MANET graph. Unfortunately, the construction of an MCDS for a connected graph is an NP-Complete problem. There are two ways to face this problem, namely using an approximation algorithm or making use of a heuristic to reduce the problem into one solved in polynomial time. We chose to undertake the heuristic approach when creating the MB. Apart from that, and in favour of simplicity and timeliness, we opted towards establishing any CDS and not the minimum one. We use two heuristics to discover the CDS, the computational capabilities of the MNs (the most resourceful nodes) and their prospective, relative location stability (the nodes that are less likely to affect the network topology and thus do not lead to frequent MB re-formations).

The nodes that will be part of the MB should therefore have sufficient resources to handle the additional requirements, such as communicating with other MB-members to reach to management decisions. The nodes forming the MB are collectively the set of nodes with the highest computational resources in the MANET. Every node is calculating a value that denotes its capability to become a member of the MB. The value of this property is then used in the selection process for the dominating set.

Our proposed capability function (CF) exploits the following attributes: memory requirements (MEM), processing power (PP), battery power (BP), mobility ratio (MR) and current load (CL). These 5 variables need to be combined in a single equation, the Capability Function (CF). MEM, PP and BP are obviously proportional to CF while MR and CL are inversely proportional. By assigning weights to these variables in accordance to their significance, we have the initial CF equation (1).

$$CF\ (x) = \frac{(w_1 \times MEM\ (x)) \times (w_2 \times PP\ (x)) \times (w_3 \times BP\ (x))}{(w_4 \times MR\ (x)) \times (w_5 \times CL\ (x))}$$

(1)

where, $\sum_{i=1}^{5} w_i = 1$, and x is the MN.

The main requirement for the CF is to lead to comparable results among MNs. For this reason the various attributes must be demoted in common range values. Space limitations do not allow us to delve into more details on how to achieve this. Equation 1 is used to derive a value for every MN that is proportional to its capability of being part of the MB.

Obviously, one should not expect the MANET topology to be known. Distributed approaches to construct the MB are thus adopted. The distributed construction of CDS has been intensely researched [12], [13], and [14]. We decided to take a similar approach. Details of the algorithm we have used to derive the CDS of the MANET are not presented due to space limitations. Our approach is based on building a relatively stable CDS with the "thickest" nodes according to the CF mentioned, but also takes into consideration the need for maintenance of the CDS due to the inherently unstable MANET nature.

3.2 Context Management

Autonomic communications solutions currently available have focused on monitoring device specific characteristics and network conditions in order to infer configuration adjustments on the devices or the network as whole. We differentiate our approach by extending the sensed environment to also consider user-specific information (i.e. user profiles and user explicit information) that can have an effect on the underlying network, as well as physical environment attributes with the same property (i.e. device location and vicinity information). Cross-layer context gathering is the basis of our middleware platform that exploits this information in order to allow for MANET self-configuration. The collection of context from the surroundings of the mobile nodes is handled by a series of interfaces that communicate with the available sensors, constituting the monitoring component of our platform. We consider the term context in a generic fashion, incorporating both computational and physical resources.

Each MN is responsible for collecting its own context information and processing it to higher-level context information that has an impact on the management plane of the MANET. For example a MN might collect its current location and monitor this through a GPS receiver installed on it, but this information is not useful for the MB. Useful information for the MB would be the mobility prediction for each MN, since having this can be used for proactive configuration changes, as it will be shown in Section 4. Other higher-level context information can refer to QoS requirements, security requirements and prospective network load. This set of elaborate context information is in effect aggregated from simpler context information. The advantages of this approach are obvious. By aggregating the context information available to a MN to a set of "advanced" contexts that are passed to the MB, less control load is imposed on the MANET in terms of traffic. It also distributes the processing and storage load of handling all the context information among the MNs of the MANET. The alternative would be to pass all this information to the MB, which would then be responsible for processing it, storing it and infer configuration changes based on it. The set of advanced MN contexts that are passed around from MNs to the MB are predefined and their processing occurs using the functionality of our middleware platform as described later.

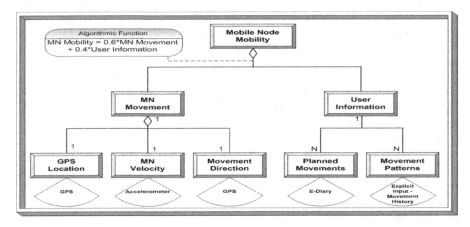

Fig. 2. Mobile node mobility context as derived from simpler contexts

Figure 2 presents an example of how the aggregated context of MN mobility can be derived from simpler contexts collected from device sensors. The analysis of elaborate contexts to simpler ones is based on the sensors used, while it should also be noted that semantic metadata information and algorithmic functions describe the way this analysis occurs in a human-understandable and a formal way respectively (e.g. in the example of Figure 2, the MN mobility is more dependent on the movement metrics rather than the user specific information since we deem the former as more credible). We represent the context using an XML-based model that takes all this information into account and allows hits lightweight processing, specific details though lie outside the scope of this paper.

3.3 MANET Self-configuration

The proposed middleware platform builds on the aggregated context information that is collected from all nodes to reach to management decisions for the MANET as a whole. These decisions are then implemented as (re-) configuration changes. Only this context information is transported across the MANET, limiting thus the traffic requirements. It also relieves the MB from a series of resource- and time-consuming processing operations, which are handled individually by every node, distributing thus the processing load. We have already mentioned that the set of aggregated context information is prespecified. The same stands for the rules that are used to establish the need for configuration changes in the MANET. The MNs forming the MB of the MANET know these rules in the form of policies. When certain preconditions are met, the rules are activated and the corresponding configuration changes are deployed on the MNs. One such example that will be elaborated in the next section is monitoring MN mobility. When the relative mobility of the MNs is changing, it might be beneficial to change the routing protocol used in the MANET. These rules in our platform are currently static and predefined. We are working towards a more dynamic and adaptive scheme based on higher-level policies, so as to increase the degree of autonomy of our system.

The configuration changes are deployed on the MANET through software plugins that carry the corresponding functionality. These plugins can be any software module, from a simple set of commands, e.g. a script, to complex applications, as long as they conform to the defined interface. All plugins should conform to standard interfaces regarding activation, deactivation and reconfiguration.

One question that arises is how the MB members collaboratively monitor and act upon the aggregated context of all MNs. For each aggregated context there is a function used to calculate its value as far as the related rule is concerned. Every MB member calculates this value collectively for the MNs it dominates and floods this information within the MB. At the end of this process every member of the MB will have a MANET-wide understanding of the rule-specific value for every aggregated context. In the previous example, relative mobility is the rule variable for routing protocol selection. Every MB member calculates its relative mobility to that of the MNs it dominates, floods this information to the rest of the MB members and receives relevant information from them. The new values it receives are used to update its relative mobility so as to include those of the rest of the nodes in the MANET.

This MANET-wide value for every aggregated context is compared against the rules in the MB nodes to establish if the need for a configuration change occurs. If so, then the appropriate action is passed from every MB member to the nodes it manages through a particular plugin. The fact that all MB members have the same values for the context and the same predefined rules ensure that the same action, if any, will be employed on the MANET, achieving a uniform self-configuration scheme.

3.4 Middleware Architecture

Figure 3 depicts the proposed system's architecture from a high-level perspective. This middleware platform is installed on every MN of the MANET, empowering it with the necessary functionality. As it will be seen at the experimentation phase, the architecture proposed is relatively lightweight. We will describe the platform and provide justification for our design choices regarding monitoring, context handling and self-configuration.

Context Monitoring. The sensed environment is accessed by means of sensors. These sensors are diverse in the way they provide the sensed information to whoever needs it. We designed a generic interface for that purpose, the Sensor Communication Interface (SCI), to which all communication protocols with the sensors conform. Every device is equipped with the SCIs for the sensors it carries and we consider them supplied as software modules bundled with the sensors. Realizing that a device might require accessing a sensor for the first time (i.e. a new positioning device) and does not have the particular SCI, we have implemented the SCI Manager. This is responsible for advertising the SCIs the device holds and discovering and retrieving SCIs from other MNs by communicating with their respective SCI Managers.

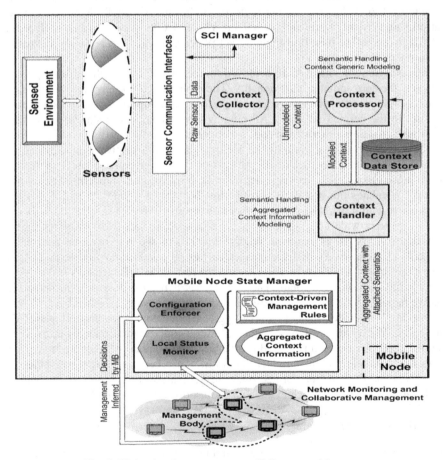

Fig. 3. Higher-level context-aware middleware architecture

Sensors do not produce context information but raw data that has to be translated into meaningful information i.e. context. For this reason semantics regarding the data the sensors produce are included in the various SCIs so that the raw data gains some semantic meaning before it is passed to the Context Processor. The Context Collector is responsible for this task. Another task that this module is in charge of is the pruning of the abundant context information. Sensors produce a plethora of data that are not all useful. For example GPS receivers inform for every single location change, even in the scale of some meters. This amount of detail might not be needed to be collected. The Context Collector retains custom filters for each context collected that states which changes in values are deemed significant to be stored and which should be discarded.

Context Handling. The Context Processor and the Context Handler are the two modules that collectively manage locally the context information for a MN. The former is responsible for modelling the primitive context information collected from the sensors to the generic context model we have devised. Semantic information is tagged to the context in order to allow for semantic operations to be performed. The Context Collector comprises 3 entities, namely the Processing Interfaces, the Context Modeller and the Semantic Handler. The Processing Interfaces entity is used to provide different interfaces for the handling of various data types provided by sensors. One sensor might for example produce binary data and another scalar. This entity provides the generic feature for the platform to be able to respond to every possible input. The Context Modeller then is instantiated with its main activity being the translation of the simple data to the model representation proposed. The Semantic Handler enriches the semantics of the context, with metadata more specific to the uses of the platform. The sensors provide some metadata about their collected data to give an understanding of what they are monitoring. For example a GPS might yield that it is collecting MN location through a "location" value. The Semantic Handler builds on this and provides more semantics like "latitude-longitude/positioning" etc. The purpose of this is to ensure that the platform is not explicitly bundled with sensors, i.e. the "location" metadata but it is rather bundled with the general notion described by more than one words. The Context Processor stores context information in the local data store created for this reason.

The Context Handler is responsible for the task described earlier: collecting simple contexts and aggregating them to higher-lever contexts that are going to be sent to the MB. To do that in a generic fashion it exploits Context Handlers and Aggregated Context Modelling. These two entities collaborate with the Semantic Handling entity to infer useful knowledge on the aggregated context. The modelling of this higher-level context is based on predefined models that are hard-coded on the platform. The platform is open enough though to support new aggregated context models that may be required from the MB. The MB might decide for example upon using a context of MN QoS requirements. The MNs are not aware of the model to be followed to infer this context from simpler contexts. The MB members then transfer the model properties to the MNs and acting upon it the MNs respond to the MB with the desired QoS requirements context.

Self-configuration. We consider that the functionality the MNs, regardless of the heterogeneity of the available platforms, is manipulated and altered through software plugins. For instance, a routing protocol used by mobile nodes, is as far as our platform is concerned a loadable plugin that has open interfaces to allow its activation, de-activation or reconfiguration according to management demands. The self-configuration aspects of our context-driven middleware platform are thus implemented through the use of these software plugins that can be implemented simple scripts or Java, C/C++ or any other programming language objects in our experimental prototype.

Self-configuration is handled through the MN State Manager module. The main responsibility of this module is to collect and advertise the aggregated context information to the MB. Communication with the MB (through XML-RPC as will be elaborated later) is handled by the State Manager, as is communication with other MNs. Hard-coded into this module are the general Context-Driven Management Rules that are used by the MB to examine if necessary conditions are met and configuration changes are necessary. The Local Status Monitor has the obvious functionality of retaining and making available the information on the current local status of a MN. The Configuration Enforcer receives "orders" from the MB regarding configuration changes through software plugins. When such "orders" are given, the Configuration Enforcer imposes them on the platform by acquiring the required plugin if it does not have it and activating it.

The plugins are considered to be owned by at least some nodes of the MANET, since we cannot consider them being generated at runtime. For example, if the plugin is a routing protocol like the case study in Section 4, this must exist in some of the MANET nodes. The nodes that have the required plugin are informed by the MB to distribute it within the MANET by means of efficient flooding to their neighbours and so forth. The flooding is efficient in two ways: i) the receiving MN is first queried to establish it does not have the plugin already and ii) the plugin is flooded only to MNs that share the same platform with the owner of the plugin (this is necessary for heterogeneous environments with multiple platform configurations, such as our experimental testbed).

4   Usage Scenario and Testbed Evaluation

For purposes of validation and experimentation we have implemented the proposed programmable middleware platform and deployed it in our experimental testbed. After reviewing the specific implementation details, we present the results obtained when testing our implementation in the testbed.

4.1   Testbed Configuration and Platform Implementation

To test the platform's performance and efficiency and also examine its operation in a real environment, we deployed it in our experimental MANET testbed that comprises 2 laptops and 4 PDAs (see Table 1 for configuration details). The testbed is a 6-hop MANET and is considered as a relatively reliable environment so that the results can be extrapolated and general conclusions can be drawn.

Table 1. Testbed hardware configuration

Platform	Configuration Attribute	Description
PDA	Processor	400 MHz Intel XScale
	Memory	48 MB ROM, 128 MB RAM
	Operating System	Familiar Linux 2.4.19
	Wireless interfaces	Integrated wireless LAN 802.11b
Laptop	Processor	1,7 GHz Intel Centrino
	Memory	512 MB RAM
	Operating System	Debian Linux 2.6.3
	Wireless interfaces	Integrated wireless LAN 802.11 a/b/g

The platform is implemented using the Java 2 Micro Edition (J2ME). This version requires a much smaller memory footprint than the standard or enterprise edition, while at the same time it is optimized for the processing power and I/O capabilities of small mobile devices. We also used the Connected Device Configuration (CDC) framework instead of the limited one (CLDC), as the latter lacks support for required advanced operations. We chose to use Java because of its ubiquity and platform independence. Our platform caters also for both Java and C/C++-based plugins. The use of Java requires MNs to have the Java Runtime Environment (JRE) installed. Although this is relatively memory-hungry, our hands-on experience confirms that even the resource-poor PDAs can comfortably support the execution of the JRE.

The communication between MNs uses the lightweight XML-RPC protocol [17]. XML-RPC is a subset of the Simple Object Access Protocol (SOAP) with only basic functionality enabled. It allows software running on different operating systems and hardware architectures to communicate through remote procedure calls (RPCs). XML-RPC uses the HTTP protocol as transport and XML encodings for the RPC protocol itself. We chose an XML-based approach because we also use XML to represent contextual data collected by MNs. We could have possibly chosen Web Services, but this approach would have certainly been more heavyweight. In addition, Web Services, in the same fashion with distributed object technologies such as CORBA, necessitate object advertisement and discovery functionality, which is not required in our platform that relies on simple message passing modelled through RPCs. Given our recent performance evaluation of XML-RPC and other management approaches [16], we believe that XML-RPC provides a useful blend of functionality and performance.

Trivial FTP (TFTP) [18] was used for the distribution of the plugins. It is less complex than FTP and consumes less network resources. TFTP has no password-based user authentication, which saves both time and traffic in a trusted environment; as already mentioned, security in an ad hoc environment is an important issue but is outside the scope of the current work. In addition, TFTP uses only one connection, contrary to FTP that requires two connections, one for control and one for data traffic.

4.2 Autonomic Routing Protocol Selection

The scenario we chose to test on our experimental testbed includes the dynamic change of the routing protocol used in the MANET. MANET routing protocol performance is dependent on the stability of the network itself. Reactive routing protocols are better suited for very volatile network topologies, while proactive approaches for more static MANETs. The scenario implemented was that of the dynamic routing protocol change according to contextual information regarding the mobility of MNs. MNs use initially the reactive AODV routing protocol [19] for their ad hoc communication over 802.11b, while at some point indicated by the change in the mobility pattern they switch to the proactive OLSR protocol [20] as the network becomes close to stationary. This decision is derived and imposed by the MB. Both these routing protocols are realized as C-based user space daemons. Practical problems during this experiments included wireless link interference given that the wireless interfaces were in a confined space. In addition, since testing for various network topologies was necessary, we used a MAC address filter tool to emulate broken links or unreachable destinations.

The scenario serves the purpose of presenting both the self-configuration and self-optimizing aspects of the platform, as well as the platform functionality. The self-configuration aspect is apparent from the scenario itself, while in this case the self-optimizing aspects refer to the fact that by changing the network protocol we achieve better performance of the MANET by means of bandwidth consumption (proactive and reactive routing protocols consume different amount of bandwidth and work better in different network states).

We experimented with many different topologies, routing protocols and other plugins to get a concrete understanding of the platform's operation. In the following subsections we present experimental results regarding the routing protocol switch scenario for three different yet representative network topologies: *star*, *random* and *line*. The star topology models a centralized approach, with the MB conveniently located in the centre and comprised of one node, having a 1-hop distance from other nodes. The line topology is the one that performs worse than the others, and models a sparse MANET with 6-hop diameter (the MB in this case is comprised of 4 nodes in a total of 6). The random topology models a middle-ground situation between the previous topologies and models the most common case real-world scenario (2 nodes form the MB). Although we have implemented context processing and dissemination in our platform, getting mobility information requires sensors MNs such as accelerometers, GPS support, etc. Given the practical difficulty of sensing real mobility changes, we chose to generate them artificially, through pre-specified timers and mockup context information. As we were mostly interested to assess the performance in terms of the plugin dissemination and activation, this approach is adequate. We plan though to focus on context-based performance issues in future work. Finally, it is essential to emphasize that the results have derived by a number of identical experiments and mean values are presented. Table 2 presents the results regarding the three described topologies as far as incurred traffic is concerned and convergence time.

Results from testbed measurements prove first of all that the platform functions properly, since the routing protocol dynamic change performs smoothly and in accordance with the network mobility, while the situation can revert to the original configuration if the necessary conditions are met. The platform as evaluated in our testbed seems to fulfil its goal as being lightweight and deployable on devices with limited resources, such as PDAs. The time needed for the initialization of the base functionality is 26 msec for the laptops and 741 msec for the PDAs, while the memory utilization was 3788 bytes and 4208 bytes respectively. The differences in time are attributed to the significantly different processing capabilities, while memory consumption is almost identical, which was expected since the platform is the same for both configurations.

Table 2. Experimental testbed results under various MANET topologies

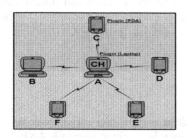

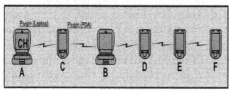

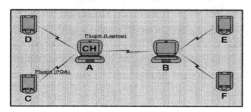

Star Topology
Time required for convergence: 41.96 sec
Routing related traffic: 7736 bytes
Inter-MN traffic: 41742 bytes
TFTP traffic: 1064880 bytes
The MB is formed of 1 node, solely A
Line Topology
Time required for convergence: 47.94 sec
Routing related traffic: 14332 bytes
Inter-MN traffic: 83145 bytes
TFTP traffic: 1530924 bytes
The MB is formed of 4 nodes, C, B, D, E
Random Topology
Time required for convergence: 44.43 sec
Routing related traffic: 12068 bytes
Inter-MN traffic: 51491 bytes
TFTP traffic: 1366896 bytes
The MB is formed of 2 nodes, A and B

The other parameters of the testbed experimentation prove the efficiency of the platform. From the moment the management body identifies the need to alter the routing protocol, up until the activation of the new routing protocol the time required is at acceptable levels, being dependent on the size of the routing plugin and the network size. The OLSR routing plugin has a size of 450 KB for the laptops and 98,1 KB for the PDAs. The convergence time required for the alignment of nodes capabilities depends on the distributed plugin. In our test case the plugin size is significant, and thus requires considerable time for its deployment throughout the network. The measured time takes into account the fact the wireless links are not stable throughout the experiment due to interference reasons. In a number of experiments, link breakages occurred without any external intervention, and we attribute these to the inter-MN interference. Given these link breakages, the time measured in our experiments includes the additional latency introduced for route reconstruction.

Another important observation is the fact that the inter-MN traffic is rather limited with a maximum of 83145 bytes for the line topology, which is attributed to the fact that this is the sparsest one and the MB is composed of many nodes due to the specific node location. Even so, the inter-MN traffic is not large enough to make our hybrid management approach inapplicable. The inter-MN traffic includes the traffic required to construct and maintain the MB, the aggregated context advertisements from the MNs to the MB and other platform specific MN calls. Regarding the TFTP traffic this includes the transfer of the routing protocol plugin to the MNs that do not have it. This noteworthy traffic size is justified if one considers the significant size of the plugin and the fact that two versions are disseminated in the MANET (laptop and PDA versions).

5 Conclusions

We presented the foundations and major design principles of a context-aware, programmable middleware platform that enables self-configuration in MANETs. The platform has been implemented and successfully deployed on our experimental testbed, with encouraging initial results. Our future work focuses on further expanding the architecture to take into account more elaborate management policies that conform and adapt to the dynamic nature of the MANETs. We have limited our experimental evaluation of the platform to include only results from actual deployment on our testbed. We plan though to test its performance, scalability and its effect on MANET optimization using also simulation tools, complementing those MANET simulations with real-world practical experiments as suggested in [15]. Understanding the major security implications that may arise from the deployment of software modules on mobile nodes, we plan to expand our framework to incorporate advanced security mechanisms using possibly "sandbox" techniques for controlled execution in a failsafe environment and authenticated remote activation of software modules.

References

[1] Haas, R., Droz, P. and Stiller, B., "Autonomic service deployment in networks", *IBM Systems Journal*, Vol. 42, No 1, 2003

[2] Kephart, J.O. and Chess, D.M., "The Vision of Autonomic Computing", *IEEE Computer*, January 2003

[3] Ganek, A. G. and Corbi, T.A., "The dawning of the autonomic computing era", *IBM Systems Journal*, Vol. 42, No 1, 2003

[4] Crawford, C.H. and Dan, A., "eModel: Addressing the Need or a Flexible Modeling Framework in Autonomic Computing", *10th IEEE International Symposium on Modeling, Analysis and Simulation of Computer and Telecommunications Systems (MASCOTS02)*, October 2002

[5]   Dong, X., Hariri, S., Xue, L., Chen, H., Zhang, M., Pavuluri, S. and Rao, S., "AUTONOMIA: An Autonomic Computing Environment", *IEEE International Conference on Performance, Computing and Communications*, April 2003

[6]   Chadha, R., Cheng, H., Cheng, Y.-H., Chiang, J., Ghetie, A., Levin, G., and Tanna, H., "Policy-based mobile as hoc network management", *5th IEEE International Workshop on Policies for Distributed Systems and Networks (POLICY04)*, 2004

[7]   C. Tschudin, H. Lundgren, H. Gulbrandsen, "Active Routing for Ad hoc Networks", *IEEE Communications*, Vol. 38, No. 4, April 2000.

[8]   C. Bohoris, A. Liotta, G. Pavlou, "Evaluation of Constrained Mobility for Programmability in Network Management", *11th IEEE/IFIP Int. Workshop on Distributed Systems: Operations and Management (DSOM'00)*, December 2000.

[9]   G. Goldszmidt, Y. Yemini, "Evaluating Management Decisions via Delegation", *Proc. of IEEE Integrated Network Management III*, pp. 247-257, Elsevier, 1993.

[10]   A. Malatras, G. Pavlou S. Gouveris, S. Sivavakeesar and V. Karakoidas, "Self Configuring and Optimizing Mobile Ad Hoc Networks", *to appear as a short paper in the Proceedings of the IEEE International Conference on Autonomic Computing (ICAC 2005)*, June 2005

[11]   S. Gouveris, S. Sivavakeesar, G. Pavlou and A. Malatras, "Programmable Middleware for the Dynamic Deployment of Services and Protocols in Ad Hoc Networks", *to appear in the Proceedings of the IEEE/IFIP Integrated Management Symposium (IM 2005)*, May 2005

[12]   P.-J. Wan, K. M. Alzoubi and O. Frieder, "Distributed construction of connected dominating set in wireless ad hoc networks", *IEEE Infocom 2002*

[13]   R. Friedman, M. Gradinariu and G. Simon, "Locating cache proxies in MANETs", ACM MobiHoc 2004

[14]   U. Kozat and L. Tassiulas, "Network layer support for service discovery in mobile ad hoc networks", *IEEE Infocom 2003*

[15]   Tschudin, C., Gunningber, P., Lundgren, H., Nordstrom, E., "Lessons from experimental MANET research", *Ad Hoc Networks*, Vol. 3, Issue 2, pp.221-233, March 2005, Elsevier, 2005

[16]   G. Pavlou, P. Flegkas, S. Gouveris, A. Liotta, *On Management Technologies and the Potential of Web Services*, IEEE Communications, special issue on XML-based Management of Networks and Services, Vol. 42, No. 7, pp. 58-66, IEEE, July 2004.

[17]   XML-RPC specifications web site, http://www.xmlrpc.com/spec, accessed April 2005

[18]   K. Sollins, The TFTP Protocol, IETF RFC 1350, July 1992

[19]   C. E. Perkins, E. M. Belding-Royer, and S.R. Das, *Ad hoc On-Demand Distance Vector (AODV) Routing*, draft-ietf-manet-aodv-13.txt.

[20]   Clausen, T., Jacquet, P., *Optimized Link State Routing Protocol (OLSR)*, RFC 3626, October 2003.

Autonomous Self-deployment of Wireless Access Networks in an Airport Environment

Holger Claussen

Bell Laboratories, Lucent Technologies,
The Quadrant, Stonehill Green, Westlea, Swindon,
SN5 7DJ, United Kingdom
claussen@lucent.com

Abstract. In environments with highly dynamic user demand, for example in airports, high over-dimensioning of wireless access networks is required to be able to serve high user densities at any possible location in the covered area, resulting in a large number of base stations. This problem is addressed with the novel concept of a self-deploying network. Distributed algorithms are proposed, which autonomously identify the need of changes in position and configuration of wireless access nodes and adapt the network to its environment. It is shown that a self-deploying network can significantly reduce the number of required base stations compared to a conventional statically deployed network. In this paper, this is demonstrated in a specific test scenario at Athens International Airport, simulating a moving user hotspot after the arrival of an airplane.

1 Introduction

In an airport environment, the arrival and departure of airplanes results in a highly dynamic environment. User demand and positions are changing rapidly with the result that high over-dimensioning of wireless access networks is required to meet the need of high bandwidth services at any possible hot-spot location. In this paper, autonomous adaptation of base station positions is investigated as a possible means to reduce the total number of required base stations in such environments. Such self-deploying network [1] would be able to identify the need for changes in both base station positions and configuration, and implement these changes without human intervention. The potential reduction of required base stations is investigated in a specific scenario at Athens International Airport, simulating a moving user hotspot after the arrival of an airplane. Mobile base stations are considered which are deployed on a rail at the ceiling of the terminal building, and are able to move autonomously along this rail, as illustrated in Fig. 1. Instead of over-dimensioning the network for the highest expected user density at any possible location, mobility of base stations allows it to adapt autonomously to changes in user locations and demand with the result of a significantly reduced number of required base stations.

While base station mobility might seem futuristic for commercial wireless communication systems (due to the costs involved in providing base station mobility), this concept has near-term applications in the field of military and emergency communications, where fast network deployment is required in high-risk areas or in environments

I. Stavrakakis and M. Smirnov (Eds.): WAC 2005, LNCS 3854, pp. 86–98, 2006.

that are difficult to access. The airport scenario was chosen for its simplicity in order to minimise the computational complexity of the environment simulation and to demonstrate the proposed algorithms, which are not limited to such a one-dimensional self-deployment implementation.

Base station positioning has been studied extensively in the past, using simulated annealing [2,3], evolutionary algorithms [4], linear programming [5], and greedy algorithms [6,7]. Other work has explored the trade-offs between coverage, cell count and capacity [8]. It has been shown that the identification of the globally optimum base station locations in a network of multiple base stations is an NP-hard problem, far too complex to solve computationally [4-6]. Further difficulties are that most of the system parameters required to find an optimal solution are unknown, and the optimal positions change constantly due to the changes in user demand, user positions, and base station positions.

The objective is the development of algorithms that are able to find near-optimum solutions for self-deployment and self-configuration, based only on limited local system knowledge. To achieve a high robustness and scalability, radically distributed processing which results in self-organising behaviour is investigated. An additional objective is to avoid or minimise direct communication between base stations in order to reduce the signalling overhead and allow technology independent operation. In this way, the network may consist of base stations with different access technologies such as UMTS or 802.11.

This paper is organised as follows. In Section 2, the use of stigmergy [9] for indirect communication between base stations is investigated in order to achieve a globally self-organising behaviour of base station locations in a network. In Section 3 the difficulties involved in finding the optimal locations of base stations in a network are discussed. Globally and locally optimal solutions are presented and modified, to allow self-deployment with limited local system knowledge. Simulation results in an airport scenario are presented in Section 4 and finally, conclusions are drawn in Section 5.

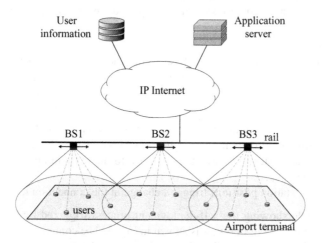

Fig. 1. Autonomous, self-deploying wireless access network. Mobile base stations are mounted on a rail in an airport environment.

2 Self-deployment and Stigmergy

Avoiding direct communication between base stations makes the optimisation problem very challenging since some means of communication is necessary to optimise the network globally. This problem may be addressed by using indirect communication, where each base station modifies its surrounding environment, and these changes then influence the behaviour of neighbouring base stations. In the field of biology, such interaction is known as stigmergy and is widely used by social insects to coordinate their activities by means of self-organisation (e.g. ants use decaying pheromone trails to find shortest paths).

In wireless communication systems, the environment in the network relates to the connections to the mobiles. When mobiles connect to the base station with the strongest received control pilot power, these connections provide information on the coverage of neighbouring cells. One possible driver for a change in the network environment is the modification of base station positions. Other possibilities are, for example, changing user demand or the adaptation of the pilot powers to achieve load balancing (either equal transmit power, or equal capacity) in each cell. The modification of the network environment through re-positioning or load balancing provides an indirect way of communication between the base stations.

One advantage of the proposed indirect communication is that it can be considered as a universal language which allows interoperability of heterogeneous systems (i.e. systems with different access technologies) since base stations do not need to be able to exchange data directly with other base stations in the network.

Examples:
An example of the self-organisation process, resulting from indirect communication between base stations and local optimisation of each base station location is illustrated in Fig. 2. Base stations are shown as solid squares and mobiles are shown as circles with a line to the connected base station. The optimal base station positions are shown as squares.

Start condition:
   All mobiles are connected to the base station dependent on the connection rule (strongest received control pilot power). This defines the current network environment.

Continuous self-deployment process:
   • In each step, the optimal positions for all base stations are calculated, based on the current network environment (i.e. connections) seen by each base station.
   • In each following step, all base stations move to the optimum positions predicted in the previous step.
   • The new base station positions trigger a change in the connection to the mobiles.

A further example showing the self-deployment process triggered by load balancing via modification of the pilot powers is shown in Fig. 3. The contour plots illustrate the received control pilot power. When BS2 reduces its pilot power for load balancing (Step 1), BS1 takes over several connections (Step 2). As a result, both base stations optimise their positions for their changed connections (Step 3).

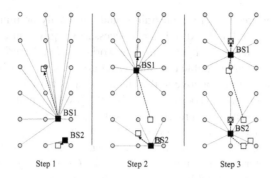

Fig. 2. Self-deployment using stigmergy

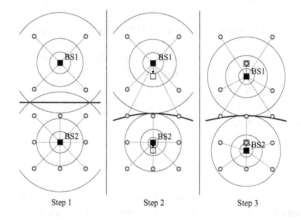

Fig. 3. Load balancing through re-positioning of base stations

3 Base Station Positioning Algorithms

The optimal position of a base station can depend on a variety of factors. While optimisation of the resource efficiency is an obvious criterion, other factors or constraints such as suitable locations, costs, or legislation also play an important role. This investigation, is focussed on the optimal use of resources (i.e. transmit power and available frequency spectrum), within constraints such as maximum transmit power levels of single base stations or possible locations. From this standpoint, rules for optimal positioning of individual base stations, and base stations in a network, can be stated as follows:

Rule1: Local optimisation of individual base stations
The optimal position for an individual base station allows it to sustain all requested connections with the minimum possible transmit power.

Rule 2: Global optimisation of base stations in a network
The optimal positions of all base stations in a network allow the network to sustain all requested connections with the minimum possible transmit power.

Note that both rules are subject to constraints, and the locally optimum position of a single base station according to Rule 1 is not necessarily equivalent to the position of the same base station in a globally optimised network based on Rule 2.

To satisfy the minimum possible transmit power criterion for an arbitrary small bit-error rate, the receivers must operate at the Shannon capacity limit. In fact, recent advances in coding theory (turbo codes, LDPC codes) allow communications very close to the capacity limit even in the presence of fast fading. Therefore, the capacity limit itself may be targeted as the optimisation point for the wireless access network.

The following assumptions are made: In order to use simple capacity equations, the intra- an inter-cell interference is modelled as a white Gaussian random variable with zero mean. This can be justified by arguing that for a large number of interferers, the total interference becomes Gaussian. In addition, only the slow fading components of the channel are taken into account for the base station positioning.

3.1 Minimum Power Requirement for a Link with Given Capacity

The channel capacity C for a channel perturbed by additive white Gaussian noise is a function of the average received signal power $P_{Rx} = E\{s(t)s(t)^*\}$, the average noise power $N = E\{n(t)n(t)^*\}$ and the bandwidth B, where $s(t)$ and $n(t)$ denote the signal and noise values at the time instant t. The well known capacity relationship (Shannon-Hartley theorem [10]) can be expressed as

$$C = B \log_2 \left(1 + \frac{P_{Rx}}{N}\right). \tag{1}$$

In order to write (1) in terms of transmitted power P_{Tx}, the impact of the channel loss $L = L_p \cdot L_s$, characterised as a combination of attenuations resulting from path loss L_p and shadow fading L_s and must be taken into account. Note that this requires knowledge of the positions of the connected mobiles and knowledge of the environment (i.e. shadow fading properties). In addition, gains at the base station and the mobile, G_{BS} and G_{UE}, can be included. Then, the channel capacity can be rewritten as

$$C = B \log_2 \left(1 + \frac{P_{Tx} \cdot G_{BS} \cdot G_{UE}}{N \cdot L}\right). \tag{2}$$

Finally, the minimum required transmit power for a radio link of capacity C for given values of bandwidth B, channel attenuation L and received noise N (including interference) operating a factor of α from the capacity limit, can be determined as

$$P_{Tx} = \frac{\alpha \cdot N \cdot L}{G_{BS} \cdot G_{UE}} \left(2^{C/B} - 1\right). \tag{3}$$

Here, the capacity C represents the requested data rate and the bandwidth B of the radio link is known.

3.2 Globally Optimum Positioning

For joint optimisation of the whole network, the optimal positions of all base stations minimise the total transmitted power for all requested links (Rule 2). The optimum set of coordinates for all M base stations and all K_m requested links to the mth base station can be written as

$$(\mathbf{x}_{opt}, \mathbf{y}_{opt}) = \arg\min_{(\mathbf{x},\mathbf{y})} \left\{ \sum_{m=1}^{M} \sum_{k=1}^{K_m} P_{Tx,m}^{(k)}(x_m, y_m) \right\}, \tag{4}$$

where $(\mathbf{x}, \mathbf{y}) = (\{x_1...x_M\}, \{y_1...y_M\})$ is the set of possible base station position coordinates. The indices for the base station and the link are denoted by m and k, respectively. $P_{Tx,m}^{(k)}(x_m, y_m)$ denotes the required transmit power from (3) for the kth link of the mth base station at the coordinates (x_m, y_m) within the possible region of deployment.

Alternatively to using specific connections for the calculation of the required transmit power $P_{Tx,m}^{(k)}(x_m, y_m)$, the above problem may be solved for a given user and demand distribution. Then, for each potential user location the expected value $E\{ P_{Tx,m}^{(k)}(x_m, y_m) \}$ may be used instead. Each base station can collect the required user statistics during operation. This approach results in the average optimum position and can be used to optimise the positions of non-mobile base stations that require human intervention to move.

The optimisation of (4) implies a search over a very large number of candidates, which grows exponential with the number of base stations. Therefore, an exhaustive search for jointly optimal positions for more than a few base stations in a limited area is impractical due to prohibitive computational complexity (i.e. NP-hard problem). In addition, centralised processing is necessary and complete system knowledge is required. However, in reality most of the required parameters (e.g. channels and interference at new positions) are unknown. Therefore, even if the computational complexity were manageable, it would still be impossible to compute the globally optimum positions due to incomplete system knowledge.

3.3 Locally Optimum Positioning

For each individual mth base station, the position can be optimised locally, by searching for a position, which minimises its transmitted power for all K_m requested links (Rule 1). Then, the locally optimum coordinates of each mth base station may be calculated as

$$(x_{opt}, y_{opt}) = \arg\min_{(x_m, y_m)} \left\{ \sum_{k=1}^{K_m} P_{Tx,m}^{(k)}(x_m, y_m) \right\}. \tag{5}$$

Again, the optimisation problem may be solved for a given user and demand distribution instead of for specific connections by using the expected value of the transmit power, required at each potential user location.

In contrast to the global optimisation, the local optimisation can be solved in a decentralised manner, based only on local system knowledge. However, as before, not all of the required system knowledge is available. At each potentially new base station position, the channel conditions (i.e. L_s), and therefore also the interference at both, mobiles and base stations, are unknown.

3.4 Positioning with Limited System Knowledge

As shown in Section 3.2, the globally optimal positioning of networks is a challenging task due to limited knowledge of the constantly changing system parameters and the prohibitive computational complexity. The locally optimum solution of Section 3.3 is of manageable computational complexity, but suffers from the same problem of incomplete system knowledge. As a consequence, other solutions based on partial system knowledge are required that provide results close to the optimum solution.

Current values for shadow fading and interference levels seen by each node can be easily measured. However, when the base station positions change relative to the interference sources, both, the shadow fading values and also the interference, can change unpredictably. Therefore, the shadow fading values L_s, and the interference levels, which dominate N in (3), at any new potential base station position can be considered as unknown. Under this assumption, the local optimisation criterion of (5) may be modified to

$$(x_{opt}, y_{opt}) = \arg \min_{(x_m, y_m)} \left\{ \sum_{k=1}^{K_m} \varphi_m^{(k)}(x_m, y_m) \right\}, \tag{6}$$

with

$$\varphi_m^{(k)}(x_m, y_m) = \frac{\alpha \cdot L_p}{G_{BS} \cdot G_{UE}} \left(2^{C/B} - 1 \right). \tag{7}$$

The strategy is to take any knowledge available into account, and ignore (or replace with their expected value) all unknown contributions. Here, L is replaced with L_p, since $E\{L\} = L_p$ and N is ignored. Alternatively, N could be estimated by calculating intercell interference based on path-loss only, and assuming constant intra-cell interference.

Equation (6) represents a convex optimisation function that can be solved using either an exhaustive search, or less complex approaches such as steepest descent or conjugate gradient methods [11].

4 Simulation Results

In order to evaluate the impact of autonomous self-deployment on the required number of base stations, both conventional and self-deploying wireless access networks were simulated for a specific test scenario in the terminal building at Athens International Airport. The scenario is illustrated in Fig. 4, where base stations are shown as solid squares and mobiles are shown as circles with a line to the connected base station. The arrows indicate the movement of the user hotspot.

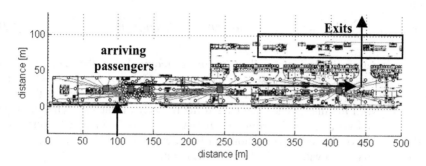

Fig. 4. Test scenario in the terminal building of Athens International Airport

Simulation steps:

(a) Start condition: uniform user and base station distribution along the terminal corridor.
(b) An airplane arrives and the passengers create a user hot-spot (shown in Fig. 4)
(c) The arriving passengers move along the corridor in direction of the airport exits (indicated by arrows).
(d) The arriving passengers leave the airport, and the user hotspot disappears.
(e) Finally, the user distribution becomes uniform again.

System level simulations were performed for the downlink of a generic wireless system to identify both the required number of base stations and the network performance, in terms of total required transmit power, for self-deploying and conventional networks. The evaluation was performed in an iterative manner until a convergence point for the link transmit powers was reached. In this way it is possible to take into account that the transmit power of each link depends on the powers of all other links in the system, and vice versa. It is assumed that each mobile connects to the base station with the highest received control pilot power. Load balancing via modification of the control pilot power is employed such that all base stations try to stay within both power and capacity limits. An additional pilot for channel estimation is assumed to require 10% of the transmit power used for data at each base station. For each simulation step, the evaluation was performed as follows, using the parameters shown in Table 1.

$P_{BS}(0) = \text{zeros}(M)$ *% initialise base station powers with zeros*
for $i = 1 \ldots I_{max}$ *% for a maximum of I_{max} iterations*
    for $m = 1 \ldots M$ *% for all M base stations*

$$P_{BS,m}(i) = \sum_{k=1}^{K_m} P_{Tx,m}^{(k)}(i-1) + P_{pilot} \quad \textit{\% calculate BS powers} \tag{8}$$

$$\delta_m(i) = \left| P_{BS,m}(i) - P_{BS,m}(i-1) \right| / P_{BS,m}(i) \tag{9}$$

    end
    if $\max(\delta_m(i)) < 0.01$ *% convergence criterion*
        break;   *% break iterations when BS powers are converged*
    end
end

$$P_{\text{Network,DL}} = \sum_{m=1}^{M} P_{\text{BS},m}(i) \qquad \% \text{ network performance metric} \qquad (10)$$

In each iteration i, the inter-cell interference required for the calculation of $P_{\text{Tx},m}^{(n)}(i-1)$ can be calculated based on the transmit power of each mth interfering base station as $P_{\text{I,inter}} = L\,P_{\text{BS},m}(i-1)$ from the previous iteration, where L is the channel loss between the interference source and the receiver of interest. When multiple links are served simultaneously from a single base station, the intra-cell interference for the nth link of the mth base station can be calculated as $P_{\text{I,intra}} = L[P_{\text{BS},m}(i-1) - P_{\text{Tx},m}^{(n)}(i-1)]$, based on values from the previous iteration.

Table 1. Simulation parameters

Parameter	value
Maximum BS transmit power	0.25 W
Maximum number of users per BS	32 users
Channel bandwidth B	3.84 MHz
Link capacity C	64 KBit/s
BS antenna gain – cable loss $G_{\text{BS,[dB]}}$	5 dB
UE antenna gain – cable loss $G_{\text{UE,[dB]}}$	0 dB
Operation point (from channel capacity) $\alpha_{\text{[dB]}}$	7 dB
UE noise figure $NF_{\text{[dB]}}$	10 dB
Shadow fading standard deviation	6 dB
Shadow fading spatial correlation	$r(x)=e^{-x/20}$
Path loss $L_{\text{p,[dB]}}$	$37+30\log(d)$ dB
Maximum BS speed	5 m/s

For the optimisation of the base station locations, the positioning algorithms based on limited local system knowledge of (6) and (7) are employed and solved by using a simple steepest descent algorithm. It is assumed that each base station has knowledge of the path loss, but the shadow fading variations are unknown. A spatially correlated shadow fading environment was generated as described in [12].

The simulations indicate that the self-deploying network requires at least five mobile base stations to serve all user requests of the simulated scenario. The user and base station locations, and the control pilot power during the autonomous self-deployment process are depicted in Fig. 5. As start condition, all base stations are uniformly distributed to provide service to a uniform user distribution of 75 mobiles (a). Then a plane arrives and the passengers create a user hotspot of additional 75 mobiles (b). The capability of autonomous repositioning allows the base stations to adapt to the changing user and demand distributions and move to the user hotspot to increase the capacity in this region. When the users move in direction of the airport exits, the base stations follow their movement and hand the users over to their neighbouring base stations (c). In this way, a small number of base stations have the ability to serve a large number of users in highly dynamic scenarios. Arriving at the exits the users leave the airport and the hotspot disappears (d). As a consequence, the base stations spread out again to serve the remaining users (e).

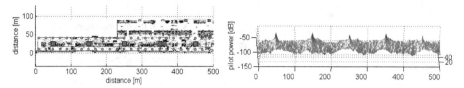

(a) Uniform user distribution along the gates

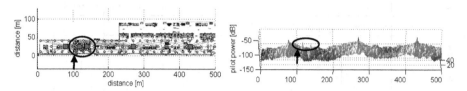

(b) A plane arrives and the passengers create a user hotspot. Base stations move to the user hot-spot to provide the required capacity.

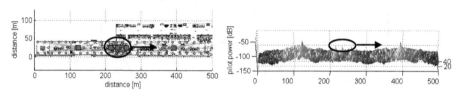

(c) The users move along the corridor to the airport exits. Base stations follow their movement.

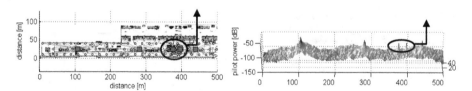

(d) Arriving at the exits, the users leave the terminal building (hotspot disappears).

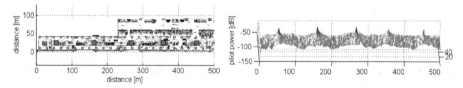

(e) The base stations spread out uniformly to serve the remaining users

Fig. 5. Simulation steps of a self-deploying network in an airport environment. A minimum number of five mobile base stations is required for this scenario.

In the same scenario, a conventional wireless access network with fixed base station deployment requires at least nine base stations to achieve similar performance as the self-deploying network. This over-dimensioning is required to allow the network

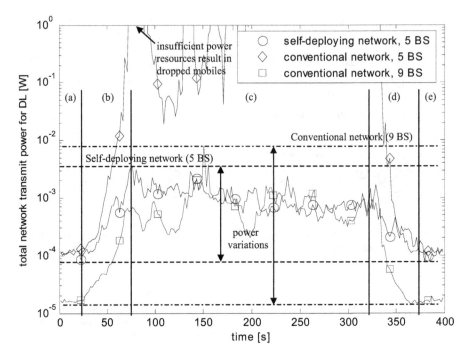

Fig. 6. Performance comparison of self-deploying and conventional networks

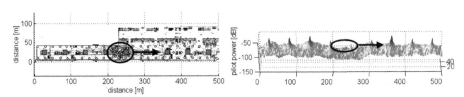

(c) The users move along the corridor to the airport exits.

Fig. 7. Simulation step of a conventional network with fixed base station deployment in an airport environment. A minimum number of nine base stations are required to achieve a similar performance as the self-deploying network with five base stations.

to cope with the moving user hotspot, without having the ability of base station repositioning. Therefore, it must be dimensioned for the highest expected user density at any possible location.

Figure 6 depicts a performance comparison of conventional and self-deploying networks. It is shown that a self-deploying network with only five base stations is able to outperform a conventional network with nine base stations. In addition, the self-deploying network shows much less variations in the required transmit power. A conventional network with five base stations exceeds the maximum base station power resources, and therefore is not able to provide all requested services in the test scenario.

The results confirm that self-deploying wireless access networks are able to significantly outperform conventional networks, since they are able to adapt effectively to changing user demand and user locations, and therefore do not require high over-dimensioning as conventional networks to cope with dynamic network environments.

5 Conclusions

In this paper, the concept of a self-deploying wireless access network was used to reduce the required number of base stations in highly dynamic environments. Distributed algorithms based on the channel capacity were proposed that are able to autonomously identify required changes in position and configuration of wireless access nodes, dependent on the demand and locations of users. It was shown that self-deploying networks using the proposed algorithms are able to significantly outperform conventional networks with fixed base station positions. For the investigated test scenario at Athens International Airport, this resulted in a reduction of the required number of base stations from nine, for the conventional network, to only five self-deploying base stations with improved network performance. This promising result demonstrates the potential advantages of autonomous, self-deploying wireless access networks. Future research will have to investigate both, technical robustness and economic viability of such self-aware and self-designing networks, critical for the widespread adoption in next-generation wireless access architectures.

Acknowledgements

A portion of this work was part-supported by the EU commission through the IST FP5 Project IST-2001-39117 ADAMANT. The author wishes to thank Reza Karimi, Lester Ho, Francis Mullany, and David Abusch-Magder for helpful discussions.

References

[1] F. J. Mullany, L. T. W. Ho, L. G. Samuel, and H. Claussen, "Self-deployment, self-configuration: Critical future paradigms for wireless access networks," in *Lecture Notes in Computer Science 3457 – Autonomic Communications*, pp. 58-68, 2004.
[2] S. Hurley, "Planning effective cellular mobile radio networks," *IEEE Transactions on Vehicular Technology*, vol. 51, no. 2, pp. 243-253, 2002.
[3] S. U. Thiel, P. Giuliani, L. J. Ibbetson, and D. Lister, "An automated UMTS site selection tool," in *Proc. 3G Mobile Communication Technologies*, pp. 69-73, 2002.
[4] N. Weicker, G. Szabo, K. Weicker, and P. Widmayer, "Evolutionary multiobjective optimization for base station transmitter placement with frequency assignment," *IEEE Transactions on Evolutionary Computation*, vol. 7, no. 2, pp. 189-203, 2003.
[5] R. Mathar and T. Niessen, "Optimum positioning of base stations for cellular radio networks," *Wireless Networks*, vol. 6, no. 6, pp. 421-428, 2000.
[6] R. Chandra, L. Qiu, K. Jain and M. Mahdian, "Optimizing the Placement of Integration Points in Multi-hop Wireless Networks," in *Proc. IEEE ICNP*, 2004.

[7] K. Tutschku, "Demand-based radio network planning of cellular mobile communication systems," in *Proc. 17th Annual INFOCOM*, pp. 1054-1061, 1998.

[8] D. Abusch-Magder and J. M. Graybeal, "Novel algorithms for efficient exploration of the trade-off between cell count and performance in wireless networks," *BLTJ*, vol. 10, is. 2, 2005.

[9] E. Bonabeau, M. Dorigo, and G. Theraulaz, *Swarm intelligence – from natural to artificial systems*, Oxford University Press, 1999.

[10] C. E. Shannon, "A mathematical theory of communication," BSTJ, vol. 27, pp. 379-423, 623-656, 1948.

[11] W. H. Press, S. A. Teukolsky, W. T. Vetterling, and B.P. Flannery, *Numerical recipes in C++ - The art of scientific computing,* Cambridge University Press, 2002.

[12] H. Claussen, "Efficient modelling of channel maps with correlated shadow fading in mobile radio systems," in *Proc. IEEE International Symposium on Personal Indoor and Mobile Radio Communications PIMRC*, 2005.

Knowledge Networks

Maurice Mulvenna[1], Franco Zambonelli[2], Kevin Curran[1], and Chris Nugent[1]

[1] School of Computing and Mathematics, University of Ulster, Newtownabbey, BT37 0QB, UK
{md.mulvenna, kj.curran, cd.nugent}@ulster.ac.uk
[2] DISMI - Università di Modena e Reggio Emilia, Italy
franco.zambonelli@unimore.it

Abstract. For future network scenarios to exhibit autonomic behaviour, both networks and application components and services need to be aware of their computational and environmental context, and must tune their activities accordingly. In this position paper, we propose an abstract architecture for knowledge networks that addresses the key issues of how both physical contextual knowledge and social knowledge from the users of communication networks can be used to form a knowledge space in support of autonomic agents dealing with network elements and applications. We discuss that the availability of raw contextual data is not enough to achieve meaningful autonomic behaviours. Rather, contextual information should be properly organised into 'networks of knowledge', to be exploited by both network and application components as the basic 'nervous system' in which situational stimuli reify into digital knowledge, and by means of which components can properly orchestrate their activities in a globally meaningful way. Here we firstly discuss the fundamental role of knowledge networks, and try to sketch what actual form and position such knowledge networks could assume. Then, we analyse some simple scenarios of use, showing how it is possible for the components of an autonomic communication system to build such knowledge networks autonomously; and, at the same time, to exploit them for orchestrating their activities in a type of stigmergy-based knowledge-rich system. Eventually, we sketch a rough research agenda and discuss the relations with other research areas.

1 Introduction

We envision that future networks will be able to provide composite, highly distributed, pervasive services in a situated and fully autonomic way. In other words, they will be made up of components capable of [KepC03, Zam05] understanding the general context – physical, technological, social, user-specific and request-specific – in which they operate; and spontaneously aggregating with each other and orchestrating their activities accordingly to that context, so as to support a range of activities and services activities that are simply not possible or impractical now, with the important addition of requesting no configuration efforts from users.

In particular, we expect services to be able to:

(i)   Improve our interactions with the physical world by providing us with any needed information about our surrounding physical environment and

I. Stavrakakis and M. Smirnov (Eds.): WAC 2005, LNCS 3854, pp. 99–114, 2006.

exploiting such information to adapt/enrich their behaviour on such basis (e.g., consider adapting the behaviour of a tourist service network on the basis of the location from which the service is invoked and of the current weather and traffic conditions) [Est02];

(ii)   Get the best of the network infrastructure and resources upon which they operate, being able to ensure sufficient quality of service adaptively and independently of the actual network characteristics (e.g., independently of the fact that we require them from a Wi-Fi PDA, from a GPRS phone, or from whatever connectivity and connected devices will be available at that time) [MikM04];

(iii) Facilitate our social interactions, by properly reflecting and exploiting the social context in which we are currently employing a service, e.g. for mere entertainment, or socialisation, or in the context of business activities. Such social possibilities could be particularly appreciated in an increasingly open and multicultural environment such as the EU [ChoP03, Pen05].

A central challenge for the above vision to become real is the promotion of suitable solutions for enabling the components of an autonomic communication infrastructure (whether network-level or application-level components) to become situation-aware. Assuming that mechanisms exists to produce all necessary "situational" knowledge (e.g., sensors and monitoring mechanism [Est02, Gel02], user and social profilers [Pen05], etc.), for components to exploit the knowledge properly it is necessary that all the available knowledge (which can be in a dramatic amount, can be distributed, decentralized, and can come from a multitude of sources) is organised for utilisation.

Organising all available situational information implies that any relations between information is properly represented and correlated (according to well-defined ontological constructs), so as to facilitate their retrieval and their understanding. To promote accessibility, it is necessary that information produced locally at one place is properly diffused in the network whenever this may be of a more global relevance. Also, it may be important that such information can be exploited for mediated (i.e., stigmergic) interactions among the components of the infrastructure, so as to promote both robust self-organising behaviours [DiM04] and fruitful cross-layer interactions.

These needs lead us to the general concept of *knowledge networks*, intended as a form of overlay – distributed in a network scenario and being an integral part of the overall infrastructure – in which all the information about the context is properly represented, organised, and correlated, and around which semantically-enriched stigmergic interactions among the components of the autonomic infrastructure can take place [Par97]. That is, a distributed knowledge infrastructure representing a sort of nervous system for the autonomic communication system, across which all information and stimuli needed for the coordinated functioning of the system flow and get organized.

This position paper aims at unfolding the idea of knowledge networks and it is organised as follows. Section 2 details on the need for knowledge networks, and tries to identify what actual role and position they could assume in future autonomic communication scenarios. Section 3 elaborates on the potentials of knowledge networks in future scenarios, also with the help of a few examples. Section 4 sketches a rough research agenda and discusses related work in the area. Section 5 concludes.

2 Knowledge Networks

We are now witnessing an age of computing ubiquity where our work and home environments are increasingly enveloped by computing resources. This comes at a cost, which is the significant problem of configuration and complexity of these resources. If computing power is to serve us, and the converse is to be denied, then these resources and their rich panoply of services must be able to carry out their increasingly complex functions without significant intrusion into our lives.

These services, with underlying technology network entities encompassing autonomic computing and communication systems, require a high degree of contextual knowledge, including knowledge about the social, computational, and physical environments in which they are situated, as well as self-knowledge about their own functioning. There is a requirement for future autonomic networks to provide meaningful knowledge-based decision making, and ultimately to infuse pervasive systems and improve our human experience of interaction. This is what Weiser [Wei91] describes as the notion of *calm*, where the computing resources quietly modify themselves to suit the needs of the user.

2.1 Why Are They Needed?

Autonomic communications networks (both the network resources and the application components and services exploiting them) need to reason about their situation and to understand their own behaviour. To do this they are required (both at the level of individual components and as a whole) to be introspective and reflective, and to feed back the results of these processes to be used to improve performance. This is the *raison d'etre* to make networks smarter, to make them more self-aware, and to provide the knowledge with which they can manage themselves. In order to manage themselves, the network and its entities and services need some form of "*knowledge networks*" through which all available knowledge is properly represented, correlated, and accessed. The reasons that lead to that concept of knowledge networks are synthesised below.

Firstly, there is a basic need for expressive and flexible means to promote context-awareness. Networks, their entities and services need to have an awareness of situations with differing degrees of granularity [Ste05]. There is a requirement for some form of computational model of context processing as in [Bal00] that orchestrates context stimuli and components in a coherent representation. We also need some way to gauge the quality of our contextual information objectively as it is gathered, as from the Quality of Context mechanism of Buchholz *et al.* [Buc03], in which any contextual information comes associated with parameters including precision of information, correctness probability, trust worthiness, resolution and regency. Simply said, contextual information cannot reduce to a trivial set of data to be accessed by components, but requires some higher-form of organization.

Secondly, contextual information cannot be simply considered as local and locally available to components and services. For a satisfactory adaptive orchestration of distributed activities (whether this is intended to be the orchestrated configuration of network components or the coordination of distributed service components), the exploitation of local knowledge only may not be enough. Nor can one think of concentrating in a single site or of replicating anywhere all available knowledge, especially when this knowledge represents dynamically evolving situations, i.e., it is subject to obsolescence. The compromise solution is to enable components which need more than simply local knowledge to organise and correlate distributed knowledge into sorts of networks that enable distributed components to "navigate" through the available knowledge to attain, on demand, the required degree of contextual awareness.

Third, there is a recognised need for future autonomic communication scenarios to promote cross-layer interactions [SAC05]. This means that the service level and the network level cannot work as separated universes, each towards its own goals. Rather, a continuous exchange of information must occur between the service and the network level, and vice-versa, so as to ensure that the overall activities of the system, at each level, will contribute towards the achievement of a satisfactory functioning. For this coordination and exchange of information to occur without significant interoperability issues, there must be some place where common information can be stored and can be properly organised so as to be accessible and understandable by both the network and the application levels, and accordingly to the means proper of each level.

Fourthly, it is known that a reasonable and effective way to promote self-organization and self-adaptation (i.e., autonomic behaviour) in distributed systems is via stigmergy, i.e., by indirect interactions occurring via a computational environment in which components can spread and sense information [Par97]. The presence of a distributed network of knowledge, to be accessed for sensing and effecting by both network and application level components, can act as the computational environment to enforce stigmergic self-organization. Moreover, if such space other than simple digital pheromones can contain properly represented and correlated situational knowledge, one can think at leveraging stigmergy to more sophisticated forms of cognitive self-organization.

2.2 What Form Could Knowledge Networks Take?

Knowledge networks are *reflective spaces* for autonomic communication systems. Being capable of storing distributed, heterogeneous, dynamically constructed, sophisticated knowledge, they can form a conceptual middle layer across which network components as well as application-level components can access information and can coordinate with each other. They act as a form of network memory, in which knowledge may be replenished continually as the network and its entities evolve and reflect introspectively. But what form do knowledge networks take, and where does the knowledge reside?

The schematic in Figure 1 illustrates how we conceive knowledge networks as a conceptual layer, positioned between the physical network level (there included the physical level, reified in the forms of the environmental information that can be produced by sensors) and the application level (there included the social level, reified in the form of social information produced by social/user profilers).

In general, the knowledge generated by both levels reaches the same conceptual knowledge level, and here it is properly put in context of extant knowledge. This means that the knowledge has to be:

- Dynamically generated and represented in proper ontological relations;
- Properly correlated, i.e., networked with existing knowledge on the basis of what it represents and of what use it and related knowledge may be to the application or the network level.

We assume that each entity in the network, whether a software agent or a network component, has the capability of accessing the knowledge network layer for reading the knowledge in it, understanding the ontological relations between different pieces of knowledge and navigating links that relate distributed knowledge. By this, components can also properly understand where newly produced knowledge can be inserted in a knowledge network, and how this has to relate with existing knowledge. That is, components have the capability of dynamically shaping the knowledge networks to have it always reflect the current overall situation of the network.

In general, we do not consider the presence of specific computational entities in charge of maintaining and updating knowledge networks [Cla03]. Such a solution would be too heavyweight to be general-purpose, and would introduce additional complexities. Rather, we consider that components at both the application and the network level will be directly in charge of populating, storing, and maintaining

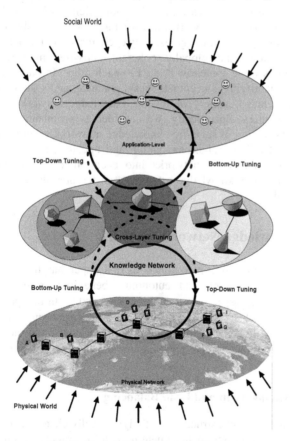

Fig. 1. Knowledge Networks in Context

portions of fully distributed knowledge networks, arguably with the help of reactive code fragments associated to knowledge pieces and aimed at automating their update and maintenance upon changing conditions.

For the actual production and update of the network of relations in a knowledge network, we may consider an ontological construct at both network and application level that is replenished continuously via the introspective process described earlier. This behavioural feedback loop, in essence, is the knowledge generator that dynamically populates the ontology. The ontological construct must be designed to be very flexible, even to the extent of facilitating self-revision [Hef01]. It must also be capable of fusing contexts [May03] and knowledge from different ontologies as networks and devices interconnect in an *ad hoc* manner.

Given the above considerations, the knowledge networks can act as the mean via which intra-level tuning of activities may occur (see Figure 1). However, they can also act as the mean via which the application-level can tune its activities to reflect events occurring at the network-level, and vice-versa. In addition, given that the activities of a component may reflect in some change/update in the knowledge network, that some other components can sense and for which its own behaviour can be affected, the possibility of stigmergic interaction is intrinsically promoted.

In general, we consider the possibility of a multiplicity of knowledge networks to co-exist in the same overall network infrastructure, each possibly serving different application-level or network-level goals. However, the need for achieving effective cross-layer interactions and globally coherent activities may require different knowledge networks to be somehow related to each other. In particular we envision the possibility of identifying conceptually easy, practical, and scalable ways by which to compose and relate a variety of diverse knowledge networks and the diverse knowledge they contain. Specifically, we consider as promising the possibility of enforcing the construction of scale-free knowledge networks, exhibiting a self-similar structure that can facilitate robust navigation and update. Also, this could promote nesting of various knowledge networks into each other and, accordingly, could tolerate an exploitation of knowledge networks at different scales (zooming in and out depending on needs).

3 Putting Knowledge Networks to Work

The knowledge network concept outlined in this paper has the potential to make possible a sophisticated degree of autonomic behaviour in future networks by providing them with introspective, cross-layer knowledge. In this position paper, we do not have the clear visibility to help describe some proof-of-concept implementation, nor do we already have crystallised ideas about how all the above ideas could be realised. Nevertheless, we can try sketch some potential applications of the concept.

3.1 Resource Management and Load Balancing

Any distributed network infrastructures with dynamically changing resource demands requires some sort of resource management tools to have its all resources effectively exploited and to provide reasonable quality of service to the application level. For the sake of simplicity, let us focus on the load balancing issue.

Traditional distributed load balancing tools consider the presence of system level processes, devoted to handle specific local resources (whether computing, communication, or memory resources) capable of monitoring the current load on local hardware resources [ShiKS92, Xu95]. Whenever they perceive specific resources are overloaded (or under utilized), these processes engage other processes in some sort of negotiation, aimed at re-distributing the load on the system resources (e.g., by re-allocating some application-level process or by establishing different routing paths).

A variety of strategies have been proposed for distributed load balancing. Different strategies can be conceived for having local processes understand if they are overloaded or underloaded: this can rely on static non-adaptive load thresholds, or they can be based on some sorts of load information exchange with other nodes to comparatively estimate the local load. Different strategies can also be conceived for negotiation, depending on which nodes (overloaded or underloaded) initiate negotiations, and on which nodes in the system (all nodes or a limited number of "close" nodes) have to be involved in it.

However, for all the above traditional approaches, the strategy rely on local system processes to perceive local load information, possibly acquire more global information by requesting it to colleague processes on different nodes, and act on the basis of this information. Nothing is traditionally said about the possibility of organising distributed load information to promote more informed decisions without having processes to explicitly coordinate with each other every time a decision has to be taken. Nothing is traditionally said about the possibility of exploiting application-level information to enforce load re-distribution patterns that, other than satisfying the hardware viewpoint, can also accommodate specific application-level needs.

The idea of knowledge networks lets us envision a radically different approach to load balancing. Rather than having processes elaborate local information, we could think of having local load information be injected (and updated upon significant changes in value) in a knowledge network to contribute to the dynamic formation of a distributed "load field", representing in a sort of virtual landscape of the distribution of load over the network. The local value of load field and its local gradients can then be perceived by system level processes to understand "where" in the network/landscape load increases or decreases, and to somehow understand not only what is the local load, but also how such local load relates to the overall load in the network. Also, such fields can be enriched with semantic information describing e.g., the types of resources involved and any additional resource-specific information to could serve the load balancing purpose.

Given the availability of the load field, one can think of having load distribution occur by simply imposing load (i.e., the entities that actually produce such load, such as data packets for communication load and application processes for computational load) to distribute in the network by "rolling down" the load field to reach underloaded zone. This eventually achieves a satisfying (sub-optimal) balance of resource exploitation, without involving any negotiation among system-level load balancer processes. Also, provided that the load field is promptly updated upon any significant change in the load of some resources (which can be achieved via simple reactive code fragments associated to information in the knowledge network), the resulting dynamic distribution of load is made self-adaptive, in that any allocation of load to resources will automatically reflect the current global load situation. To some extent, we can consider this way of achieving load balancing as a sort of stigmergic interaction occurring via the load field.

From the application viewpoint, the above approach makes it possible for application components to play an active role in load distribution, other than the passive role of being distributed here and there. Firstly, since they too have access to the knowledge network, they can somehow "bias" the structure of the knowledge network to have it reflect their own needs. For instance, they can artificially "heighten" the shape of the load field in some zones to be ensured that they will have a specified amount of resources devoted to their execution without having other application-level components roll down to these zones. Secondly, they can enrich the load field with any type of application-specific knowledge, to be connected (via proper ontological construct) to the available load and resource information available at the lower level. In this way, one can put to work fruitful cross-layer interactions, where: on the one hand, application-level components can fruitfully exploit both types of information towards the achievement of their application goals; on the other hand, network-level components can direct application-level components towards those part of the system where their needs can be better satisfied without negatively affecting the overall systems functioning, which the availability of a semantically enriched load field enable to effectively evaluate in an introspective way.

3.2 Pervasive Computing

An application scenario which can strongly take advantage of our knowledge networks approach is pervasive computing, here intended as the support of individual and collaborative human activities in an environment densely populated by embedded computers (e.g., sensor networks and computer-based cameras), computer-enriched objects (e.g., smart furniture), and personal computing systems.

Such pervasive computing scenarios are typically open and dynamic: new computers join the scenario at any time (as carried on by humans getting in the environment or brought in via computer-enriched objects) the same as some can leave or being dismissed. The need of exploiting at the best all the available computing resources requires spontaneous inter-operability, i.e., the capability of all computing-based devices to be found in the environment (a priori unaware of each other and never explicitly configured to work together) to start interacting with each other towards the achievement of some application goals. Also, the scenario intrinsically involves situated computational and communication activities, in that the distributed computing infrastructure is put to service for improving humans interactions with the surrounding environment and with the current "situation" of the environment.

To solve the above problems, the pervasive computing research community recognises that middleware infrastructures based on active spaces are necessary [Rom02]. These considerations about interactions in a pervasive computing scenario occur via kinds of shared memory spaces, where to store and by which to access all information about the context/situation, properly organised by the space itself according to shared ontologies, and in which uncoupled (data-mediated) interactions among components (application-level and network-level) may occur without having components to know each other.

Our concept of knowledge networks leverage the active spaces approach by suggesting organising contextual/situational information into fully distributed networks of knowledge. As in active spaces, networks of knowledge can be used for uncoupled interactions among components. However, the knowledge networks approach provides for a more dynamic and lightweight perspective, in that it does not suggest that the organization of knowledge should take place by specific processes devoted to this, but rather suggest an approach in which all components contributed to the building and maintaining of the available situational knowledge. Also, it provides a better support for distributed self-organizing and self-adaptive activities, being a fully distributed knowledge network on which to rely for effective stigmergic coordination.

4   Research Agenda and Related Work

For our idea of knowledge networks idea to become a practical approach, several open research problems have be unfolded. This section analyses some of the most relevant issues – defining a broad research agenda for knowledge networks researches – and discuss how related research thrusts can somewhat contribute to it.

In general, these issues can be all generally related to the following problems (Figure 2), each of which analysed in the following sub-sections: *(i)* how to represent knowledge using proper ontological constructs; *(ii)* how to generate, compose, and relate distributed knowledge; *(iii)* how to have knowledge networks evolve and according to which structure;*(iv)* how to exploit this knowledge to achieve autonomic behaviour at both the network and the application levels.

Before continuing, we emphasise that our approach here is clearly distinguished from the 'knowledge plane' approach [Cla03], and thus introduces different research issues. The knowledge plane approach considers an additional network layer between the network and the application layer, as the place in which nearly all network control activities take place. The knowledge plane is populated by heavyweight intelligent agents [ZamJW03], managing and exchanging knowledge about the current state of the network, and that directly enact forms of control over both network and application components. In our idea, instead, knowledge networks are not intended to be populated, handled, and managed, by additional knowledge-level components. Rather, to avoid the burden of an additional distributed computational layer, and to more fruitfully promote cross-layer interactions, we consider knowledge networks as managed by existing components at the application and network levels (at least supported by some simple reactive code fragments). Thus, while the research issues in the knowledge plane approach relate to how have agents in the knowledge plan interact with each other to properly control the network, the research issues in our knowledge approach relates to how components can generate, maintain, and exploit knowledge.

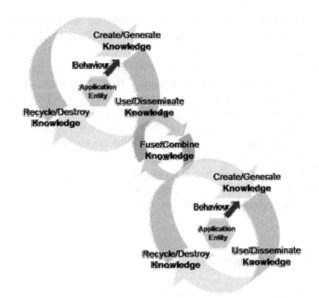

Fig. 2. Knowledge Network Lifecycle

4.1 Defining Ontologies for Knowledge Networks

Ontological constructs [Usc96] can enable the modelling of contextual information semantically. They provide a general model which is independent of programming language, underlying operating system or middleware. Other knowledge 'consumers' in the network must be able to access and use the ontological formalisms developed. Accessing information stored in a network of distributed contextual knowledge requires the specification of information locators, e.g. in the form of an addressing scheme as well as request routing procedures. The relation between knowledge representation and addressing scheme (i.e. how can information be mapped deterministically or probabilistically to locators) as well as request routing schemas are important aspects.

One approach within the ontology category has been proposed as the Aspect-Scale-Context Information (ASC) model [Str03a]. In this model, using ontologies provides an uniform way to specify the models core concepts as well as an arbitrary amount of sub-concepts and facts, together enabling contextual knowledge sharing and reuse in an ubiquitous computing system [DeB03]. These implementations build up the core of a Context Ontology Language (CoOL), which is supplemented by integration elements such as scheme extensions for Web Services and others [Str03b].

The CONON context modelling approach by Wang *et al.* [Wan04] is based on the same idea of the ASC/CoOL approach, namely to develop a context model based on ontologies because of its knowledge sharing, logic inferencing and knowledge reuse capabilities. Wang *et al.* created an upper ontology which captures general features of basic contextual entities and a collection of domain specific ontologies and their features in each sub-domain. The CANON ontologies are serialized in OWL-DL which has a semantic equivalence to well researched description logics. This allows for consistency checking and contextual reasoning using inference engines developed for concept languages.

A promising emerging context modelling approach based on ontologies is the CoBrA system [Che03]. This system provides a set of ontological concepts to characterize entities such as persons, places or several other kinds of objects within their contexts.

All the above, may be of help to characterize the ontological constructs to be put at work for the production of knowledge in the context of autonomic communication scenarios.

4.2 Building Knowledge Network Ecosystems

How may situated knowledge networks be put to use, and how can knowledge be combined/split in differing scales of use by network entities and entity aggregates? There is need to experiment with ensembles of knowledge to facilitate knowledge consumption and use at all differing scales in our networks.

Knowledge is generated from the behaviour and behavioural analysis of individual and aggregated autonomic network- and application-level entities. This behavioural knowledge floods into an ontological construct at entity scale. We need a mechanism that distributes our behavioural knowledge dynamically. To retrieve particular knowledge, it must be possible to address information. Instead of using a fixed structure e.g., by assigning unique identifiers, we can experiment with path languages for topologies taking the 'semantic proximity' of the information in account. The goal is to be able to give directions in a fuzzy way (e.g., "follow this street, turn left on the second traffic light and walk till you see the red building"), which still offers promise in yielding an accurate and unambiguous addressing schema.

These issues have some relations with overlay networks in P2P computing [RowD01, Bab02, Rat02, Bab02, And04]. Indeed, autonomic knowledge networks will be sorts of overlays. However – unlike traditional overlays approaches in P2P computing – knowledge networks are not intended to simply support navigation of data and messages in a dynamic network of components. Rather they are intended to provide components with a local representation of the situation, that can then be used by them to adapt their behaviour e.g. to enforce properties of self-preservation, self-aggregation, and self-organisation in general. With this regard, some recent proposals for semantic overlay networks may be of great relevance [Loe04]. Semantic overlay networks are created by network nodes in P2P systems using content metrics to relate entities. Network queries are routed via the semantic overlay network, reducing the load on nodes with non-related content. Semantic overlay clusters, cluster P2P super-peers by their characteristics, enhancing search and integration significantly. Although guided by policies defined by human experts, this approach shows merit in flood reduction in overlay networks, with potential of application to overlay knowledge networks and especially knowledge network research. Thus, the study of semantic overlays may be of some relevance for the finalization of our knowledge networks concept.

Some additional source of inspiration for knowledge networks could come from some modern middleware proposals for mobile and ubiquitous computing, which consider exploiting forms of distributed data structures – to be dynamically built and self-adapting – to act as the basic mean via which adaptive coordination activities can be promoted. Such middleware proposals include among others LIME [Pic01] and TOTA [MamZ04], Smart Messages [Bor04], Limbo [Dav02]. These approaches, by having distributed data structures typically represent some application-level knowledge, definitely shares something with our knowledge networks approach. However, so far, very little has been said on the possibility of building scalable global distributed data structures in accord with some semantic relations and ontological constructs.

4.3 Enforcing Self-similarity and Robustness

Three considerations must be made when thinking at the possible structure of knowledge networks:

- *(i)*   they should somehow reflect the structure of those networks whose work they are intended to support, i.e., the application/social networks and the technological networks;
- *(ii)* they must evolve over time in an adaptive way yet preserving their properties; and
- *(iii)* they must be scalable and promote composability.

These three issues, though, are strictly related with each other.

Both social (and application-level) networks and technological networks (e.g., the Internet and the Web) tend to evolve towards "scale-free" topologies [AlbB02]. These classes of networks, also found in biological and physical systems, exhibit neither completely random nor completely regular connection topologies [Wat98]. They are characterised by the small-world phenomenon [Mil67]; highly clustered like regular lattices, yet preserving small characteristic path lengths. Dynamical systems models with small world coupling display enhanced signal propagation speed, computational power and synchronisability, properties which can be of great importance for the effective propagation of knowledge in autonomic communication scenarios. In addition, the scale-free characteristic tends to enforce robustness and scalability in the network structure: the same overall structure is preserved as the network evolves over time; and the network exhibit the same structural properties at different levels of observation. Again, these properties would be very important for representing evolving distributed knowledge in a robust way, and for enabling a scalable way with which to structure and compose knowledge.

In summary, it will be interesting to explore how to structure knowledge networks into scale-free structures, so as to reflect the structure of the social and technological networks they support, to support robust adaptive evolution, and to support scalability and scale-free composability at different scales of observation, and to analyse the implications of this structuring.

4.4 Promoting Cognitive Stigmergy

Swarm intelligence approaches consider that global self-organizing and self-adapting behaviour can be made to emerge in systems of a large number of lightweight agents that indirectly interact via the mediation of an environment [Par97, Bon99, ParB04].

Agents, by depositing and by sensing "pheromones", and by having the environment properly diffuse pheromones according to specific laws, can – to most extent unconsciously – self-organize their global activities into robust and adaptive patterns.

Our concept of autonomic network knowledge could potentially act as a form of computational environment via which indirect, stigmergic interactions, may take place to promote self-organization and self-adaptation of activities. Still, this requires leveraging the traditional concept of stigmergy into a concept of *cognitive stigmergy*. Self-organizing and self-adaptive coordinated activities at both the network and the application level should be enforced not simply by reacting to a local concentration of meaningless pheromones. Rather, they should be driven by the actual meaning of the knowledge represented within knowledge networks.

Clearly, to preserve the advantages of swarm intelligence approaches, this should occur without requiring ants to become heavyweight agents, and a proper trade-off between the purely reactive behaviours promoted by traditional stigmergy and the purely cognitive behaviour promoted by artificial intelligence approaches have to be found.

Similar considerations can be made for those approaches to self-organisation based on indirect interactions such as the morphogen gradients of amorphous computing [Nag02, MagM04] and the field-diffusion in teams of mobile robots [McL04].

5 Conclusions

The ambitious scoping of this position paper and of the associated research road map focus on the development of sophisticated knowledge representational schema for next-generation autonomic networks. Our research should deliver knowledge representation schemes and ontologies for situated and autonomic communication-intensive services, structural mapping of knowledge ensembles to network and aggregated network entities, software interfaces for programming interaction with knowledge networks, and tools, metrics and algorithms for the evaluation and monitoring of knowledge networks.

We acknowledge that the scale of research outlined in this paper is very large and that work on developing mediated network knowledge requires us to address significant *stages* of challenges. In addition to those outlined already in the paper, challenges include managing the ontology lifecycle, in particular automated knowledge acquisition for dynamic ontology construction, the use of knowledge-level techniques to address provable, correctness-preserving transformations and adaptive algorithms, and working to understand the role of planning knowledge, including understanding and changing global and local goals. It is also important to consider that the protection of use of sensitive security and privacy information raised by applying such a shared knowledge space to a highly distributed application is addressed at the design stage of a research programme such as this.

We still have no stable ideas about how these knowledge networks will look, and to which extent they will be effectively able to deliver the promise of acting as the nervous system of a future autonomic communication infrastructure. Nevertheless, this appears indeed a challenging and fascinating research topic, involving a number of related research issues likely to impact on future autonomic communication scenarios and worthy of investigation. Bringing together network and knowledge engineering to address the problems in pervasive computing shows promise, and should open up new research directions, in particular once we begin to design and implement for real-world issues using this paradigm.

Acknowledgments. Work supported by the Italian MIUR in the context of the "Progetto Strategico ICT: IS-MANET: Infrastructures for Mobile Ad-Hoc Networks", and by UK Invest Northern Ireland Networking Support using EU Structural Funds. All support is gratefully acknowledged.

References

[And04]   S. Androutsellis-Theotokis, D. Spinellis, "A Survey of P2P Content Distribution Techniques", *ACM Computing Surveys*, 36(4):335-371, Dec. 2004.

[AlbB02]   R. Albert, A. Barabasi, "Statistical Mechanics of Complex Networks", *Reviews of. Modern. Physics*, 74(47), 2002.

[Bab02]   O. Babaoglu, H. Meling, A. Montresor, "Anthill: a Framework for the Development of Agent-Based Peer-to-Peer Systems", *Proceedings of the 22nd IEEE Conference on Distributed Computing Systems*, Vienna (A), May 2002.

[Bal00]   C. Balkenius, J. Moren, "A Computational Model of Context Processing", *6th International Conference on the Simulation of Adaptive Behaviour*. The MIT Press, 2000.

[Bon99]   E. Bonabeau, M. Dorigo, G. Theraulaz, *Swarm Intelligence*, Oxford University Press, 1999.

[Bor04]   C. Borcea, "Spatial Programming Using Smart Messages: Design and Implementation", *24th Int.l Conference on Distributed Computing Systems*, Tokio (J), May 2004.

[Buc03]   Buchholz, T., A. Kupper, and M. Schiffers, "Quality of context: What it is and why we need it", *Workshop of the HP OpenView University Association 2003 (HPOVUA 2003)*, Geneve (CH), 2003..

[Che03]   Chen, H., Finin, T., Joshi, A. "Using OWL in a Pervasive Computing Broker". *Proceedings of Workshop on Ontologies in Open Agent Systems (AAMAS 2003)*, Melbourne (AU), July 2003.

[ChoP03]   T. Choudhury, A. Pentland, "Modeling Face-to-Face Communication Using the Sociometer", *ACM Conference on Ubiquitous Computing*, Seattle, WA, USA, 2003.

[Cla03]   D. Clark et al., "A Knowledge Plane for the Internet", *Proceedings of the 2003 ACM SIGCOMM Conference*, Karlsruhe (D), ACM Press, 2003.

[Dav02]   N. Davies, et al, "L2imbo: A distributed systems platform for mobile computing", *ACM Mobile Networks and Applications*, 3(2):143-156.

[DeB03]   J. De Bruijn, "Using Ontologies - Enabling Knowledge Sharing and Reuse on the Semantic Web", *Tech. Rep. Technical Report DERI-2003-10-29*, Digital Enterprise Research Institute (DERI), Austria, Oct. 2003.

[DiM04]   G. Di Marzo, A. Karageorgos, O. Rana, F. Zambonelli (Eds.), *Engineering Self-organizing Systems: Nature Inspired Approaches to Software Engineering*, LNCS No. 2977, Springer Verlag, May 2004.

[Est02]   D. Estrin, D. Culler, K. Pister, G. Sukjatme, "Connecting the Physical World with Pervasive Networks", *IEEE Pervasive Computing*, 1(1):59-69, 2002.

[Gel02]   H.W. Gellersen, A. Schmidt, M. Beigl, "Multi-Sensor Context-Awareness in Mobile Devices and Smart Artefacts", *Mobile Networks and Applications*, 7(5): 341-351, Oct. 2002.

[Hef01]   Heflin, J., *Towards the Semantic Web: Knowledge Representation in a Dynamic, Distributed Environment*, PhD., Thesis, University of Maryland, College Park, 2001.

[KepC03]   J. Kephart, D. Chess, "The Vision of Autonomic Computing", *IEEE Computer*, 36(1), 2003.

[Loe04]   A. Loeser, F. Naumann, W. Siberski, W. Nejdl, U.Thaden, "Semantic Overlay Clusters within Super-Peer Networks", *International Workshop on Databases, Information Systems and Peer-to-Peer Computing (DBISP2P)*, Berlin (D), 2003.

[MamZ04]   M. Mamei, F. Zambonelli, "Programming Pervasive and Mobile Computing Applications with the TOTA Middleware", *2nd IEEE Conference on Pervasive Computing and Communications,* Orlando (FL), IEEE CS Press, March 2004.

[May03]   R. Mayrhofer, H. Radi, A.Ferscha. Recognizing and predicting context by learning from user behavior. 2003: Austrian Computer Society (OCG).

[MikM04]   M. Mikic-Rakic, N. Medvidovic, "Support for Disconnected Operation via Architectural Self-Reconfiguration", *International Conference on Autonomic Computing*, New York, NY, USA, 2004.

[Mil67]   Milgram, S., The Small World Problem. *Psychology Today*, 1967. 2: p. 60-67.

[McL04]   J. McLurkin, J. Smith, "Distributed Algorithms for Dispersion in Indoor Environments using a Swarm of Autonomous Mobile Robots", *Proceedings of the 7th International Symposium on Distributed Autonomous Robotic Systems*, Toulouse (F), 2004.

[Nag02]   R. Nagpal, "Programmable Self-Assembly Using Biologically-Inspired Multi-agent Control", *1st Int.l Conference on Autonomous Agents and Multi-agent Systems*, Bologna (I), July 2002.

[NagM04]   R. Nagpal, M. Mamei, "Engineering Amorphous Computing Systems", in *Methodologies and Software Engineering for Agent Systems: the Handbook of Agent-Oriented Software Engineering*, Kluwer Academic Publishing (New York, NY), 2004.

[Par97]   V. Parunak, "Go to the Ant: Engineering Principles from Natural Agent Systems", *Annals of Operations Research*, 75:69-101, 1997.

[Par04]   V. Parunak, S. Brueckner, J. Sauter, "Digital *Pheromones for Coordination of Unmanned Vehicles"*, *Workshop on Environments for Multi-agent Systems (E4MAS)*, LNAI 3374, Springer Verlag, 2004.

[Pen05]   A. Pentland, "Socially-Aware Computation and Communication", *IEEE Computer*, 38(3):33-40, March 2005.

[Pic01]   G. P. Picco, A. L. Murphy, G. C. Roman, "LIME: a Middleware for Logical and Physical Mobility", *22nd IEEE Intl. Conference Distributed Computing Systems*, 2001.

[PicM04]   G. P. Picco, A. L. Murphy, "Using Coordination Middleware for Location-Aware Computing: A Lime Case Study", *Proceedings of the 6th International Conference on Coordination Models and Languages, LNCS No. 2949*, Feb. 2004.

[Rat02]   S. Ratsanamy et al., "GHT: A Geographic Hash Table for Data-Centric Storage", *1st ACM Int.l Workshop on Wireless Sensor Networks and Applications*, Atlanta, Georgia, USA, September 2002.

[Rom02]   M. Roman et al., " Gaia : A Middleware Infrastructure for Active Spaces", *IEEE Pervasive Computing*, 1(4):74-83, Oct.-Dec. 2002.

[RowD01]   A. Rowstron, P. Druschel, "Pastry: Scalable, Decentralized Object Location and Routing for Large-Scale Peer-to-Peer Systems", *18th IFIP/ACM Conference on Distributed Systems Platforms*, Heidelberg (D), Nov. 2001.

[Str03a]   Strang, T. Service Interoperability in Ubiquitous Computing Environments, *PhD thesis, Ludwig Maximilians University Munich*, Oct. 2003.

[Str03b]   Strang, T., Linnhoff-Popien, C., Frank,K. CoOL: A Context Ontology Language to enable Contextual Interoperability. In LNCS 2893: *Proceedings of 4th IFIP WG 6.1 International Conference on Distributed Applications and Interoperable Systems (DAIS2003)*, J.-B. Stefani, I. Dameure, and D. Hagimont, Eds., vol. 2893 of Lecture Notes in Computer Science (LNCS), Springer Verlag, pp. 236–247, 2003.

[SAC05]   EU IST Commission, "Situated and Autonomic Communication Initiative", *Future and Emerging Technologies Report*, www.cordis.lu/ist/fet/comms.html, 2005.

[ShiKS92]   N. G. Shivaratri, P. Krueger, M. Singhal, "Load Distributing for Locally Distributed System", *IEEE Computer*, 25(12):33-44, Dec. 1992.

[Ste05]   Sterritt, R., M.D. Mulvenna, and A. Lawrynowicz, "Dynamic and Contextualised Behavioural Knowledge in Autonomic Communications", *Proceedings of the 1st IFIP Workshop on Autonomic Communications*. Berlin: Springer-Verlag, 2004.

[Usc96]   Uschold, M. and M. Grueninger, Ontologies: Principles, methods, and applications. Knowledge Engineering Review, 1996. 11(2): p. 93–155.

[Wan04]   Wang, X. H., Zhang, D.Q., Gu, T., Pung, H.K., Ontology Based Context Modeling and Reasoning using OWL. *In Workshop Proceedings of the 2nd IEEE Conference on Pervasive Computing and Communications (PerCom2004)*, (Orlando, Fl.,USA, March 2004), pp. 18–22.

[Wat98]   Watts, D., J., Strogatz, S.H., Collective Dynamics of 'Small-World' Networks. *Nature*, 1998. 393: p. 440-442.

[Wei91]   M. Weiser. "The Computer for the 21st Century", *Scientific American*, 265(3):94–104, September 1991.

[Xu95]   M. Xu, B. Monien, R. Luling, F. C. M. Lau, "Nearest Neighbour Algorithms for Load Balancing in Parallel Computers", *Concurrency: Practice and Experience*, 7(7):707-736, Oct. 1995.

[Zam05]   F. Zambonelli, M.P. Gleizes, R. Tolksdorf, M. Mamei, "Spray Computers: Explorations in Self-organization", *Journal of Pervasive and Mobile Computing*, 1(1):1-20, May 2005.

[ZamJW03]   F. Zambonelli, N. Jennings, M. Wooldridge, "Developing Multiagent Systems: the Gaia Methodology", *ACM Transactions on Software Engineering and Methodology*, 12(3):410-470, July 2003.

Towards a Reliable, Wide-Area Infrastructure for Context-Based Self-management of Communications

Graeme Stevenson*, Paddy Nixon*, and Simon Dobson

Systems Research Group,
School of Computer Science and Informatics,
UCD Dublin, Ireland
graeme.stevenson@ucd.ie

Abstract. In this paper we describe *ConStruct*, a distributed, context-aggregation based service infrastructure which supports the development of context-aware applications. ConStruct operates by automatically generating and maintaining directed context-processing graphs which connect applications to the sources of data they require at a relevant level of abstraction. The infrastructure also supports the dynamic creation of context processing elements to bridge gaps between available and requested information. ConStruct provides a reliable, scalable infrastructure; focused on self-maintenance in order to alleviate developer workload. We describe the infrastructure design and implementation, the associated programming model, and our planned extensions to the infrastructure.

1 Introduction

A defining trait of pervasive computing environments is the presence of large numbers of sensors, embedded into the physical surroundings, which provide information on a variety of characteristics of the environment in which they operate. This *context* information can be utilised by applications to modify their behaviour in response to changes in their execution environment, or to convey such changes to users.

The information required by an application is rarely at the same level of abstraction as that provided by individual data sources. For example, a sensor may be able to indicate that a person has been detected in a room, but an application may be interested in the current occupancy of the room, or whether a meeting is currently taking place in the room. In order to obtain such information, an application will frequently have to aggregate data from multiple sources. Such sources may differ in many respects. For example, the level of accuracy they provide, and the data formats and communication protocols they use. As a result, application developers are required to spend the majority of their time on the details

---

* This work was undertaken while Stevenson and Nixon were at The University of Strathclyde, Glasgow, Scotland.

I. Stavrakakis and M. Smirnov (Eds.): WAC 2005, LNCS 3854, pp. 115–128, 2006.

of obtaining the information they require, rather than on their primary goal of defining the behavioural changes in the applications that use the information.

The aim of our research is to provide a service infrastructure which will carry out the task of obtaining and processing sensor data from a variety of disparate sensor technologies, and deliver it to applications at the level of abstraction they desire. This allows developers to focus on the task of specifying application behaviour with respect to that information.

There are four challenges in the areas of *flexibility, maintainability, scalability* and *inter-operability* that must be met. Firstly, developers cannot anticipate at development time the physical sources of data that will be available to their applications. Mechanisms must therefore be provided which are flexible enough to support runtime binding between an application and viable data sources. Infrastructure support must also be provided for automatically locating potential data sources, and for bridging the gap between available information and required information using data-aggregation techniques. Secondly, as sensors and data-processing components may be prone to failure, the infrastructure should automatically detect and repair faults wherever possible. Thirdly, an infrastructure should scale to provide support for large numbers of devices, sensors and applications. Finally, as sensors and applications may be deployed on a wide range of heterogeneous devices, standard data formats and communication protocols should be used to provide independence from hardware, operating system, and programming language. Detailed analysis of these requirements can be found in [1].

We have designed and implemented ConStruct, a middleware infrastructure for context-aggregation, with the goal of meeting these challenges. ConStruct draws from several concepts of state-of-the-art context processing research (see Section 3) and extends from this by introducing the automatic synthesis of data sources to bridge gaps between available and requested context information, and by providing mechanisms to reuse components across concurrently executing applications with different data requirements. Work on ConStruct is still in progress, with several strands of research planned to move towards the provision of context in an autonomic computing setting.

The rest of the paper is organised as follows. Section 2 presents an overview of ConStruct. Section 3 presents a discussion of related work. Section 4 discusses Context Entities, the building blocks of the infrastructure. Sections 5 and 6 describe how Context Entities are dynamically composed to provide answers to application queries. In Section 7 we discuss application mobility and communication between multiple instances of the infrastructure services. Section 8 describes the programming model used for developing Context Entities and Applications. We discuss the work still required to meet the goal of context provision in an autonomic computing environment in Section 9, and conclude with a summary in Section 10.

2   Overview of ConStruct

ConStruct is comprised of a number of execution environments called *Ranges*, which self-organise to form a partially connected overlay network referred to as

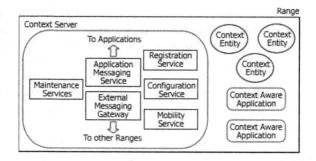

Fig. 1. The set of components which make up ConStruct

the *ConStruct-NET*. Each Range is functionally equivalent and contains a set of services that are used for the management of *Context Entities*, independent units of execution which provide and process context information, and *Context-Aware Applications*, which use the Range services to request and consume the context information produced by entities. Any entity or application which utilises the services provided by a Range is referred to as being a part of that Range. The infrastructure places no restrictions on the physical placement of Range components within the network.

The ConStruct-NET is formed using a self-organising, self-repairing peer-to-peer protocol [2], and provides functionality for dealing with applications which may move between Ranges during their lifetime, and for managing the interactions required to obtain context information from data sources in remote Ranges.

The services provided by a Range are grouped together to form the Context Server. There are six services in total, as shown in Figure 1.

When an entity starts, it sends a request to the Registration Service, advertising the type of information it supplies. This information is used by the Configuration Service, to compose and instantiate graphs of Context Entities, called configurations, which are capable of answering application queries. The External Messaging Gateway is used in this process to obtain context from other Ranges via the ConStruct-NET, whilst the Maintenance Services monitor the status of all the entities and applications in a Range, performing repairs to configurations as required. The Mobility Service is responsible for supporting applications relocating to other Ranges. Finally, the Application Messaging Service provides an additional mechanism for message based entity to application communication outwith the confines of a configuration.

The current implementation of ConStruct is built using the Java Message Service (JMS) [3], a standard, asynchronous messaging API, which supports communication between loosely coupled, distributed components.

3   Related Work

Whilst a lot of work has been undertaken in the field of context delivery over the last decade, the projects closest in spirit to our own are those that provide support for context aggregation. [4] introduces the Contextor, an extension of a Context Widget [5]. Contextors can be composed, and recomposed, into colonies, typed functional units which perform data-aggregation. iQueue [6], provides similar support by automatically combining composers, data aggregation functions written using iQL [7], a purpose built specification language. The iQueue runtime attempts to resolve queries by selecting appropriate data sources using application provided criteria. Finally, Solar [8] allows applications to compose operators using operator-graphs which are instantiated at runtime using available resources. Applications may also specify policies defining how to discard or summarize data flows wherever buffers overflow. Runtime support is provided for load balancing operators across execution environments (planets), for restarting failed operators, and for client mobility.

We note that although existing infrastructures have looked at the problem of automatically generating context-processing graphs (iQueue, Contextors), and context processing across distributed environments (Solar), no project has yet looked at the combination of these elements in tandem with the runtime synthesis of context-processing elements to bridge the gap between available and requested information when only approximate matches are available. This is one of the features of ConStruct.

4   Context Entities

A Context Entity (analogous to a Contextor [4], or Operator [9]), is a lightweight software component which represents either a source of data or a function which operates on the data produced, or computed, by other Context Entities. Each entity has its own thread of execution, and may consume and publish events, which represent context information. This section describes Context Entity meta-data, entity architecture, and the different types of entity supported by the infrastructure.

4.1   Entity Profiles and Naming

It is impractical to require application developers to identify the physical sources for the information they require at development time. Not only would this be a time consuming process for anything more than a trivial application, it would also lead to the development of applications which were inflexible in the face of device failures and changes in the resource pool - two prominent characteristics of their operating environment. To overcome this challenge we require that data sources be identified by the properties of the information they supply rather than by their network location or unique identifier [10].

In order to achieve this, each Context Entity is associated with a profile - XML formatted meta-data which describes the properties of the information supplied by the entity. An entity profile consists of four parts: a classification (see Section 4.3), a location, a description of the output generated by the entity, and descriptions of any inputs the entity requires.

The location parameter describes the logical placement of the Context Entity in the network (based on the Intentional Naming System [11]). For example, an entity representing a coffee machine in room 10 on level 11 of the Livingstone Tower would have the location [LivingstoneTower/L11/R1110].

The description of the output supplied by the entity consists of two parts: a reference to an ontology which describes the format of the events published by the entity, and a set of attribute-value pairs which describe the static properties of the events published by that entity. The property names correspond to those outlined in the ontology.

The combination of entity location and output event description is used to identify resources within the network. This is similar in spirit to Solar [9] and iQueue [6]. It is this format which entities use to express any input requirements they may have.

There is an ongoing effort in the research community towards developing ontologies which can describe the data supplied by a multitude of diverse data sources (e.g., [12]). We assume the existence of such ontologies, although their provision is outside the scope of this work.

We are currently looking at ways in which we can improve the expressiveness of our context specifications. There are two extensions of particular interest. The first allows for the input requirements of a Context Entity to be derived from the output required of it. For example, an entity which will compute the distance in metres between two people (specified at runtime) given their GPS coordinates. The second extension allows Context Entities to define the properties of one input requirement based on the runtime output of another. For example, an entity that provides a list of all the occupants collocated with a given person. This entity requires two inputs: the location of the person we are interested in, and events that describe the detection of people within that location. In order to correctly set up the latter input, we must first have access to the data from the former.

4.2   Entity Architecture

Context Entities consist of three main parts: a control channel, an event channel, and a functional core. Context Entities can send messages to the control channels

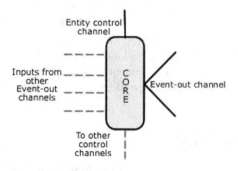

Fig. 2. The architecture of a Context Entity

belonging to other entities (or infrastructure services), and may also receive events from the event channels belonging to other entities. This is illustrated in Figure 2.

The functional core of a Context Entity defines how the value of its output events are calculated from its input events, while the control channel of an entity may receive events from the infrastructure services in order to check its status, or from other Context Entities asking it to calculate a new value. These functions discussed in more detail later.

In our current implementation, the control channel corresponds to a JMS Queue - which has one-to-one delivery semantics, whilst the output event channel corresponds to a JMS Topic - which has one-to-many delivery semantics.

4.3   Entity Classification

Context Entities may use data from a wide variety of sources to perform a number of different computations. Influenced by the work in [4], we provide support for seven different flavours of Context Entity. A *source* represents any physical or computational component from which data originates (e.g., a door sensor, or an entity which delivers user preferences). A *fusioner* obtains input from multiple entities which supply events of the same type (X), and outputs events (also of type X) whose quality has improved over that of the input events (e.g., a more accurate estimation of the location of a person based on events produced by RFID and IR sensors). An *aggregator* outputs an event of arbitrary type based on one or more input events, also of arbitrary type (e.g., detecting the activity taking place in a room based on the time, the identity of the people in that room, and the associated noise level). A *transformer* takes an input event of type X and recasts the data into another format without altering its level of abstraction. The output event may be of the same type (e.g., converting a temperature reading from Celsius to Fahrenheit) or a different type (converting data from one event ontology to another). A *generaliser* takes in, and outputs data of type X, where the output data is at a lower level of granularity than that of the input data. We envision the generaliser being used to implement privacy policies, where users may wish to restrict the accuracy of any personal data which is made available to other users (e.g., reducing the accuracy of a location measurement from a room name to a building name). A *filter* takes a single input of type X and outputs a subset of its input events based on some criteria (e.g., to filter out location events about a specific person from a general location service). Finally, a *merger* takes in multiple inputs of type X and outputs each event received without alteration. The purpose of the merger is to aid reuse of event streams and operators (see Section 6.2).

5   Configuration Model

As we described earlier, ConStruct uses the functionality provided by Context Entities to generate answers to queries submitted by Context-Aware

Applications. This is achieved by connecting Context Entities together to form directed, acyclic graphs which produce the required context information as a result. We call these graphs *configurations*. This section describes the architectural style used as the basis for configurations, and describes the interaction model which controls communication between Context Entities.

5.1 Architecture of a Configuration

The architectural style we use for configurations is based upon *Chiron-2* (C2) [13], which was originally devised to support component reuse in GUI based systems. The style consists of a number of components (Context Entities), which are connected together to form a hierarchy using message routing devices (control and event channels). The key property of this style is that components are only aware of other components which reside directly above it in the hierarchy, and have no knowledge of those components which reside lower down. The C2 style supports two forms of communication: notifications, which are passed down the architecture, and requests, which are passed up the architecture. In our case notifications (events) are passed using event channels, while requests are communicated using control channels. Applications represent the lowest level of the hierarchy and form the sinks of the graph.

5.2 Entity Interaction

The interaction model used by ConStruct supports both *active* and *passive* Context Entities. Active entities are characterised by the fact they automatically publish new context information when it becomes available, while passive entities wait until data is requested from them before supplying it. In order to accommodate both types of Context Entity, we use the following interaction model, based on [6]:

- When an Context Entity receives an event from one of its input sources, it will send an event-request message to the entities which lie in the level of the hierarchy directly above it (with the exception of the entity which sent the original event). Once it has received a new event from each of these sources it will calculate and publish a new value.
- When a Context Entity receives an event-request message from an entity (or application) in the level of the hierarchy directly below it, it will send an event-request message to each of the entities which lie in the level of the hierarchy directly above it. Once new values have been returned, the entity will calculate and publish a new value.

We wish to extend our configuration model to provide support for cyclic graphs. This would allow us to support applications and services which employ feedback techniques. We would also like to support the provision of services where entities require to coordinate their efforts with their peers (such as the traffic monitoring/route planning application described in [10]).

6   Query Resolution

The Configuration Service employs *Automatic Path Creation* (APC) techniques in order to generate configurations that are capable of satisfying application queries. This section describes three aspects of this process: the APC mechanism itself, the techniques implemented to reuse existing configurations and entities where possible, and the process of maintaining configurations during their lifetime.

6.1   Query Processing

Restricting, for now, discussion of the resolution process to a single Range, the process carried out by the Configuration Service upon receipt of a query is as follows:

1. First, the Configuration Service searches for Context Entities which match the desired location and output event type requested by the application.
2. The attribute-value pairs describing the output supplied by each candidate entity are then compared to the application's requirements. Entities are classified into one of four categories: *no match, partial match, exact match*, and *over match*. The no match category contains entities which have conflicting attribute-value pairs to that of the application request. The partial match category contains entities who's attribute-value pairs are a subset of those required by the application. The exact match category includes entities who's attribute-value pairs have exactly one-to-one correspondence with the application request. Finally, the over match category contains entities who's attributes form a superset of those required by the application.
3. If any exact match category contains at least one entity, the next step is to examine each of their input requirements (if any) in turn, and determine if they can be satisfied (using this procedure). This is a recursive process which continues until physical sources of data are found (i.e., Source entities). If there is a choice to be made among multiple entities, the one with the classification that provides the higher quality of data is chosen (e.g., Fusioner > Source > Aggregator).
4. If there are no exact matches, the next step is to examine the input requirements for any partial matches in a similar manner. If a complete configuration can be formed, a filter is automatically generated and configured to bridge the gap between the output of the configuration and the requirement of the application.
5. Should the previous two groups fail to yield a positive result, the final option is to evaluate the group of entities in the over match category. The results of all successfully evaluated configurations can then be merged together to provide the application with the best possible match available.

If the above procedure succeeds in generating a configuration, the Configuration Service sends messages to each entity involved, detailing the identity of the event streams that each should subscribe to. On completion, the Configuration

Service then sends a message to the application informing it about the identity of the event channel representing the end point of the configuration, to which the application will subscribe.

6.2 Reuse of Event Streams and Context Entities

In the process of generating a configuration, the Configuration Service will try to reuse active event streams and Context Entities (i.e., those which are part of an existing configuration) wherever possible. In the case of event streams, this is a straightforward process. If the output of an active entity is found to satisfy either the application query itself, or one of the inputs required by a entity within the new configuration, the Configuration Service will utilise the existing event stream, thus satisfying that particular branch of the configuration completely.

In the case of Context Entities, the process is slightly more involved. When we talk about reusing a Context Entity, we refer to the functionality of the entity, rather than the role it plays in an existing configuration. For example, an entity that converts context information from one ontology to another can perform the same role in multiple configurations which require information from different sources. The process of reusing an entity in multiple configurations involves merging the desired event streams, tagging each stream with an identifier, passing the event stream through the reusable entities, and finally filtering out the original event streams. This is illustrated in Figure 3.

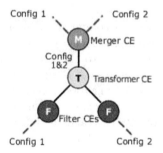

Fig. 3. Example showing the reuse of a transformer entity across two configurations

6.3 Runtime Maintenance

Pervasive computing environments are considered to be dynamic with respect to the resources available within them at any one time. Another tenet of such environments is that the failure of computational devices should be treated as commonplace. To deal with these features, ConStruct provides a suite of maintenance services that: monitors Context Entities and Applications for failure; performs repairs to configurations where possible; and re-evaluates configurations when new resources become available.

Application and Context Entity monitoring takes the form of periodic passing of ping/pong style messages. If the control channel of an application/entity has been closed or if a response has not been received within a set number of iterations the application/entity is assumed to have failed. In the case of an application failure, any configurations to which they were the sole subscriber can be removed, and the involved entities told to deactivate. In the case of an entity failure, the Configuration Service will be used to try and repair the branches of any configurations which utilise the entity. If a configuration can be repaired successfully and the end point of the configuration is unchanged there is no need to inform the application. Otherwise, the application is told to change their subscription, or that their requirements can no longer be satisfied.

Periodic re-evaluations of queries are performed in order to take advantage of additions to the resource pool. Should a preferable configuration be found to one already in use, the affected branches of a configuration are altered, old branches deactivated and applications informed as above if necessary. We are currently working on providing support for runtime configuration adaption based on changing Quality of Service parameters (e.g., accuracy, confidence, error, and bandwidth). Although this information has always been available (should an entity choose to provide it), its interpretation was previously left to data consumers. As different data types have different QoS parameters associated with them, we aim to develop an model which is extensible, allowing us to perform informed analysis of the QoS parameters of new data types as they emerge.

7   ConStruct-NET

The ConStruct-NET facilitates the communication of context information over a wide-area by connecting distributed Ranges. This allows applications (by way of the infrastructure services) access to the context information they require, irrespective of their network location and Range they are part of. This section gives a brief overview of the communication mechanisms employed to form the ConStruct-NET, the extension of the configuration model to communicate with data sources located in remote Ranges, and the infrastructure mechanisms that support application mobility between Ranges.

7.1   Inter-range Communication

The *External Messaging Gateway* (EMG) component of a Range is responsible for initialising (or joining) the ConStruct-NET, and for all communication between remote Ranges and its own. The ConStruct-NET is implemented using *Pastry* [2], which provides the basis for communications, and message routing within peer-to-peer applications. Details on the message routing algorithm employed by Pastry, and the self-organising and self-repairing characteristics of a Pastry network are described in [2].

7.2   Extending the Configuration Model

In order that the context information supplied by entities in remote Ranges can be utilised, we impose two additional requirements on the single Range

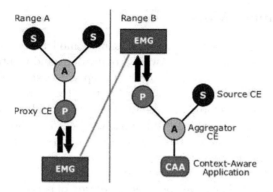

Fig. 4. Example showing the use of a proxy to communicate information across the ConStruct-NET

model described above. Firstly, that the query resolution algorithm incorporates searches across multiple Ranges. Secondly, that the model used for executing configurations is extended to provide support for obtaining information from and sending requests to entities across the ConStruct-NET.

If part (or all) of a configuration cannot be resolved locally, the Configuration Service will route a request through the EMG to the Configuration Services belonging to other Ranges, which will attempt to complete the configuration. Should a remote Configuration Service be able to contribute to the configuration, the process will recurse from that point in a similar manner until the configuration is completed, or the process fails.

In order that that our configuration model remains consistent, we have introduced *proxy* Context Entities, which bridge the gap between Ranges, serving as a local representation of a remote entity. Proxy entities use the EMG to communicate with the entity they represent. This is illustrated in Figure 4.

7.3   Application Mobility

Mobile applications may use the Mobility Service to retain their configurations whilst they relocate to another Range, or during periods where they experience temporary loss of connectivity (e.g., out of range of a Wi-Fi access point). The Mobility Service acts as a proxy between an application and the end point of its configurations. Should message delivery to the application fail, the Mobility Service will cache events on the applications behalf. When an application rejoins the ConStruct-NET (either in the same or a different Range), it uses the infrastructure to route a message to the Mobility Service, asking it to resume message delivery. Should an application fail to reappear within a reasonable time period (set by the administrator of a Range), the assumption is made that the application will not return. At this point the Mobility Service will stop acting on the application's behalf, and the maintenance procedures will perform cleanup operations as normal.

8   Programming Model

ConStruct provides a simple two-step programming model that allows developers to easily create their own Context-Aware Applications and Context Entities.

To create a new Context Entity, the first step is to extend from the entity base class which provides all the functionality required to interact with the infrastructure services. Developers are required only to provide an implementation for the *evaluate()* method, which returns an XML encoded String representing the event produced by that entity. Access to any inputs required by the entity are achieved by calling the *getSource("sourceName")* method. Updated events for these inputs are obtained automatically (see Section 5.2) before the *evaluate()* method is called. The developer has the option of specifying how often the entity should evaluate as part of its constructor. If no value is given, the entity is treated as being passive.

Similarly, the base class from which applications extend only requires developers to provide an implementation for the *eventHook()* method. This is called automatically when an application receives a new event.

The second step in the development process is to write the XML context specification for the entity (its profile) or application (its queries). Profiles include the entity location, input, and output details of a entity as discussed in Section 4.1, whilst the format used for a specifying each application query includes a local name (used as a parameter when the *eventHook()* method is called, signalling which query the event is associated with), location, event type, and associated attribute-value property list. Context specifications are stored in an external text file, and identified to the entity/application through the object's constructor. The process of verifying and using the data provided by the profiles and queries is handled automatically.

We have build several applications using our programming model, including an In/Out Board, a Context-Aware Coffee Break Notifier, and a Smart To-Do List. Details of these can be found in [1].

9   Towards Context Provision in an AC Environment

Up to this point, the focus of our work has been on investigating the necessary abstractions to allow us to decompose high-level services into low-level building blocks, and on the mechanisms to facilitate their communication and reuse. The techniques we use in this process have allowed us to place all the maintenance complexity and communication logic into the software, minimising the effort required of developers to build mobile applications which can source data from any location within the ConStruct-NET.

In addition to the ongoing work we have described throughout this paper, there are several issues which we must address before we reach the stage of providing a context delivery mechanism which is suitable for an autonomous networking environment.

Whilst the infrastructure has self-organising and self-healing properties at the Range level, both in terms of the Pastry network protocol used to form the

ConStruct-NET and the fact that it provides automatic creation and maintenance of configurations within a Range, at the macro level we have a reliance on centralised services. This single point of failure, whilst effective at lessening the processing burden on individual devices, is a fair criticism of our work from an autonomic communications perspective.

We aim look at the feasibility of decentralising our protocols. This raises a number of issues, such as: the efficiency of the discovery protocol; the time required to construct a configuration; memory footprint; CPU load, which will be of critical importance for battery powered devices; and preserving the facility to source data from remote locations. Another key issue involves the synthesis and reuse of data sources - where we currently use the infrastructure services to do this work, another approach will be required. The concept of a domain, as we have with a Range is useful, and retaining this concept when decentralising our protocol is something to consider.

Finally, security of data is also an important issue - primarily in terms of access control, although encryption may be a requirement in some cases. Providing access control mechanisms for dealing with context is a complex issue. Challenges include the need to determine ownership of the data; to resolve conflicting privacy preferences (between users and/or administrative domains), and to provide mechanisms for permitting access to information at different levels of granularity (e.g., granting access to your location information at room, building, or city level depending on the identity of the interested party).

10   Summary

In this paper we have described ConStruct, a service infrastructure designed to enable the collection and aggregation of context information from a myriad of distributed data sources, and the distribution of that information to the applications that require it at an appropriate level of abstraction. We detailed the mechanisms which we use to allow runtime synthesis of new data sources to bridge processing gaps, and the techniques we use to support the reuse of processing elements across multiple configurations. We also described how ConStruct facilitates context processing and dissemination over a wide-area using multiple deployments of the infrastructure services. We concluded this paper by discussing some of the issues that we still need to address in order to apply our technology to the autonomic computing domain.

References

1. Graeme Stevenson. A Service Infrastructure for Change-Tolerant Context-Aware Applications. Master's thesis, University of Strathclyde, Glasgow, Scotland, 2005.
2. Antony Rowstron and Peter Druschel. Pastry: Scalable, Decentralized Object Location, and Routing for Large-Scale Peer-to-Peer Systems. *Lecture Notes in Computer Science*, 2218:329–350, 2001.
3. R. Monson-Haefel and D. Chappell. *Java Message Service*. O'Reilley & Associates, December 2000.

4. Jolle Coutaz and Gatan Rey. Foundations for a Theory of Contextors. In *Computer-Aided Design of User Interface (CADUI02)*, 2002.
5. Anind Dey. *Providing Architectural Support for Building Context-Aware Applications*. PhD thesis, Providing Architectural Support for Building Context-Aware Applications, November 2000.
6. Norman H. Cohen, Apratim Purakayastha, Luke Wong, and Danny L. Yeh. iQueue: A Pervasive Data Composition Framework. In *Third International Conference on Mobile Data Management (MDM'02)*, pages 146–153, Singapore, January 2002.
7. Norman H. Cohen, Hui Lei, Paul Castro, John S. Davis II, and Apratim Purakayastha. Composing Pervasive Data Using iQL. In *Proceedings of the Fourth IEEE Workshop on Mobile Computing Systems and Applications*, page 94. IEEE Computer Society, 2002.
8. Guanling Chen and David Kotz. Application-controlled loss-tolerant data dissemination. Technical Report TR2004-488, Dartmouth College, Computer Science, Hanover, NH, February 2004.
9. Guanling Chen and David Kotz. Context Aggregation and Dissemination in Ubiquitous Computing Systems. In *The Fourth IEEE Workshop on Mobile Computing Systems and Applications.*, pages 115–114, Callicoon, New York, June 2002.
10. Norman H. Cohen, Apratim Purakayastha, John Turek, Luke Wong, and Danny Yeh. Challenges in Flexible Aggregation of Pervasive Data, 2001.
11. William Adjie-Winoto, Elliot Schwartz, Hari Balakrishnan, and Jeremy Lilley. The design and implementation of an intentional naming system. In *Symposium on Operating Systems Principles*, pages 186–201, 1999.
12. Harry Chen, Filip Perich, Tim Finin, and Anupam Joshi. SOUPA: Standard Ontology for Ubiquitous and Pervasive Applications. In *First Annual International Conference on Mobile and Ubiquitous Systems: Networking and Services (MobiQuitous'04)*, Boston, Massachussets, USA, August 2004.
13. Richard N. Taylor, Nenad Medvidovic, Kenneth M. Anderson, E. James Whitehead Jr., Jason E. Robbins, Kari A. Nies, Peyman Oreizy, and Deborah L. Dubrow. A Component- and Message-Based Architectural Style for GUI Software. *Software Engineering*, 22(6):390–406, 1996.

Semantic Interoperability for an Autonomic Knowledge Delivery Service

David Lewis, Declan O'Sullivan, Ruaidhri Power, and John Keeney

Knowledge and Data Engineering Group,
Trinity College Dublin, Ireland
{Dave.Lewis, Declan.OSullivan, Ruaidhri.Power,
John.Keeney}@cs.tcd.ie

Abstract. The development and deployment of interconnected networks is being increasingly limited by their complexity and the concomitant cost of managing the operational network. Autonomic Communication aims to reduce this cost, by migrating management intelligence towards the network elements and empowering operational support staff to specify what network behaviour in terms of goals and constraints. Towards this aim we propose a key infrastructural service that enables the efficient delivery of network operations knowledge to, and only to, nodes that have expressed an interest in that knowledge. This Knowledge Delivery Service mediates operational network knowledge in an open, ontological form, thereby promoting the graceful evolution of network management applications from contemporary to fully autonomic. To cope with the inevitable heterogeneity of knowledge across the population of network nodes, the service provides a level of semantic interoperability that will be transparent to the nodes providing and consuming knowledge. The Service will be based on content-based networking principles. This paper describes work towards supporting semantic interoperability in such a Knowledge Delivery Service.

1 Introduction

The development and deployment of interconnected networks is being increasingly limited by their complexity and the concomitant cost of managing the operational network. Autonomic Communications aims to reduce this cost, by migrating management intelligence towards the network element and empowering users and operational support staff to specify network behaviour in terms of goals and constraints, rather than specifying how that behaviour should be achieved.

Autonomic principles are targeted at reducing the cost of handling the complexity of distributed computing systems by making them self-managing, i.e. self-configuring, self-healing, self-optimising and self-protecting [kephart]. This requires monitoring and analysing the operational knowledge in systems so that it can be used to plan and execute corrective measures, typically using some artificial intelligence techniques. This relieves the human manager from performing these tasks while

I. Stavrakakis and M. Smirnov (Eds.): WAC 2005, LNCS 3854, pp. 129 – 140, 2006.
© IFIP International Federation for Information Processing 2006

allowing the human to guide the decisions made by the autonomic manager through the definition of high-level policy rules defining goals for and constraints on the desired system behaviour.

More recently, momentum has been growing to apply autonomic principles to network operations, i.e. Autonomic Communications. An early articulation of the use of operational network knowledge by intelligent applications was proposed by David Clarke et. al. in a proposal for a Knowledge Plane for the Internet [clark]. Operational network knowledge is defined as network operations or management data accompanied by its meta-data, typically expressed as a management information model. These approaches present a major challenge in obtaining the relevant operational network knowledge. Difficulties arise because the network elements that possess this knowledge are widely distributed, they are purchased from different vendors, they perform different functions, they possess a wide range of knowledge meta-data and, perhaps most challenging of all, they are operated by different organisations.

Current approaches to Autonomic Communication typically involve distributed intelligence, such as multi-agent systems, swarm intelligence, or cellular automata [mullany], operating at the network element level, adapting to changes in the knowledge that is gathered on the network and application context. This adaptation is constrained by policies representing the operational goals and constraints of network operators and users. To date, however, there has been no movement towards an inter-working consensus for these technologies or on how the knowledge required to make autonomic decisions is gathered from across a heterogeneous network, and particularly across administrative domains. This work tackles head-on the interoperability short-comings of current Autonomic Communication proposals, but in a way that ensures a smooth, commercially viable transition from contemporary network management systems to fully autonomic ones. We therefore propose a Knowledge Delivery Service (KDS) as an infrastructure that accurately and efficiently delivers autonomic network knowledge to nodes that have expressed an interest in that knowledge. Here we focus on how this service might adapt the knowledge delivered between the semantics used by the producer and those expected by the consumer. This paper describes the major technical challenges in developing semantic interoperability in a KDS and then presents initial results on how semantic interoperability knowledge can be captured and then distributed and used within such a service.

2 Background

The proposed Knowledge Delivery Service presents a demanding set of challenges that intersect Semantic Web, Content-based Networking and attribute based access control research.

Communication service operator concerns about the sensitivity and security of operational data is reflected in the hierarchical nature of the manager-agent paradigm and the intra-domain focus of architectures such as TMN (Telecommunications Management Network) [TMN]. The fragmentation of manager-agent protocols at the element layer, and the lack of a dominant interoperation technology at the higher layers has led to problems exchanging management knowledge between the vertical silos of interoperability (both syntactic and semantic) within operators' Operational

Support Systems (OSS) [adams]. When exchanging operational knowledge between operators, this is compounded by commercial confidentiality concerns, which result in bilateral agreements and inflexible custom gateways. This spells disaster for the vision of Autonomic Communications where intelligent agents operating at or near the network element level must be able to freely gather contextual knowledge about the state of the network end-to-end and adapt to changes in this context to achieve administrator-specified goals and maintain their constraints. The challenge of cross-ownership sensitivities is addressed elsewhere [feeney05], while here we focus on the interoperability issues in gather network context to guide the behaviour of autonomic network elements. These present an even more extreme case of the conditions that led to interoperability silos in conventional OSS and thus the looming prospect of Autonomic Communications silos must be urgently addressed. Proposals for end-to-end delivery of operation network knowledge are either constrained to individual protocol layers [thaler] or to following existing signalling paths [schulzrinne]. These approaches are, however, insufficient as the wider network state increasingly forms the context for intelligent decision making in network elements [karmouch]. Any scalable solution will reply on loose coupling between the producers and consumers of autonomic knowledge. In this work we focus on late semantic binding through the encoding and mapping between heterogeneous models using the existing Web Ontology Language (OWL) [owl]. As a W3C standard OWL represents a broadly applicable mean for capturing semantics with basic language primitive for capturing semantic mapping, as well as providing the basis for richer mapping languages. This is coupled with loose binding between producers and consumers of operational network knowledge using publish-subscribe communications.

Several attempts to address management model heterogeneity have been made by defining new management information modelling languages to act as a canonical model providing lingua franca between other models. Notable amongst these are the Distributed Management Task Force's (DMTF) Common Information Model (CIM) schema [cim] and the TeleManagement Forum's NGOSS technology neutral architecture [tmf053]. However, a lack of a strong semantic interoperability mechanism and reliance on conformance to poorly subscribed industrial agreements effectively render these as yet more management knowledge formats with which other schemes needed to interoperate. Recent pioneering work by Vergara and Villagra [lopezdevergara] has shown directly the value of modelling management information models in the OWL ontological format, and how this can be used to ease the interoperation between models originally conceived in different management information languages, i.e., GDMO, SMI, CIM. Our approach follows the underlying philosophy of the semantic web, where semantics, including mappings, are captured where applications require it and the necessary expertise is present. Our aim, however, is to ensure what mappings exist are made available as automatically as possible to the management applications that can use them. This also points to the adoption of the OWL-S service semantic language [martin] for defining management services knowledge as well as for supporting to dynamic management service composition.

Publish-subscribe systems provide an efficient mechanism for delivering information from its source to one or more interested parties (known as subscribers). It also allows the timely notification of events or changes to information, when compared to polling approaches, but requires publishers and subscribers to agree on the message types before interacting. Content-Based Networks (CBN) extend this approach to allow the subscriber to specify conditional filters on message properties, effectively allowing the subscriber to define the type of message in which they are interested [carzaniga][segall][strom]. By delivering operational knowledge only to those knowledge consumers who register a specific interest, while multicasting messages to consumers who share interests offers the potential for scaling knowledge delivery to Internet dimensions [crowcroft]. We propose a KDS that will be implemented as a Knowledge Delivery Network (KDN) structured along CBN principles. Thus, network elements may advertise the type of knowledge they possess to the KDS while an intelligent autonomic network element may place a subscription for the knowledge they need for the task at hand, cancelling the subscription when the task is finished. Producers of operational network knowledge express this capability using the ontological representation of the relevant management information models and while consumers express subscriptions as simple semantic queries. The advertisement-subscription mapping and subscription aggregation algorithms used in the KDN may therefore exploit ontology-based reasoning mechanisms, such as class subsumption. In this paper we will focus on how ontological mappings between management information models will be used by the KDN nodes to supplement these algorithms providing a level of semantic interoperability between the models of the autonomic context knowledge sought by the consumer and that used by its producer.

In the rest of this paper we focus on how mappings between managed object concepts on heterogeneous systems can be captured and then injected into the KDS to support the dynamic semantic interoperability of management knowledge.

3   Semantic Interoperability for Knowledge Delivery

Semantic Interoperability is a key element of the KDS. Where the semantics of information emitted by a notification producer does not immediately conform to the semantics sought by a particular consumer's subscription, a match may still be possible if the knowledge exists of how one set of semantics may be mapped to the other. Mapping one knowledge domain onto another typically requires human comprehension of both sets of semantics, though tools are increasingly able to produce mappings by extrapolating from a few human supplied mapping anchors. More automated collaborative identification of semantic mapping is being researched using intelligent agent technologies.

In the context of the KDS, the service must operate in a framework that supports the discovery, injection and interpretation of known mappings, regardless of whether they are human-designed or auto-generated. Mapping interpretation is then undertaken at runtime to route information appropriately and to aid the transformation of information between two different formats.

3.1   Discovering Semantic Mappings

A key challenge is how the mapping information between the ontologies can be derived. Automatically deriving ontology mapping information at runtime without the involvement of a human is generally considered impossible [klein] due to diversity of domains and lack of encoded semantics. The network management domain has better semantic homogeneity than many due to its standardised information models that represent the common resource semantics needed for control and user plane interoperability. However, differing management standards and proprietary extensions, which representing competitive differentiations inherent in the industry, combined with the frequent need to integrate arbitrary elements within increasingly complex network still present difficult semantic interoperability challenges. The challenge in our work, therefore, has been to identify an integrated software and process framework which will minimise the amount of design time work involved, devolve as much work as possible to a runtime algorithm, and share the mappings as much as possible such that human involvement is reduced. This devolution is crucial for the uptake of this approach in autonomic computing environments where reduction of human intervention is key. Equally important is maximising the applicability of human generated ontology mappings by ensuring that they maximise the chances of a successful runtime mapping between information conforming to concepts from the two ontologies concerned.

The process of the resultant OISIN (Ontology Interoperability for Semantic Interoperability) framework [vanderMeer] is overviewed in Figure 1 and illuminated further with an example. This example involves interoperability between intelligent agents that are resident on the management agents of different printer types and are aware of the local printer information models. A useful application here may entail agents monitoring neighbouring printers in an office for out-of-paper notifications, and conserving its paper (e.g. by automatically switching to double sided of two up) to maintain office-wide print services until notification of fresh supplies being loaded have been detected. In this example we assume printer agents have knowledge of either a local SNMP printer MIB or the DMTF CIM printer device model.

In the first phase of the OISIN process the ontologies from each party are characterised. These ontologies represent the core concepts that would be used in self-management functions in the two agents types, that is a CIM based ontology and a Printer MIB based ontology. Of course it is assumed in future that these ontologies would be pre-existing, perhaps through the use of a conversion tool such as presented in [lópezdevergara].

The tools of the Characterisation Phase transform the ontology (in Ontology Web Language (OWL) format or relational database format, etc.) into a common internal format. The software tools in this phase characterise:

- The nature of the terms, whether simple or composite. This can be helpful in determining whether finding mappings will be straightforward for the mapping algorithm and the complexity of the human driven confirmation process of the algorithm suggested mappings;
- How many terms are known/unknown by WORDNET the online dictionary developed in Princeton. This can be helpful in evaluating the degree of domain or acronym specific terminology;
- The quality of the ontologies according to the online Semantic Web search engine SWOOGLE developed at the University of Maryland. The information from SWOOGLE can provide an indication as to how widely referenced a particular ontology is, which provides one measure of quality;

- The number of candidate matches. Class and property names of the ontologies are compared (with support of WordNet and an encoded telecommunications domain specific thesaurus) to identify potential matches (through exact or synonym matches) of ontology classes and their properties. In the initial implementation a lexical matcher is used, but this is being extended to type and range matchers;
- The number of potential mappings arising from the candidate lexical matches. This information can provide an indication of the amount of overlap between the ontologies, and also provides information about the potential difficulty in finding/confirming mappings from suggested matches.

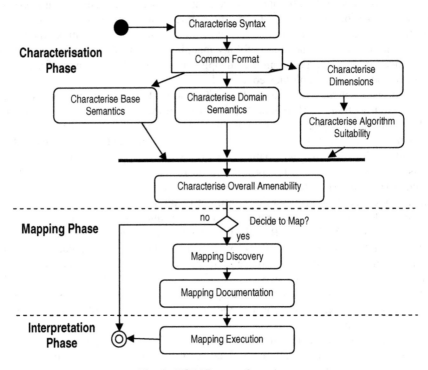

Fig. 1. OISIN Process Overview

The characterisation information generated by the **Characterisation Phase** is presented to the user via numeric and graphical charts to the user, so that a decision to map or not can be made. If it is decided to map then in the Mapping Phase the matches are presented to the user in a graphical manner. In Figure 2 for example the M identifies exact lexical matches (e.g. Printer) and the P identify partial matches on a lexical or synonym basis (e.g. Person partially matches on a partial synonym basis to Operator, Manager and User).

The user then identifies the "anchors" which correspond to key partial mappings between the ontologies. This involves examining the two ontologies to try to identify equivalent concepts. Typically during this examination the properties of the concepts are examined to identify equivalence as well. In the example shown in Figure 2 the `MaxNumberUp` property of the CIM Printer class can be seen to be equivalent to the `prtOutputMaxCapacity` property concept of the MIB Printer class. Once an anchor is chosen it is annotated with an E (e.g. `Printer` in Figure 2). In addition, XSL based transformation code/bridge can be associated with a mapping in order to provide the ability to translate from one value range to another.

During this process "anchor paths" are also identified. The concept of "anchor paths" was first introduced by [noy01]. The idea is that if two anchors are specified in a hierarchy of Ontology *A* it is likely that the classes which appear in the intervening path may correspond with those on the path of the corresponding anchors in Ontology *B*. Unlike the PROMPT/Anchor tool [noy00], we do not actually require the user to enumerate every mapping in these paths at design time but devolve the determination of what is or is not a match within these paths to the runtime algorithm.

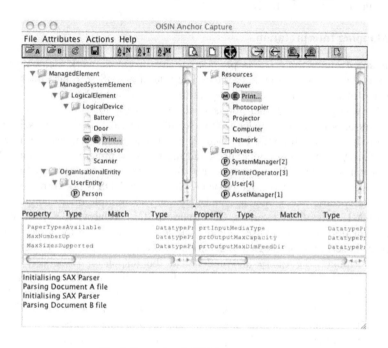

Fig. 2. Example of OISIN capture tool

A key differentiation of our approach in general is our belief that the determination of what is or is not considered class equivalence can only be undertaken in the context of the applications involved in using the mappings and what they are trying to achieve. For this reason the original matching information as well as the equivalence annotations provided by the user are made available to an application that uses the mapping information. The output of this phase is mapping information consisting of a set of anchor mappings (expressed using the `owl:equivalentClass` and `owl:equivalentProperty` XML elements) and their corresponding XSL based transformation bridges. In [osullivan] we have shown how XSL based transformation bridges can be automatically generated given a set of ontology mappings, and so we will not provide further detail here on the bridge creation. In addition the mapping information output contains the matches information and the anchor path information that has been generated, which are used at runtime by a Semantic Matching Utility when no default transformation bridge exists. The Semantic Matching Utility returns the anchor path and matching information for a term requested and this can be used by an application to dynamically create a transformation depending on the context. In the **Mapping Interpretation** phase, the mapping information is injected into the KDS for appropriate mapping execution. The transformation bridges and the Semantic Matching Utility are used by the Knowledge Discovery Service Node (KDSN) during query resolution (see next section). As discovery of such mappings is likely to be a decentralised task, the distribution of mappings to points in the KDN where they are needed for interpretation purposes is itself performed using the content-based networking feature of the KDN.

3.2 Mapping Interpretation

In our architecture, the KDS is provided to intelligent network elements to make autonomic knowledge available to them. The role of a KDS is to take queries from a client and to resolve those queries by acting as a mediator between the client and other knowledge sources that the service has access to. As well as acting as consumers of autonomic knowledge (by executing queries), intelligent NEs can also act as producers of autonomic knowledge. In our current implementation interaction between KDS client agent applications and the KDN is implemented as an API. The implementation of the API operates in the same memory space as the application, but can be considered as a KDS Edge Node (KDSEN). The KDS is not currently designed to define a specific query language and may support multiple query language styles, such as CMIS scope/filer requests, CIM queries, SQL, XQuery, RDQL or SPARQL. Our current implementation uses XPath. The terms involved in the query must be a subset of those in the ontologies understood by the KDS client, for the service to be able to formulate the query and understand the response. Associated with each KDSEN there is a repository of ontologies that describes the domain of knowledge that this agent application currently understands, i.e. the OWL version of the NE's local MIB and other models the agent uses in a manager role. In more intelligent agents this domain may extend during operation as new concepts are used in downloaded policies or as the agent learns concepts from peer agents. These ontologies are provided to the KDSEN by the application agents, either when registering notifications it is able to provide or making query subscription. The KDS aims to resolve query subscription immediately and then update the subscribing application with any matching updates until the subscription is cancelled.

The internal architecture of a KDSEN is shown in Figure 3. The registration interface allows intelligent NE agents to register with a Knowledge Discovery Service Node. During the registration the application will provide a reference to an ontology (O) that defines the domain information in which the agent will couch its queries. The registration interface is also used to inject mappings and bridges (M) from the Mapping Phase of the OISIN process into the KDSEN. The ontology and mapping repositories store the ontologies and bridges that have been registered and the mappings that have been injected at a particular KDSN. The Internal Query Interface takes in queries (Q) using terms from the ontology (O) and passes them to the Query Resolver for resolution. If the application does not have the knowledge required to resolve the query locally, the KDSEN directs it via the internal query interface and a "context connector" is introduced. This essentially is an interceptor which intercepts queries normally destined for the repository that the application normally queries.. The Query Resolver takes this query expressed using terms from the local ontology, passes it to the KDN, and resolves the responses received for that query. The KDN takes the query and routes the query to KDSENs which have directly produced information using the terms involved. In addition, it routes the query to KDSENs where the mapping information injected previously has indicated that relevant information exists. Mappings are themselves distributed by the KDN, with KDSEN's subscribing to any mapping that include concepts form locally registered ontologies.

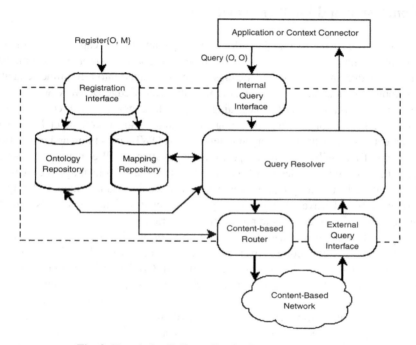

Fig. 3. Knowledge Delivery Service Edge Node Architecture

The External Query Interface of a KDEN receives a query and undertakes a response. If some transformation has to take place due to the fact that the query arrived as a result routing due to mappings, then the corresponding dynamic bridge to handle the appropriate query and response translations (automatically created during the mapping injection phase) is invoked.

In order to illustrate the above, we return to the printer agent example described earlier. First a semantic mapping is undertaken between the CIM model and SNMP model, one such mapping being the equivalence between the CIM `Printer` and the MIB `Printer` classes. This mapping is then injected into the Knowledge Discovery

Service, resulting in nodes which have a MIB based printer attached having a mapping and associated bridge registered. A context connector is introduced so that the application does not need to be altered to take advantage of the KDN. Thus when the application seeks to discover the output capacity of all the printers being managed, the connector poses the query locally via CIM and also passes the query to the KDSEN. The KDN then distributes this query to all nodes which either directly supports CIM based printers or those which have been mapped (e.g. via the MIB mapping in our case). Any node which has a MIB based printer attached then receives the query via the external query interface, applies the bridge to transform the query from CIM to SNMP based and transform the responses if necessary into CIM format. The Query Resolver of the KDSEN that originally received the query then takes all the responses and returns them.

4   Conclusion and Further Work

This paper presents some initial work in the semi-automated capture of semantic interoperability mappings for use in the run-time translation of Autonomic Communication Knowledge and how those mappings can be integrated into a run-time semantic interoperability system suitable for integration with a Knowledge Delivery Service. We are currently working towards integrating this with a conventional content-based network (Elvin [sutton]). This is being integrated with a commercial element management system, which is being augmented with OWL conversion and translation bridge support that will allow initial performance evaluations We then aim to extent our investigations into the design of the core KDN. We aim to leverage recent work showing that perfect routing can be achieved in a scaleable manner independently of subscriber joins and leaves though subscription aggregation [chand]. We also aim to address consumption of composite notifications [courtenage].

In any multi-domain scenario those responsible for any knowledge resource must be able to impose access control over who is able to access that knowledge. Access control policies have been demonstrated to work with CBN sources [belokosztolszki]. This employed role-based access control which is the predominant approach to defining access control policies. However, in a fluid, multi-domain scenarios the detailed business modelling that underpins the identification of the roles used will make this approach very brittle. Instead we will adopt a community-based policy approach, which has been shown to track more easily and accurately the dynamic organisational grouping within and between organisations [feeney04]. This approach will be adapted to support access control of knowledge available to the KDN, including a mechanism to identify possible access control conflicts and suggest resolutions. Such KDN access control requires new mechanisms for matching subscriptions to producer access control rules, without strongly binding subscriptions to consumer attributes and thereby reducing the routing efficiency gained by subscription aggregation. This will be combined with a trust-based access control mechanism for determining community membership in ad hoc organisational situations [feeney05].

Acknowledgements

This work was supported partially by the Irish Higher Education Authority through the M-Zones programme and partly by Science Foundation Ireland and Enterprise Ireland through the Centre for Telecommunication Value Chain Research. The authors would like to thank Simon Courtenage, Thanassis Tiropanis and Victor Villagra for their help in developing these ideas.

References

[adams]   Adams, E., Willetts, K., "The Lean Communication Provider: Surviving the Shakeout through Service Management Excellence", McGrw-Hill, 1996

[belokosztolszki] Belokosztolszki, A., Eyers, D.M., Pietzuch, P.R., Bacon, J., Moody, K., "Role-based access control for publish/subscribe middleware architectures". In International Workshop on Distributed Event-Based Systems (DEBS03), ACM SIGMOD, San Diego, CA, USA, 2003. ACM

[carzaniga]   Carzaniga, A., Rosenblum, D. S., and Wold, A. L., The Design and Evaluation of a Wide-Area Event Notification Service, ACM Transactions on Computer Systems, Vol. 19, Issue 3, August 2001

[chand]   Chand, R., Felber, P.A., "A Scalable Protocol for Content-Based Routing in Overlay Networks", Second IEEE International Symposium on Network Computing and Applications, April 2003, Cambridge, MA

[cim]   Common Information Model v2.5, DMTF 2000: http://www.dmtf.org/spec/cim_schema_v25.html

[clark]   D. Clark, C. Partridge, J.C. Ramming, J. Wroclawski, "A Knowledge Plane for the Internet", SIGCOMM'03

[courtenage]   Courtenage, S. "Specifying and Detecting Composite Events", In 1st International Workshop on Discrete Event-Based Systems, Vienna, 2002

[crowcroft]   Crowcroft, J., Jean Bacon, J., Pietzuch, P., Coulouris, G., Naguib. H., "Channel Islands in a Reflective Ocean: Large-Scale Event Distribution in Heterogeneous Networks", IEEE Communications Magazine, 40(9):112-115, September 2002

[feeney04]   Feeney, K., Lewis, D., Wade, V., "Policy based Management for Internet Communities", Proceeding of the 5th IEEE International Workshop on Policies for Distributed Systems and Networks, 7-9 June 2004, York Town Hieght, NY, USA

[feeney05]   Feeney, K., Quinn, K., Lewis, D., O'Sullivan, D., Wade, V. "Relationship-Driven Policy Engineering for Autonomic Organizations", to appear in proc. of 6th IEEE International Workshop on Policies for Distributed Systems and Networks, 2005

[karmouch]   Karmouch, et al "Contextware Research Cahllenges in Ambient Networks", in proc of 1st Intl workshop on Mobility Aware Technologies and Applications (Mata 2004), Florianopolis, Brazil, Oct 2004, Springer LNCS 3284, pp 62-77

[kephart]   J. Kephart, D. Chess "The Vision of Autonomic Computing" Computer, Jan 2003

[klein]   Michel Klein, "Combining and relating ontologies: an analysis of problems and solutions". In Workshop on Ontologies and Information Sharing, IJCAI'01, Seattle, USA, August 4-5, 2001

[lópezdevergara]   López de Vergara, J.E., Villagrá, V.A., Berrocal, J., "Applying the Web Ontology Language to management information definitions", IEEE Communications Magazine, Vol. 42, Issue 7, July 2004, pp. 68-74

[martin]   David Martin, Mark Burstein, Jerry Hobbs, Ora Lassila, Drew McDermott, Sheila McIlraith, Srini Narayanan, Massimo Paolucci, Bijan Parsia, Terry Payne, Evren Sirin, Naveen Srinivasan, Katia Sycara, "OWL-S: Semantic Markup for Web Services", W3C Member Submission 22 November 2004

[mullany]   Mullany, F., Ho, L., Samuel, L., Claussen, H., "Self-Deployment, Self Configuration: Critical Future Paradigms for Wireless Access Networks", in Proc of the 1st IFIP International Workshop on Autonomic Communications, Dec 2004

[noy00]   Noy, N.F., Musen, M.A., "PROMPT: Algorithm and Tool for Automated Ontology Merging and Alignment". Seventeenth National Conference on Artificial Intelligence (AAAI-2000), Austin, TX, 2000

[noy01]   Noy, N.F., Musen, M.A., "Anchor-PROMPT: Using Non-Local Context for Semantic Matching". Workshop on Ontologies and Information Sharing at the Seventeenth International Joint Conference on Artificial Intelligence (IJCAI-2001), Seattle, WA, 2001.

[osullivan]   O'Sullivan, D., Lewis, D., "Semantically Driven Service Interoperability for Pervasive Computing", Proceedings of the 3rd ACM International Workshop on Data Engineering for Wireless and Mobile Access, San Diego, CA, USA, 19th September 2003, pp 17-24

[owl]   OWL Guide, W3C, http://www.w3.org/TR/owl-guide/

[schulzrinne]   Schulzrinne, H., Hancock, R., "GIMPS: General Internet Messaging Protocol for Signaling", Internet Draft, draft-ietf-nsis-ntlp-05, February 2005

[segall]   Segall, B. et al, "Content-Based Routing in Elvin4", In Proceedings AUUG2K, Canberra 2000

[strom]   Strom et al., "Gryphon: An Information Flow Based Approach to Message Brokering", In International Symposium on Software Reliability Engineering 1998 [smirnov] Smirnov, M., "Autonomic Communications: A Research Agenda for a New Communication Paradigm", November 2004, http://www.autonomic-communication.org/publications/doc/WP_v02.pdf

[sutton]   P. Sutton, R. Arkins, and B. Segall, "Supporting Disconnectedness – TransparentInformation Delivery for Mobile and Invisible Computing," in Proceedings of IEEECCGrid 2001. Brisbane, Australia. pp. 277-285. 2001

[thaler]   Thaler, D. G., Ravishankar, C.V., "An Architecture for Interdomain Trouble Shooting", Journal of Network and Systems Management, 12(2), pp 155-189, June 2004

[tmf053]   NGOSS Architecture, Technology Neutral Specification, Membership Evaluation Version 1.51, TeleManagement Forum, July 2001

[TMN]   ITU-T Recommendation M.3010 (1992), Principles for a TMN

[vanderMeer]   van der Meer, S., O'Sullivan, D., Lewis, D. Agoulmine, N., "Ontology Based Policy Mobility for Pervasive Computing", in proc 9th IFIP/IEEE International Symposium on Integrated Network Management (IM 2005), May 2005, Nice, France

Autonomic Communication Security
in Sensor Networks

Tassos Dimitriou and Ioannis Krontiris

Athens Information Technology
{tdim, ikro}@ait.edu.gr

Abstract. The fact that sensor networks are deployed in wide dynamically changing environment and usually left unattended, calls for nomadic, diverse and autonomic behavior. The nature of security threats
in such networks as well as the nature of the network itself raise additional security challenges, so new mechanisms and architectures must
be designed to protect them. In an autonomic communication context
these mechanisms must be based on self-healing, self-configuration and
self-optimization in order to enforce high-level security policies. In this
work we discuss the research challenges posed by sensor network security
as they apply to the autonomic communication setting.

1 Introduction

During the past few years there has been an explosive growth in the research
devoted to the field of sensor networks, covering a broad range of areas, from
understanding theoretical issues to technological advances that made the realization of such networks possible. These networks use hundreds to thousands of
inexpensive wireless sensor nodes over an area for the purpose of monitoring certain phenomena and capture geographically distinct measurements over a long
period of time. Nodes employed in sensor networks are characterized by limited
resources such as storage, computational and communication capabilities. As an
example, Figure 1 shows a sensor node designed at UC Berkeley, along with its
processor and radio characteristics. The power of sensor networks, however, lies
exactly in the fact that their nodes are so small and cheap to build that a large
number of them can be used to cover an extended geographical area.

Even though originally research on sensor networks was motivated my military
applications, the availability of low cost sensors and the advances in communication networks have resulted in exciting applications [1, 2, 3] in a wide range
of fields such as counterterrorism applications, environmental and habitat monitoring, disaster management and traffic control. One reason that make such
networks attractive is the fact that they can be deployed rapidly and start operating without the need of any previous infrastructure or human intervention. For
instance, sensor networks could be deployed directly in the region of interest to
help rescuing efforts at disaster sites, or they could monitor conditions at a highly
toxic environment, along an earthquake fault, or around a critical water reservoir.

I. Stavrakakis and M. Smirnov (Eds.): WAC 2005, LNCS 3854, pp. 141–152, 2006.
© IFIP International Federation for Information Processing 2006

Processor	
CPU Clock	4 MHz
Program Memory	128K bytes
Serial Flash	512K bytes
EEPROM	4 K bytes
Current Draw	8 mA

Radio	
Center Frequency	433 MHz
Data Rate	38.4 Kbaud
Outdoor Range	1000 ft
Current Draw	25 mA (transmit)
	8 mA (receive)

Fig. 1. UC Berkeley's Mica mote and specifications

As most of the applications require the unattended operation of a large number of sensor nodes, this raises immediate problems for administration and utilization. Even worse, some times it is not possible to approach the deployment area at all, like for example in hostile environments of military applications. So, sensor networks need to become *autonomous* and exhibit responsiveness and adaptability to evolution changes in real time, without explicit user or administrator action.

Autonomic responses of sensor networks are especially important to counter security threats. Most sensor networks actively monitor their surroundings, and it is often easy to deduce information other than the data monitored. Such information leakage often results in loss of privacy for the people in the environment. Moreover, the wireless communication employed by sensor networks facilitates eavesdropping and packet injection by an adversary. The combination of these factors demands security for sensor networks [4, 5] to ensure operation safety, secrecy of sensitive data and privacy for people in sensor environments.

Nevertheless, sensor networks cannot rely on human intervention to face an adversary's attempt to compromise the network or hinder its proper operation. Neither can they employ existing security mechanisms such as public key infrastructures that are computationally expensive. Instead, an autonomic response of the network that relies on the embedded pre-programmed policies and a coordinated, cooperative behavior is the most effective way to gain maximum advantage against adversaries.

2   Limitations and Potential Attacks

Although wireless sensor networks have an ad-hoc nature, there are several limitations that make security mechanisms proposed for ad-hoc networks not applicable in this setting. In particular, security in sensor networks is complicated by more constrained resources and the need for large-scale deployments. A summary of these limitations follows below:

- *Constrained hardware*: Establishing secure communication between sensor nodes becomes a challenging task, given the limited processing power, storage, bandwidth and energy resources, as well as the lack of control of the wireless communication medium. Public-key algorithms, such as RSA [6] or Diffie-Hellman key agreement [7] are undesirable, as they are computationally expensive. Instead, symmetric encryption/decryption algorithms and hash functions are between two to four orders of magnitude faster [8], and constitute the basic tools for securing sensor network communications. However, symmetric key cryptography is not as versatile as public key cryptography, which complicates the design of secure applications.
- *Wireless communications*: Sensor networks use wireless communication which is particularly expensive from an energy point of view (one bit transmitted is equivalent to about a thousand CPU operations [9]). Hence one cannot use complicated protocols that involve the exchange of a large number of messages. Additionally, the nature of communication makes it particularly easy to eavesdrop, inject malicious messages into the wireless network or even hinder communications entirely using radio jamming.
- *Exposure to physical attacks*: Unlike traditional networks, sensor nodes are often deployed in areas accessible by an attacker, presenting the added risk of physical attacks that can expose their cryptographic material or modify their underlying code. This problem is magnified further by the fact that sensor nodes cannot be made tamper-resistant due to increases in hardware cost.
- *Large scale deployment*: Future sensor networks will have hundreds to thousands of nodes so it is clear that scalability is a prerequisite for any attempt in securing sensor networks. Security algorithms or protocols that have not designed with scalability into mind offer little or no practical value to sensor network security.
- *Aggregation processing*: An effective technique to extend sensor network lifetime is to limit the amount of data sent back to reporting nodes since this reduces communication overhead [10]. However, this cannot be done unless intermediate sensor nodes have access to the exchanged data to perform data fusion processing. End-to-end confidentiality should therefore be avoided as it hinders aggregation by intermediate nodes and complicates the design of energy-aware protocols.

All these limitations make sensor networks more vulnerable to attacks, ranging from passive eavesdropping to active interference. In particular, we distinguish attacks as outsider and insider attacks. In *outsider* attacks, the attacker may inject useless packets in the network in order to exhaust the energy levels of the nodes, or passively eavesdrop on the network's traffic and retrieve secret information. An *insider* attacker however, has compromised a legitimate sensor node and uses the stolen key material, code and data in order to communicate with the rest of the nodes, as if it was an authorized node. With this kind of intrusion, an attacker can launch more powerful and hard to detect attacks that can disrupt or paralyze the network.

3   Typical Security Requirements

Usually in sensor networks there exists one or more base stations operating as data sinks and often as gateways to other networks. In general a base station is considered trustworthy, ether because it is physically protected or because it has a tamper-resistant hardware. Concerning the rest of the network, we now discuss the standard security requirements (and eventually behavior) we would like to achieve by making the network secure.

- *Confidentiality*: In order to protect sensed data and communication exchanges between sensors nodes it is important to guarantee the secrecy of messages. In the sensor network case this is usually achieved by the use of *symmetric* cryptography as asymmetric or public key cryptography in general is considered too expensive. However, while encryption protects against outside attacks, it does not protect against inside attacks/node compromises, as an attacker can use recovered cryptographic key material [11] to successfully eavesdrop, impersonate or participate in the secret communications of the network. Furthermore, while confidentiality, when applied properly, guarantees the security of communications inside the network it does not prevent the misuse of information reaching the base station. Hence, confidentiality must also be coupled with the right control policies so that only authorized users can have access to confidential information.
- *Integrity and Authentication*: Integrity and authentication is necessary to enable sensor nodes to detect modified, injected, or replayed packets. While it is clear that safety-critical applications require authentication, it is still wise to use it even for the rest of applications since otherwise the owner of the sensor network may get the wrong picture of the sensed world thus making inappropriate decisions. However, authentication alone does not solve the problem of node takeovers as compromised nodes can still authenticate themselves to the network. Hence authentication mechanisms should be "collective" and aim at securing the entire network. Using intrusion detection techniques we may be able to locate the compromised nodes and start appropriate revoking procedures.
- *Availability*: In many sensor network deployments (monitoring fires, quality of water in reservoirs, protection against floods, battlefield surveillance, etc.), keeping the network available for its intended use is essential. Thus, attacks like denial-of-service (DoS) that aim at bringing down the network itself may have serious consequences to the health and well being of people. However, the limited ability of individual sensor nodes to detect between threats and benign failures makes ensuring network availability extremely difficult. Additionally, it is important that the network still operates under such scenarios and that its operation degrades in a predictable and stable way despite the presence of node compromises or failures.

All this discussion suggests that it is necessary to develop networks that exhibit autonomic security capabilities, i.e. be resilient to attacks and have the ability to contain damage after an intrusion.

4   Issues in Sensor Network Security Research

A security architecture for sensor networks must integrate a number of security measures and techniques in order to protect the network and satisfy the desirable requirements we have outlined. In what follows we describe some of these components (and the techniques involved) that are currently under research in sensor networks and we discuss some open challenges with respect to autonomic communication behavior.

4.1   Key Establishment and Initial Trust Setup

One important component of autonomic communication is programmable and controlled group communication. Members leave and join the group according to some membership rules and follow the same behavior pattern within the group. When setting up a secure sensor network, one must be able to embed trust rules that govern the security level of group communications as well as the self-configuration nature of the network. This includes discovering new nodes and adding them in the group as well as identifying and isolating malicious ones. Eventually this translates in establishing cryptographic keys between the members of the group.

Key establishment protocols used in traditional networks are well studied but cannot be applied here due to the inherent limited capabilities (CPU power, memory, etc.) of sensor nodes. Moreover, key-establishment techniques need to scale to networks with tens of thousands of nodes. Simple solutions such as network-wide keys [12] are not acceptable from a security point of view since compromising a single node leads to compromise of the entire network, leaving no margins for self-healing. On the other hand, having each node sharing a separate key with every other node in the network is not possible due to memory constraints (each node usually has a few KBs of memory).

Typically, the problem of initial trust setup can be solved by allocating to each sensor node a randomly selected subset from a pre-established set of keys [13, 14, 15]. Then sensors can communicate securely if they have one or more keys in common. However, these techniques offer only "probabilistic" security as compromising a node may lead to security breaches in other parts of the network. Some other techniques exist [16, 17] that are designed to restrict an adversary that compromised a node to a small portion of the network supporting in-network processing at the same time, but more research is needed in this area.

In order for sensor nodes to be able to communicate safely using established cryptographic keys, a key refresh mechanism is also needed. In an autonomic scenario, re-keying is equivalent with self-revocation of a key when the network detects an intrusion or the lifetime of the key has expired. In order to keep the desirable security level intact the network itself has to determine that rekeying is needed and initiate the appropriate mechanisms. Re-keying is thus a challenging issue, since new keys must be generated in a collaborative and energy-efficient manner, so not all security architectures can support them.

4.2   Resilience to Denial of Service Attacks

Adversaries can limit the value of a wireless sensor network through DoS attacks making it imperative to defend against them. DoS attacks can occur at multiple protocol layers [18], from radio jamming in physical layer to flooding in transport layer, all with the same goal: to prevent the network from performing its expected function. Adversaries can involve malicious transmissions into the network to interfere with sensor network protocols and induce battery exhaustion or physically destroy central network nodes. More disastrous attacks can occur from inside the sensor network if attackers compromise some of the sensors themselves. For example, they could create routing loops that will eventually exhaust all nodes in the loop.

Determining that the network is subject to a DoS attack is a very challenging problem. Especially in large-scale deployments, it is hard to differentiate between failures caused by intentional DoS attacks and nominal node failures. An autonomic sensor network must be able to *monitor* the network traffic and look for suspicious patterns that match some possibly learned rules about what is normal or abnormal behavior [19]. Then it can respond according to the type of the attack.

Potential defenses include techniques such frequency hopping, spread spectrum communication [20] and proper authentication. What is needed, however, is an autonomic coordinated response to defend against DoS attacks with a minimum latency between the detection and a coordinated response. One example could be the use of unaffected nodes to map the affected region and then route around the jammed portion of the network [21]. Further progress in this area is needed to allow for greater security against DoS attacks.

4.3   Resilience to Node Compromises

Due to the nature of their deployment, sensor nodes are exposed to physical attacks in which an attacker can extract cryptographic secrets or modify their code. In [11], the authors demonstrate how to extract cryptographic keys from a sensor node using a JTAG programmer interface in a matter of seconds. One solution to this problem would be the use of more expensive tamper resistance hardware; however, this solution would increase the cost per sensor considerably, thus ruling out deployment of sensor networks with thousands of nodes. Moreover, trusting tamper resistant devices can be problematic [22].

So, the challenge here is to build networks that operate correctly even when several nodes have been compromised and behave in an arbitrarily malicious way. One approach would be the design of proactive networks of sensors in which the sensors at regular time intervals run a protocol to update their cryptographic key material. Combined with the fact that an adversary would have to capture a large percentage of the sensors in the same time interval, security of the network would be enforced. In general, it is very difficult for an adversary to obtain global information about the entire network. Instead, an attacker only has limited information connected with the nodes she compromises. This can be turned into a defensive mechanism for the sensor network, if the compromised region can be located [23] successfully.

As a result, it would be vital to sensor network security if there was a mechanism that could effectively detect malicious code in sensor nodes and give an assurance that they are running the correct code. Lately *software-based code attestation* has been proposed as a mechanism like this. For example, SWATT [24] enables an external verifier to verify the code of a running system to detect maliciously inserted or altered code, without the use of any special hardware. This enables new intrusion-detection architectures, where other sensor nodes can play the role of the verifier and alert the rest of the network in case a compromised node is detected. We believe that this direction could offer a serious defence mechanism for sensor networks and propose it as a future research.

4.4   Routing Security

Routing and data forwarding is an essential service for enabling communication in sensor networks. Unfortunately, currently proposed routing protocols suffer from many security vulnerabilities [25] (selective forwarding, replayed messages, sinkhole and Sybil attacks, etc.), especially due to node compromises in which a single compromised node suffices to take over the entire network. Cryptographic primitives, such as encryption and authentication, are not enough to secure routing protocols; carefully re-designing these protocols with security as a goal is needed as well.

For example, multipath routing [26, 27] has been proposed as a solution. Redundant disjoint paths are used, so even if an intruder compromises a node, information can be routed by alternative paths. This strategy however provides *intrusion-tolerant* security. An autonomous communication paradigm should provide *intrusion-detection* capabilities, in order to enable self-healing processes and enable routing and other network functions to be adapted accordingly.

A closely related problem is that of secure location determination (discussed in the next subsection), which is a prerequisite for secure geographic routing. This is so because in an adversarial environment a malicious node can claim a false position to the infrastructure in order to create routing loops, or have all traffic routed through it. Nevertheless, in autonomic sensor networks, routing strategies may change in order to *adapt* to network changes [28, 29, 30]. So, for example, if location service for geographic routing becomes unavailable then a different routing strategy must be employed.

In conclusion, securing routing means providing an adaptive mechanism that secures packet flow in the network under various threats. As autonomic routing in sensor networks becomes an attractive challenge, providing security requires new design goals, like adaptability, and extensibility.

4.5   Location Aware Security

Many applications of sensor networks require location information, not only for routing purposes, but also for determining the origin of the sensed information or preventing threats against services [31, 32]. Many localization techniques have been proposed, but little research has been done in securing the localization scheme [33, 34, 35, 36]. Security in this case is twofold: Each node must determine its own location in a secure way (secure localization) and each node must verify the location claim of another node (location verification).

Since providing each node with a GPS receiver increases its cost, many localization services assume the presence of a few such nodes (usually more powerful also), which communicate their coordinates in the network and allow the rest of the nodes to estimate their position. This communication provides malicious attackers with the chance to modify measured distances and make nodes believe that they are at a position which is different from their real one. Furthermore, without location verification mechanisms, a dishonest node can cheat about its own position in order to gain unauthorized access to some services, or avoid being penalized. As more and more protocols and services are based upon location awareness, enabling sensors to determine their location in an un-trusted environment becomes essential.

4.6   Data Fusion Security

The paradigm of autonomic communication includes the *filtering* of large data feeds in order to retrieve useful information [37]. In sensor networks, thousands of sensor nodes that monitor an area generate a substantial amount of data which may be unnecessary and inefficient to be returned at the base station. Instead, certain intermediate nodes collect this data, autonomously evaluate it and reply to the aggregate queries of a remote user (Figure 2). So, data aggregation shifts the focus from address-centric approaches to a more *context* aware approach that enables sensor networks to maintain a logical view of the data.

The resource constraints and security issues make designing mechanisms for information aggregation in large sensor networks particularly challenging since aggregation nodes constitute single points of failure. An attacker upon compromising such a node may have access to valuable information and most importantly by changing the value of this information may present a wrong picture

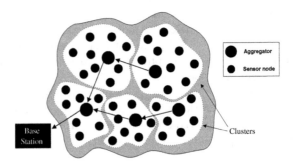

Fig. 2. An aggregation hierarchy in a sensor network. Aggregators collect information coming from the same cluster, process it and forward suitable summaries towards the base station, thus saving valuable energy resources. Although aggregators are shown bigger than simple sensor nodes, it should not be inferred that they are more powerful.

about the sensed world, thus leading to bad decisions. Several proposals for secure aggregations exist [38, 39, 40], but some open issues remain, like reorganizing the security infrastructure in case of energy depletion of an aggregation node.

4.7   Efficient Cryptographic Primitives

Because sensor nodes have limited computational and storage capabilities, traditional security solutions are often too expensive for sensor networks. More research in this domain is necessary, especially in exploring the use of efficient asymmetric cryptographic mechanisms for key establishment and digital signatures as a means for leveraging trust in sensor networks and solving some of the problems mentioned above.

Recently, elliptic curve cryptography (ECC) has emerged as a suitable public key alternative for sensor networks [41, 42], providing high security for relatively small key sizes. Since many traditional public key protocols can be turned to their EC equivalents, public-key infrastructure based on elliptic curves appears to be an attractive choice for sensor networks.

5   Autonomic Communication Challenges in Securing Sensor Networks

From the discussion in the previous sections, we see how autonomic communication behavior offers opportunities to increase security in sensor networks. We now summarize these autonomic characteristics and discuss what is needed in order to provide an integrated and complete solution for sensor networks security.

- *Self-configuration:* As the energy of sensor nodes is reduced by computation and communication, some nodes are expected to be disabled during the lifetime of the network and new ones must be deployed. Autonomic communication architectures must allow for sensor nodes to leave and join the network on-the-fly, without compromising the security level. Network configuration may also change in mobile sensor networks, resulting in new formation of groups. In all cases, the network must be able to automatically reconfigure its state, keeping the security level consistent.
- *Self-awareness:* Before a sensor network is able to respond to a security threat, it must be able to recognize it. This requires knowledge about the network's state (or more realistically, the state of neighboring nodes) and network monitoring for abnormal behaviors of sensor nodes or data traffic. To characterize normal and malicious behavior, appropriate rules must be generated, based on statistics, induction and deduction.
- *Self-healing:* Once the network is aware that an intrusion has taken place and have detected the compromised area, appropriate actions must be taken. The first one is to cut off the intruder as much as possible and isolate the compromised nodes. After that, proper operation of the network must be restored. This may include changes in the routing paths, updates of the cryptographic material (keys, etc.) or restoring part of the system using

redundant information distributed in other parts of the network. Autonomic behavior of sensor networks means that these functions must be performed without human intervention.

– *Self-organization:* Self-organization of thousands of nodes allow a sensor network to perform complex operations in a dynamic communication environment. Emphasis must be given on distributed services that allow secure location awareness, secure data fusion and implementation of complex cryptographic operations, such as access control, authentication, etc. In order to provide the needed functionality, self-organization mechanisms need to be highly scalable and adaptable.

– *Self-optimization:* Since sensor networks can be subject to unpredictable security attacks, they must be able to update their configuration on-the-fly to enable optimal behaviors in response to these changes. For example, sensor nodes should be able to function under the sudden communication load often caused by widespread security incidents, like a DoS attack, by triggering the appropriate measures.

6   Conclusion

In this paper we have presented an overview of current research challenges on sensor networks security, highlighting their autonomic communication aspects. A progress has been made in providing specialized security mechanisms, like key establishment, secure localization, secure aggregation or secure routing. While these mechanisms may protect sensor networks from specific threats, what has been lacking is a *holistic* approach that encompasses autonomic responses over a broad range of attacks. A research challenge therefore, would be the design of an adaptive security architecture that can monitor the sensor network, recognize a security threat and respond by a coordinated self-healing mechanism. In this sense, autonomic communication techniques offer opportunities for increasing sensor networks security and guaranteeing a robust and survivable solution.

References

1. Chong, C.-Y., Kumar, S.P.: Sensor Networks: Evolution, Opportunities, and Challenges. In: Proc. of the IEEE (8)**91** (2003) 1247–1256
2. Tubaishat, M., Madria, S.: Sensor networks: an overview. IEEE Potentials (2)**22** (2003) 20–23
3. Rajaravivarma, V., Yang, Y., Yang, T.: An overview of Wireless Sensor Network and applications. In: Proc. of the 35th IEEE Southeastern Symposium on System Theory (2003) 432–436
4. Hu, F., Sharma, N.K.: Secure Wireless Sensor Networks: Problems and Solutions. In: Proc. of International Conference on Computer, Communication and Control Technologies (CCCT 2003)
5. Shi, A., Perrig, A.: Designing Secure Sensor Networks. IEEE Wireless Communications (6)**11** (2004) 38–43

6. Rivest, R., Shamir, A., and Adleman, L.: A method for obtaining digital signatures and public-key cryptosystems. Communications of the ACM (2) **21** (1978) 120–126
7. Diffie, W., Hellman, M.E.: New directions in cryptography. IEEE Transactions on Information Theory **22** (1976) 644–654
8. Carman, D., Kruus, P., Matt, B.: Constraints and approaches for distributed sensor network security. Technical Report 00-010, NAI Labs (2000)
9. Hill, J., Szewczyk, R., Woo, A., Hollar, S., Culler, D., Pister, K.: System architecture directions for networked sensors. In: Proc. of the 9th International Conference ASPLOS-IX (2000) 93–104
10. Intanagonwiwat, C., Govindan, R., Estrin, D.: Directed diffusion: A scalable and robust communication paradigm for sensor networks. In: Proc. of the 6th Annual Inter. Conference on Mobile Computing and Networking (MobiCOM '00), 2000
11. Hartung, C., Balasalle, J., Han, R.: Node Compromise in Sensor Networks: The Need for Secure Systems. Tech. Report CU-CS-990-05, Univ. of Colorado (2005)
12. Basagni, S., Herrin, K., Bruschi, D, Rosti, E.: Secure pebblenets. In: Proc. of the ACM Inter. Symposium on Mobile Ad Hoc Networking and Computing, 2001
13. Eschenauer, L., Gligor, V. D.: A key-management scheme for distributed sensor networks. In: Proc. of the 9th ACM Conference on Computer and Communications Security, Washington D.C., USA (2002) 41–47
14. Chan, H., Perrig, A., Song, D.: Random key predistribution schemes for sensor networks. In: Proc. of the IEEE Symp. Security Privacy, Berkeley, CA (2003)
15. Du, W. , Deng, J. , Han, Y.S., Varshney, P.K.: A pairwise key pre-distribution scheme for wireless sensor networks. In: Proc. of the 10th ACM Conference on Computer and Communications Security, Washington D.C., USA (2003) 42–51
16. Dimitriou, T., Krontiris, I.: A Localized, Distributed Protocol for Secure Information Exchange in Sensor Networks. In: Proc. of the 5th IEEE Intern. Workshop on Algorithms for Wireless, Mobile, Ad Hoc and Sensor Networks (WMAN), 2005
17. Zhu, S., Setia, S., Jajodia, S.: LEAP: Efficient Security Mechanisms for Large-Scale Distributed Sensor Networks. In: Proc. of the 10th ACM Conference on Computer and Communications Security (CCS03), Washington D.C. (2003) 62–72
18. Wood, A. D., Stankovic, J. A.: Denial of Service in Sensor Networks. IEEE Computer **35**(10) (2002) 54–62
19. Jones, A.K., Sielken, R.S.: Computer system intrusion detection: a survey. Technical Report, Computer Science Department, University of Virginia (2000)
20. Pickholtz, R.L., Schilling, D.L., Milstein, L.B.: Theory of spread-spectrum communications - A tutorial. IEEE Transactions on Communications **30** (1982) 855–884
21. Krontiris, I., Dimitriou, T.: GRAViTy, Geographic Routing Around Voids. In preparation.
22. Anderson, R., Kuhn, M.: Low cost attacks on tamper resistant devices. In: IWSP: 5th International Workshop of Security Protocols, Lecture Notes in Computer Science **1361** Springer-Verlag (1997) 125–136
23. Deng, J., Han, R., Mishra, S.: Intrusion Tolerance and Anti-Traffic Analysis Strategies for Wireless Sensor Networks. In: Proc. of the IEEE International Conference on Dependable Systems and Networks (DSN) (2004) 594–603
24. Seshadri, A., Perrig, A., Doorn, L., Khosla, P.: SWATT: SoftWare-based ATTestation for Embedded Devices. In: Proc. of the IEEE Symposium on Security and Privacy (2004) 272–282
25. Karlof, C., Wagner, D.: Secure routing in wireless sensor networks: Attacks and countermeasures. In: Proc. of the 1st IEEE International Workshop on Sensor Network Protocols and Applications (2003)

26. Ganesan, D., Govindan, R., Shenker, S., Estrin, D.: Highly-resilient, energy-efficient multipath routing in wireless sensor networks. ACM SIGMOBILE Mobile Computing and Communications Review **5**(4) (2001) 11–25

27. Deng, J., Han, R., Mishra, S.: INSENS: Intrusion-tolerant routing in wireless Sensor NetworkS. Report CU CS-939-02, CS Dept. , University of Colorado (2002)

28. He, Y., Raghavendra, C.S., Berson, S., Braden, B.: A Programmable Routing Framework for Autonomic Sensor Networks. In: Proc. of the 5th Annual International Workshop on Active Middleware Services (AMS 2003)

29. Legendre, F., Dias de Amorim, M., Fdida, S.: Some Requirements for Autonomic Routing in Self-organizing Networks. In: Proc. of the 1st International Workshop on Autonomic Communication (WAC 2004), Berlin, Germany (2004)

30. Santivanez, C., Stavrakakis, I.: Towards Adaptable Ad Hoc Networks: The routing Experience. In: WAC 2004, Berlin, Germany (2004)

31. Liu, D., Ning, P.: Location-based pairwise key establishments for static sensor networks. In: Proc. of the ACM Workshop on Security in Ad Hoc and Sensor Networks (SASN '03), Fairfax, VA (2003)

32. Lazos, L., Poovendran, R.: Energy-Aware Secure Multicast Communication in Ad-hoc Networks Using Geographic Location Information. In: Proc. of the IEEE International Conference on Acoustics, Speech, and Signal Processing (ICASSP 2003)

33. Sastry, N., Shankar, U., Wagner, D.: Secure Verification of Location Claims. In: Proc. of the ACM Workshop on Wireless Security (2003)

34. Lazos, L., Poovendran, R.: SeRLoc: secure range-independent localization for wireless sensor networks. In: Proc. of the ACM Workshop on Wireless Security, Philadelphia, PA (2004)

35. Capkun, S., Hubaux, J.P.: Secure Positioning of Wireless Devices with Appplication to Sensor Networks. To appear in Proc. of IEEE INFOCOM (2005)

36. Du, W., Fang, L., Ning, P.: LAD: Localization Anomaly Detection for Wireless Sensor Networks. In: Proc. of the 5th IEEE Inter. Workshop on Algorithms for Wireless, Mobile, Ad Hoc and Sensor Networks (WMAN 05), 2005

37. Davide, F.: Strategic direction towards Autonomic Communication. Telecom Italia Learning Services (2004)

38. Hu, L., Evans, D.: Secure Aggregation for Wireless Networks. In: Proc. of Workshop on Security and Assurance in Ad hoc Networks, Orlando, FL (2003)

39. Przydatek, B., Song, D., Perrig, A.: SIA: Secure Information Aggregation in Sensor Networks. In: Proc. of the First International Conference on Embedded Networked Sensor Systems (SenSys) (2003) 255–265

40. Dimitriou, T., Foteinakis, D.: Secure In-Network Processing in Sensor Networks. In Proc. of the 1st Workshop on Broadband Advanced Sensor Networks (IEEE BASENETS), San Francisco (2004)

41. Malan, D., Welsh, M., Smith, M.: A Public-Key Infrastructure for Key Distribution in TinyOS Based on Elliptic Curve Cryptography. In Proc. of the 1st IEEE Inter. Conference on Sensor and Ad Hoc Communications and Networks, 2004

42. Blass, E.-O., Zitterbart, M.: Towards Acceptable Public-Key Encryption in Sensor Networks. To appear in Proc. of the 2nd International Workshop on Ubiquitous Computing (2005)

Trust Management Issues for Ad Hoc and Self-organized Networks

Vassileios Tsetsos[*], Giannis F. Marias, and Sarantis Paskalis

Pervasive Computing Research Group, Dept. of Informatics and Telecommunications,
University of Athens, Panepistimiopolis, Ilissia, GR-15784, Athens, Greece
Tel.: +302107275362; Fax: +302107275601.
{b.tsetsos, paskalis}@di.uoa.gr, marias@mm.di.uoa.gr
http://p-comp.di.uoa.gr

Abstract. Self-organized and ad hoc communications have many fundamental principles in common and also face similar problems in the domains of security and Quality of Service. Trust management, although still in its first steps, seems capable of dealing with such problems. In this paper we present an integrated trust management framework for self-organized networks. In addition, starting from our experience with the presented framework, we indicate and discuss important research challenges (among them interoperability and integration issues) for the future evolution of the trust-based autonomic computing paradigm. We argue that ontologies can address many of these issues through the semantics they convey.

Keywords: Trust management, ad hoc networking, ontologies, interoperability.

1 Introduction

Pervasive Computing envisages computing environments with ubiquitous connectivity among the deployed devices and provision of advanced "intelligent" services. As this vision comes closer to realization, new computing and communication models are deemed as necessary for the efficient handling and performance of the participating complex systems. Autonomic Communications may be one possible solution towards the next generation of large-scale networking. A case where the autonomic paradigm can be applied, is the well-known ad hoc networking paradigm. This is a special case, since ad hoc networks impose many more challenges (in all relevant computing domains) than the infrastructure-based ones. In fact, there is strong relevance between mobile ad hoc networks (MANETs) and autonomic computing and communications (ACC) which is based on their fundamentally dynamic nature. Thus, the issues addressed in this paper concern both MANET and ACC paradigms.

Security, as well as Quality of Service (QoS), issues have been well studied in existing networked systems. Such issues also arise in ACC and MANET systems, however, their handling, in general, differs from that of current systems. The special characteristics of ACC impose new security threats and risks that the future security

---

[*] Corresponding author.

I. Stavrakakis and M. Smirnov (Eds.): WAC 2005, LNCS 3854, pp. 153–164, 2006.

mechanisms should take into account. In particular, the self-optimization principle of autonomic systems, if applied in an entity/element- and not a system-basis, can lead to increased competitiveness and, thus, reduced reliability of the overall system's behavior. For example, such self-optimization could dictate the individual communicating entities to behave maliciously or in a selfish manner. In an open autonomic environment, entities, information assets, data communications and robustness are subject to the following threats:

- Attacks on the authenticity of entities, such as impersonation and *Sybil* attacks [1].
- Attacks on the privacy of communication flows, such as passive eavesdropping, or even sinkhole and wormhole [2], traffic analysis, with an objective to disclose the exchanged data and the identity of communicating entities.
- Attacks on entities or resources availability, such as denial of service attacks, materialized through sleep deprivation torture, flooding and active interfering, or even attacks on the network performance, through selfish nodes.
- Attacks on the integrity of the information assets through unauthorized alteration of distributed IT resources and of stored or exchanged data.

As it has already been identified [3], one of the most promising, though challenging, mechanisms to address both security and QoS risks in ACC is the trust establishment, evaluation and management between the cooperating entities. One of the core processes involved in the trust evaluation process is the reputation management, where cooperating entities exchange their experiences and recommendations about third parties. Trust is a soft-security method in contrast to hard-security methods such as certificate-based authentication and Public Key Infrastructures (PKI). The main advantages of soft-security are that it requires less formal information about the cooperating entities and it does not assume any infrastructure to be available. Since these are fundamental assumptions of the ACC and MANET paradigms, it seems quite reasonable to exploit trust mechanisms in such systems.

Thus, frameworks that provide self-protection of the distributed information assets, entities authenticity, and resource availability are considered essential. Since ACC specifies a self-evolving paradigm such self-protection frameworks will be situation-driven. Nodes and entities should react consistently and correctly to different situations [4] based on high level policies. In the first part of this paper, we present ATF (Ad hoc Trust Framework), a lightweight framework for trust management in self-organised networks, like MANETs. This framework is designed so as to detect selfish, malicious or unreliable behavior of communicating network nodes and provide feedback to the various services of each node on how to assess the trustworthiness of the corresponding services of other peers. ATF is lightweight in the sense that it does not perform extensive risk and behavior analysis for trustworthiness assessment, nor does it include a reasoning service capable of adapting the systems behavior with respect to other nodes' behavior and alternative strategies. Additionally, it does not involve computationally heavy security tasks, such as key generation, key agreement and cryptography. This simplicity allows its application in real systems with minor integration effort (see also Section 4). In the second part of the paper, we discuss some design aspects of trust systems with respect to the ACC vision. Based on our experience with ATF design, we indicate some challenging research areas. Among them is the interoperability between heterogeneous trust-aware ACC entities, the design of trust policies and the semantics of trust. In addition, being familiar with semantic knowledge representation and engineering we foresee many applications of ontologies in these areas and we outline some possible ways for their contribution.

2 Architecture of the ATF

ATF is based on a distributed and modular architecture. Each of the modules resides in every node and performs a well-defined set of actions to evaluate the reputation of another node or to inform others about the trustworthiness of third nodes. ATF incorporates self-evidences, recommendations, subjective judgment and historical data to evaluate the trust level of other nodes. These elements are the inputs to a trust computation model. The model also consolidates, among others, user's natural behavior, through the designation of a user-oriented Trust Policy. Such policy defines the parameters that will influence the trust computation process.

The ATF architecture is depicted in Fig. 1 and consists of the following components: *Trust Sensors (TS), Trust Builder (TB), Reputation Manager (RM) and Trust Policy (TP).* Every node implements these components. Every node also provides a number of typical *communication functions (i.e., services)* such as packet forwarding, routing, naming, etc. In general, as function can be considered any service or application provided by a node. Moreover, every node implements a special virtual service, called *Recommendation Function (RF),* which provides recommendations for specific nodes to third parties (this function is implemented by RM). *ATF* adopts the definition introduced in [5] for the reputation of a node's function, which is defined through the triplet: *Reputation = {NodeId, Function, Trust Value}.* Thus, the reputation of a function f of node i is defined as $R(i,f) = \{i, f, TV_{i,f}\}$ with $TV_{i,f}$ being the *Trust Value (TV)* for the function f of node i. This value is updated through direct evidence, recommendations and subjective criteria.

Before proceeding, it is necessary to provide the nomenclature that will be used hereafter (see Fig. 2). We use the term "detector" to denote a node that directly monitors the behavior of another node's functions, called "target". A "requestor" is a node that asks for recommendations, which are issued by "recommenders". "Neighborhood" is the set of all adjacent nodes.

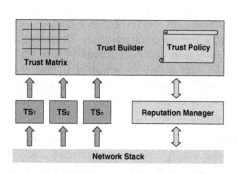

Fig. 1. ATF architecture

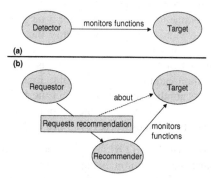

Fig. 2. Trust entity model (a) direct monitoring scenario, (b) recommendation-based scenario

Trust Sensors (TS). The majority of the proposed reputation systems agree that the most significant factor for trust building is the direct experience (or direct evidence). In the SECURE project [6], this evidence monitoring is performed independently by each node through a "Monitor" [8] which logs every activity in an "Evidence Store". In [8] a Watchdog mechanism is proposed as an observation device for routing behavior of nodes participating in ad hoc networks. The proposed mechanisms are usually function-specific. For example, monitoring the packet forwarding function of an adjacent node is different, in terms of functionality and semantics, from monitoring the routing service. ATF is based on *TSs* to detect direct evidences. A *TS* operates similarly to common sensors: translates a physical phenomenon or behavior in a machine interpretable form. In our case this phenomenon is the trustworthiness of a node. A *TS* monitors the behavior of an adjacent node, on behalf of its housed node, and compares this behavior to a predefined reference attitude, (*i.e. expected functionality*). In that sense, the *ATF* scheme uses *TSs* to assist a node to define the credibility of others. The proposed generic methodology consists of the following:

- Definition of a conceptual model of *a node's expected functionality*. This model is, generally, in close relationship with the observation methods selected (as discussed below). For example, if the observation method is pattern recognition and analysis, the expected functionality would be expressed in terms of acceptable patterns.
- Definition of the *observation methods/mechanisms*. Possible realizations include the pattern analysis of logs, messages overheard through promiscuous mode of operation, return codes of remote procedures, etc. The observation method is dependent on the function that a particular *TS* is supposed to monitor.
- *Quantification of the difference between the observations and the expected functionality*. An intuitive and easy-to-implement approach to this issue is the categorization of observations to *Successes* and *Failures* relating to the expected functionality. The number of successes and failures eventually leads to the quantification of the actual direct evidence.

At this point we should mention a special-purposed trust sensor of ATF that evaluates the trustworthiness of a node regarding its recommendation function. *RFTS* (Recommendation Function Trust Sensor), as any other *TS*, compares the observations, i.e., recommendations received for a target node, with the direct evidence of that node. If these values differ significantly the trustworthiness of the respective recommenders regarding their *RF* is decreased. In addition, a testing mechanism can decrease the impact of lying recommenders as follows: in regular intervals a requestor requests recommendations for particular functions of target nodes for which it maintains a large number of interactions in the recent past. The requestor increases or decreases the trustworthiness of the recommenders' *RF*, according to the deviation between these two values (direct evidence and recommendation).

Trust Builder (TB). This component computes the TV of other nodes' functions. In particular, it computes the TV of the nodes with which there is established interaction or intention for cooperation. All these TVs lie in the Trust Matrix, which is consulted by applications and other system/network services, and the role of the Trust Builder is to maintain and update this matrix. The actual TV computation depends on several factors and is described in Section 3.2. Factors such as, interaction history and direct evidence may introduce high certainty for a node's behavior, whereas recommendations might have less contribution. The weighting of these factors should be defined after extensive simulations and expressed through a Trust Policy.

Reputation Manager (RM). The RM role is to manage the recommendations exchange procedure (i.e., to provide recommendations from other nodes to the TB in order the latter to compute the TVs, and to give recommendations about third parties). In an on-demand schema, originator requests recommendations for a target node when it has insufficient experience with it. Thereafter, RM selects the nodes to contact (recommenders) in order to obtain requested values. These should be as close as possible to the originator in order to minimize communications overhead, but should also be rated as "good" recommenders (i.e., the originator has a high TV for their recommendation function). When recommendations arrive, the RM aggregates them and returns a single value to the TB. When a recommender receives a request for recommendation, its RM contacts TB and obtains the DE (see Section 3.2) for the requested function of target node, if any. Next, the recommender returns this value to the requestor node (see Fig. 2b). RM could be also responsible for informing other nodes whenever the TV of a function of a node is rapidly changing. In this event-driven schema, TB triggers RM upon trust value changes and RM informs the other nodes (e.g. through flooding or multicasting).

Trust Policy (TP). As already mentioned, each node maintains a Trust Policy (TP); a set of parameter values, which can fully define the functionality of its Reputation Manager and Trust Builder. In the current version of our model such policy is quite simple, but in the future will be enriched with more advanced features that enable trust strategy definition and enforcement.

3   Trust Computation Model

3.1   The Qualitative Perspective

The majority of the trust computation approaches acknowledge that two main components should be taken into consideration: *Direct Evidence* (DE) and *Recommendations (RECs)* from third parties. The *DE* is calculated based on *TSs'* feedbacks and is useful for evaluation of adjacent nodes' functionality. *RECs* are communicated between the entities participating in the trust network, according to a reputation dissemination protocol, implemented in RM.

The proposed scheme incorporates several user-defined time-dependent weights. Time-dependence is important, since it allows the modelling of temporal trust strategies, which can be followed by the participating nodes. Additionally, the weights are defined separately for each node in its *Trust Policy (TP)*. For the *ATF* scheme, time is treated as a discrete sequence of *direct* interactions between the nodes. Thus, time elapses in a different rate for every separate node. We use only direct interactions as a time reference, since they are generally regarded more important than the indirect ones (recommendations) for the trust building process. Moreover, interactions can be categorized to positive and negative (according to the success/failure model incorporated by each *TS*) to enable flexible computation of trust.

Socio-cognitive approaches to trust [9] imply that trust computation should also include a *subjective component*, since each node has a unique, subjective, way to trust others. Here, we adopt this approach, and a separate *Subjective-factor* component (*SUB*) is introduced in the trust computation model. This component is time-dependent, as well, so as to enable time-variant trusting behaviour of nodes (e.g., a node may want to trust a newcomer node only up to a point, until it establishes a specific number of successful interactions with it). *SUB* is defined in the *TP* of each individual node, and can model typical trust characters, such as unwary, suspicious, unbeliever, etc. This component provides flexibility in the trust strategy of a user, without imposing significant complexity in the overall trust computation.

History is an additional concept that has drawn attention in the trust community. Several researchers use history as an implicit component in the trust computation. Some assign specific weight to past observations or recommendations in order to provide stable and smoothed TV fluctuation [10]. Others assign minor weights to evidence (direct or indirect) received in the past to allow for reputation fading [11]. Even if the first approach seems more suitable for trust modeling, the scheme proposed here can support both policies. In the following paragraphs a detailed description of how the TB manipulates all the aforementioned factors is provided.

3.2   The Quantitative Perspective

This section describes the mathematical formulae for the proposed trust computation model. The *trust time (T)*, as already mentioned, counts the directly observable interactions. We monitor the temporal evolution of every function and node, so this time scale is represented as matrix of size NxF, where *N is the number of nodes* in the network, and *F corresponds to the overall number of supported functions.*

$$T \equiv (T_{n,f}) \in N, \ where \ n = 1...N, f = 1...F$$

Each node should have at least one time matrix. Each time the outcome of an interaction (success or failure) with a node's function is captured, the corresponding element of the detector's time matrix is increased by one unit.

Each detector maintains a NxF *Trust Matrix (TM)*, representing the *TV* that the node computes per monitored function of a target node. Each element $TM_{n,f}$ ($1 \leq n \leq N$ and $1 \leq f \leq F$) refers to a specific function f of a particular node n, and it varies with time. The formulae for *TM* and *TV* are:

$$TV(n,f,t) = TV' \cdot u(1 - TV') + u(TV' - 1)$$
$$TV' \equiv TV'(n,f,t) = [a \cdot DE_{n,f} + (1-a) \cdot REC_{n,f}] \cdot SUB_{n,f}(t), \ t = T_{n,f}$$
$$TM \equiv (TM_{n,f}), \ TM_{n,f} = TV(n,f,t) \in [0,1]$$

$$DE_{n,f} \in [0,1], \ REC_{n,f} \in [0,1] \ and \ SUB_{n,f}(t) \in [0,2]$$

Thus, $TV' \in [0,2]$, as well. In order to map the *TV* values within the [0,1] interval we use a unit step function $u(t)$ (see Eq. 1) to normalize *TV'* into the final *TV*. The range of *TV(n,f,t)* is [0,1], where 0 declares distrust for a specific function f of a target node n, and 1 declares complete trust in n for f.

$$u(t) = \begin{cases} 0, & t < 0 \\ 1/2, & t = 0 \\ 1, & t > 0 \end{cases} \tag{1}$$

DE_{nf} is the DE for a target node n and its function f, as observed by the corresponding TS of the detector. The elements of the DE matrix are defined as:

$$DE_{n,f}(t) = w \cdot TS(n, f) + (1 - w) \cdot A_H(DE_{n,f}) \text{ and } TS(n, f) \in \{0,1\}$$

A "0" value of TS indicates Failure, while "1" denotes Success. The coefficient w adjusts the weights assigned to recent and historical DE values and the A_H is an average of the last H DE values.

REC_{nf} stands for the aggregated recommendations we have for the function f on node n from third parties. These recommendations are third parties' DEs. We also keep the history of $RECs$ received. Thus, each node has a NxF matrix, with elements:

$$REC_{n,f}(t) = w \cdot NEWREC(n, f) + (1 - w) \cdot A_H(REC_{n,f})$$

NEWREC is the more recent REC.

The SUB component of the TV computation formula incorporates the node's subjectivity, as discussed in the previous subsection. SUB is a NxF matrix with elements in the $\{ f : T \to [0,2] \}$ domain. Thus, its elements are time-functions. The range [0,2] allows the detector to distrust (i.e. value 0) the target node, trust it (i.e. value 1), be enthusiastic about the target node (i.e. value 2) or develop any other intermediate form of subjective trust strategy. We have chosen the value 2 as an upper bound to prevent enthusiastic nodes from endangering the network's rationality. An example SUB time-function could be defined as:

$$SUB_{n,f}(t) = u(t - 20), \quad t = T_{n,f}$$

This function indicates that no matter what DEs or RECs a requestor node has for a target function, it will not trust it until twenty direct interactions have been observed. We should remind that all parameters involved in the present model (including SUB) are defined in each node's trust policy.

We have evaluated the performance and quality of the proposed framework in [25]. This evaluation showed that the on-demand recommendation requests and the corresponding responses introduce small communication overheads. Moreover, simulations showed that ATF enables peers to rapid identify the trust values for functions provided by peer nodes (e.g., forwarding). Thus, ATF provides sufficient means to fair nodes for rapidly identifying and isolating selfish nodes.

4   Integration and Interoperability Issues

Trust management for ACC and MANET systems raises, as happens with every new computing paradigm, questions regarding its seamless integration with current network protocol stacks. In particular, one should define how the introduction of trust affects the architecture and operation of current services and identify potential difficulties or problems during such integration.

In general, three types of modifications are needed for such integration (see Fig. 3):

1. **Introduction of a trust plane.** This is a vertical plane similar to the user/control and management planes of other networking specifications. It includes all the necessary components in order to assess trust of third entities: sensing mechanisms, policy-driven evolution of trust, interaction memory etc. Trust plane operates also as a broker since it disseminates to all other interested layers the observations of each specific layer. For example, it may give feedback to the networking layer (i.e., routing) about the physical layer operation of peer entities.

2. **Recommendation exchange through a trust protocol.** Such protocol could be implemented in the application layer and is responsible for the request and receipt of recommendations (i.e., in ATF it would be part of the Reputation Manager).

3. **Trust-aware versions of current protocols.** The observations collected by the trust plane or the recommendations collected through the trust protocol should somehow affect the operation of the network stack. This can be only performed if the protocols support trust-driven reconfigurability.

The trust plane is a distributed entity, residing in each node and dealing with trust management issues. It is similar in nature to the knowledge plane proposed in [24]. The similarities consist of the autonomic and distributed nature of the planes, and the "subjective" approach the proposed constructs follow. The trust plane, however, imposes harder design requirements. In addition, the representation and reasoning of trust entities and relationships is strict and the operation of the whole autonomic systems network is very sensitive to any weak interpretation of trust.

A good example of a trust-aware, self-adapted and reconfigurable "protocol" is the software radio [12]. Specifically, software radio serves as a radio communication system technology that uses software for the modulation and demodulation of the signal. The use of software is not only cost-beneficial; it releases the physical layer from the tight hardware integration. That is, the interface to the physical layer is no more a fixed hardware interface, but a set of interfaces provided by the deployed software. The net result is that the physical layer can be altered to any supported protocol by simple software redeployment triggered by the trust plane. In general, the protocol adaptations may be as minor as a parameter tweaking or as major as a complete operation protocol swap in cases of fully autonomic operation [13].

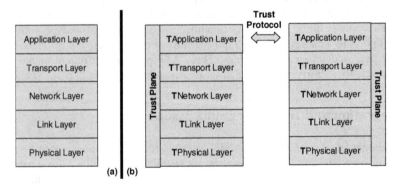

Fig. 3. (a) Traditional network stack (b) New stack elements imposed by incorporation of trust (bold text in figure)

One problem of such extended reconfigurability is the extensive management overhead required and the interoperability issues raised when nodes with different protocol configurations wish to communicate. In general, an autonomic network may sooner or later evolve to a collaboration domain of heterogeneous nodes. This is in contrast to the traditional networking paradigm, where all nodes adhere to strict standards (e.g., SSL, RSVP) in order to "stay connected". Such paradigm, although is typically simpler to implement, imposes a major restriction to the network: the nodes should not differentiate regarding their interfaces or protocols. This restriction aims, besides the obvious co-operation simplicity, at the *preservation of service and protocol semantics* that describe their messages, parameters and interaction sequences.

Obviously, this way of inter-networking is not suitable for flexible autonomic networks. In such networks, the nodes in order to continue collaborating should adhere to some common *interoperability rules*. This is a *soft-standardization* approach in opposition to the abovementioned *hard-standardization* one. A good enabler for such interoperability frameworks are the ontology-based knowledge management facilities. These have recently met wide acceptance by the research community as a means to introduce Artificial Intelligence techniques into practical applications. The most popular initiative in this discipline is the Semantic Web [15].

Ontology [16] is a terminology shared by all parties interested in an application domain (e.g., networking engineers, companies and forums). Apart from the taxonomy of the concept model of interest, an ontology also contains restrictions and formal axioms relevant to this model. Towards the vision for soft-standardized interoperability ontologies can be utilized as protocol hierarchies where a protocol is classified under a class if it satisfies its necessary (and sufficient) conditions. Such conditions may, for example, involve existence of protocol parameters or restrictions on parameter values. In general, every discrete protocol configuration can be mapped to an ontology class. The classification performed by reasoning engines on the ontology instances can infer compliance or not between the various protocols. Similar approaches for integration have been already exploited in other domains, such as network management [17].

5 Other Issues

Trust Semantics. Another aspect that is more technical but closely related to interoperability is the clear definition of trust semantics. For example, autonomic systems developed in different domains may not represent their trust values using the same representation forms. As a simple scenario consider a system A that assesses trustworthiness using real numbers in the range [0,1] (i.e., what ATF does), a system B that uses integers in the range [1,12] and system C that uses fuzzy set theory in order to translate its observations to symbolic values. Apparently, unless we map each system's values to a commonly-adopted trust value reference system, these systems will not be able to communicate their trust-related information. Such mapping entails the careful specification of semantics for each involved trust value system.

This problem can be also addressed by ontology-based knowledge representation, since ontologies are the ideal candidate for playing the role of such reference system.

The interacting systems should align their trust models with the conceptual trust model of the reference system. Furthermore, through such alignment, the semantics of the reference system are assigned to each specific trust model. This can be of great value, since the axioms and restrictions described in this reference trust ontology are automatically inherited by the specific models and can be exploited for advanced trust reasoning. Some interesting trust reasoning techniques utilizing semantic (web) technologies are presented in [18][23].

Trust Policies. From the initial research steps of ACC, policies have been recognized as a means for allowing self-organization of systems adhering to some predefined rules. Modern policy management has become quite formalized. Hierarchical policy architectures are proposed [21][22] that use distributed policies in hierarchical environments (grids, storage, ad hoc networks, etc.). The evolution of policy management divides the policy lifecycle into separate building blocks, which can be modified independently. This ability creates new opportunities to experiment and evaluate different policy schemes in the definition stage, in the enforcement stage or in any other in-between stage.

The introduction of multi-level policy architectures might play a significant role in the future of autonomic computing. Multi-level policies may bear a hierarchical or even a mesh structure, depending on the complexity of the facility and the specific need of the situation. In hierarchical policy architectures, a grand policy sets the basic rules and more specific policies apply to specific tasks.

After specifying the policy architecture, the policy definition will likely be a major field of innovation and experimentation. For example, advanced tools such as stochastic processes could be involved to randomize and, thus, conceal the trust-based decision making processes from potential enemies/attackers. Moreover, elements of game theory can be exploited to optimize the performance in such environments, where conflict of interest is apparent (in [14] trust management is described as a strategy game). Nevertheless, the complexity of the solutions will be a major factor to the acceptance or not of the final scheme.

For the implementation and enforcement of rule- and logic-based policies Semantic Web technologies (i.e., ontology languages) can be used. In fact many modern policy-based trust systems [19][20] have adopted such technologies due to their rich expressiveness, tractability (i.e., low computational complexity) and developer community adoption. However, exploiting ontologies for expressing policy rules implies that we have already established well-defined semantics for trust itself.

Finally, regarding the reasoning behind policy enforcement and trust assessment, while cognitive-evolutionary approaches may seem suitable for construct with overwhelmingly many parameters, such as the trust plane and trust policy, their relaxed reasoning could lead to security compromises in any unforeseen lapses. On the other hand, a full-knowledge, strict-reasoning approach is probably utopic given the complexity of the issue. Hence, the most likely path would be to accept the possible trust breach of a subset of the connected systems and put effort on isolating the identified malicious systems. To accomplish this self-healing task, features such as redundancy, anomalies detection and self-reconfiguration are necessary.

6 Conclusions

Trust-based communications is a key element of self-organized systems, as they enable advanced interaction models, even between unknown entities. We described a lightweight trust framework, suitable for ad hoc communications that incorporates direct evidence, recommendations, history and subjective factors, in order to evaluate the trustworthiness of peer nodes. Such model requires several modifications to the current system architectures, which in their turn affect the interoperability of nodes. We discussed such integration and interoperability issues, as well as other issues related to trust-aware self-adaptable systems. Finally, we believe that ontology-based knowledge representation can clarify the semantics of such systems and, thus, we outlined some options towards this direction. As a future work, we aim to further investigate the coupling of ontologies and trust frameworks in order to propose more specific solutions to the issues discussed in this paper.

Acknowledgement

This work was performed in the context of the project entitled "PYTHAGORAS: Support of Universities' research groups" co-funded by the "Operational Programme for Education and Initial Vocational Training" (O.P. "Education") and the European Social Fund.

References

1. Douceur, J.: The Sybil Attack. In Proceedings of the First International Workshop on Peer-to-Peer Systems (IPTPS '02), Cambridge, MA (2002)
2. Perrig, A., Hu, Y-C., Johnson, D. B.: Wormhole Protection in Wireless Ad Hoc Networks, Technical Report TR01-384. Dep. Of Computer Science, Rice University (2001)
3. EC Directorate F, IST Framework, Future and Emerging Technologies, Situated and Autonomic Communications (COMS) - Communication Paradigms for 2020, available at http://www.cordis.lu/ist/fet/comms.htm
4. Chess, D. M., Palmer, C. C., and White, S. R.: Security in an autonomic computing environment. IBM Systems Journal, Vol. 42, No 1 (2003)
5. Abdul-Rahman, A., Hailes, S.: A Distributed Trust Model. in Proc. New Security Paradigms Workshop, ACM (1997)
6. Marti, S., Giuli, T. J., Lai, K., Baker, M.: Mitigating routing misbehaviour in mobile ad hoc networks. in Proc. of Mobicom2000, Boston, USA (2000)
7. Cahill, V. et. al.: Using Trust for Secure Collaboration in Uncertain Environments. IEEE Pervasive Computing, 2(3) (2003) 52-61
8. English, C., Terzis, S., Nixon, P.: Towards Self-Protecting Ubiquitous Systems: Monitoring Trust-based Interactions. Journal of Personal and Ubiquitous Computing, (2005), to appear
9. Castelfranchi, C., Falcone, R.: Trust is much more than subjective probability. Mental components and sources of trust. HICSS33, Hawaii (2000)
10. Michiardi, P., Molva, R.: CORE: A collaborative reputation mechanism to enforce node cooperation in mobile ad hoc networks, in Proc. 6th IFIP Communications and Multimedia Security Conference, Portorosz, Slovenia (2002)

11. Buchegger, S., Le Boudec, J-Y.: A Robust Reputation System for P2P and Mobile Ad-hoc Networks. in Proc. Second Workshop on the Economics of Peer-to-Peer Systems (2004)
12. Software-defined radio forum, http://www.sdrforum.org/
13. IST End-to-End Reconfigurability Project E2R, http://e2r.motlabs.com/
14. Jøsang, A.: The right type of trust for distributed systems, in Proc. of Workshop on New Security Paradigms, California, USA (1996)
15. Berners-Lee, T., Hendler, J., Lassila, O.: The Semantic Web, Scientific American (2001)
16. Gomez-Perez, A., Corcho, O., Fernandez-Lopez, M.: Ontological Engineering: with examples from the areas of Knowledge Management, e-Commerce and the Semantic Web, Springer (2004)
17. López de Vergara, J. E., et al.: Semantic management: Application of ontologies for the integration of management information models, IM 2003 - 9th IFIP/IEEE International Symposium on Integrated Network Management, vol. 9, no. 1, March 2003, pp. 131 - 134
18. Golbeck, J., Parsia, B., Hendler, J.: Trust networks on the semantic web, Proceedings of Cooperative Intelligent Agents (CIA), Helsinki, Finland (2003) 238-249
19. Kagal, L., Finin, T., Joshi, A.: A Policy Based Approach to Security for the Semantic Web, 2nd International Semantic Web Conference (ISWC2003), Florida, USA (2003)
20. Tonti, G., et al.: Semantic Web languages for policy representation and reasoning: A comparison of KAoS, Rei, and Ponder. International Semantic Web Conference (ISWC 03), Florida, USA, 2003
21. Verma, D. C.: Policy-Based Networking–Architecture and Algorithms, New Riders (2000)
22. Neisse, R.: An Hierarchical Policy-Based Architecture for Integrated Management of Grids and Networks. Fifth IEEE POLICY Workshop, New York, USA (2004)
23. Semantic Web Trust and Security Resource Guide, www.wiwiss.fu-berlin.de/suhl/bizer/SWTSGuide/
24. Clark, D., Partridge, C., Ramming, J.C., Wroclawski, J.: A Knowledge Plane for the Internet. SIGCOMM '03, Karlsruhe, Germany (2003)
25. Marias, G., Tsetsos, V., Sekkas, O., Georgiadis, P.: Performance evaluation of a self-evolving trust building framework. IEEE SECOVAL Workshop, Athens, Greece (2005)

Multipath Routing Protocols for Mobile Ad Hoc Networks: Security Issues and Performance Evaluation

Rosa Mavropodi and Christos Douligeris*

University of Piraeus, Department of Informatics,
80 Karaoli & Dimitriou, Piraeus 185 34, Greece
{rosa, cdoulig}@unipi.gr

Abstract. The evolution of wireless network technologies and mobile computing hardware made possible the introduction of various applications in mobile ad hoc networks. These applications have increased requirements regarding security and the acceptable delays in order to provide high quality services. Multipath routing protocols were designed to address these challenges. With the use of multiple paths for the communication between a source and a destination, the autonomic user becomes almost unaware of a possible network failure, due to security attacks or link collapses. Several secure multipath routing protocols have been proposed but little performance information and extensive comparisons are available. In this paper, we briefly describe some security issues that multipath routing protocols face and we evaluate the performance of three existing secure routing protocols under different traffic conditions and mobility patterns.

1 Introduction

Mobile ad hoc networks have received great attention in recent years, mainly due to the evolution of wireless networking and mobile computing hardware that made possible the introduction of various applications. These applications have increased requirements in order to ensure high quality for the provided services. Security in such infrastructureless networks has been proven to be a challenging task. Multipath routing protocols were initially proposed in order to design robust and secure networks. The maintenance of multiple routes towards a destination prevents initiation of a new path discovery from the source node each time there is a link failure, due to a network fault or a malicious attack. In addition, the existence of multiple paths may prevent node congestion, since it may balances the traffic load through alternative routes.

Routing protocols may generally be categorized as *table driven* (often called proactive) and *source initiated* (or on-demand). In table driven protocols

---

* This work has been supported in part by the IST FET Coordination Action ACCA (006475).

I. Stavrakakis and M. Smirnov (Eds.): WAC 2005, LNCS 3854, pp. 165–176, 2006.

(*e.g.* ZRP [13]), each host continuously maintains complete network routing information. On-demand schemes (*e.g.* DSR [12]), the routing discovery process is invoked only on demand, in a query/reply approach. According to the number of paths that are discovered from a route request, the routing protocols are divided into *single* path (*e.g.* [12, 14]) and *multipath* (*e.g.* [9, 7]). The number of the discovered paths that are actually used for sending data is another feature of the routing protocols. Some protocols use only a single path for the communication, while others distribute the data through different channels. The route discovery process in the multipath protocols may be initiated either when the active path collapses (in that case communication is performed with one of the alternative paths), or when all known paths towards the destination are broken [5]. The route discovery may stop when a sufficient number of paths are discovered or when all possible paths are detected. The protocols of the second case, are also known as *complete*. Path found by routing protocols can be *node-disjoint* [11] or *link-disjoint* [6] if a node (or a link) cannot participate in more than one route between two end nodes.

The area of secure routing protocols is of particular interest in routing security, since the lack of fixed infrastructure makes routing an obvious target for malicious nodes. Several solutions for secure routing have been proposed, such as collaborative monitoring of the routing behavior between nodes [15, 16], motivating nodes to behave well with fictitious currency [18] or participation of nodes in routing paths based on quantifiable criteria [17]. However, these solutions cannot resist Denial-of-Service (DoS) attacks of malicious nodes, since they are designed for single path routing protocols.

Three secure multipath routing protocols have been recently proposed in order to resist Denial of Service (DoS) attacks of collaborating malicious nodes, which single path protocols fail to address; the Secure Routing Protocol (SRP) [8], the multipath routing protocol of [2] and the Secure Multipath Routing protocol (SecMR) [4]. In this paper for simplicity reasons the protocol of [2] will be called *Multipath*.

In this paper, we evaluate the performance of the currently proposed secure multipath routing protocols of SRP, SecMR and Multipath by simulating their behavior in various traffic scenarios and under different mobility patterns. In section 2, we briefly describe some security issues that routing protocols face in mobile ad hoc networks. In section 3, we present a short description of the compared protocols, while in section 4 we present the simulation results. Finally, in section 5 we discuss possible enhancements and we conclude this paper.

2   Security Issues in Multipath Routing Protocols

A major security issue that single path routing protocols fail to resolve is Denial of Service (DoS) attacks of collaborating malicious nodes. With single path routing protocols it is trivial for an adversary to launch a DoS attack, even if security measures are taken. A malicious node controlled by the adversary may participate passively in the routing path between two end nodes and may behave as a legitimate intermediate node. The malicious node can stop the communication at any time it seems most advantageous to the adversary. Although communication may be cryptographically protected, network characteristics (such as variation in traffic) or external factors may be used by the adversary in order to identify the proper time to disrupt communication. Even though the end nodes may initiate a new route request after the DoS attack, the time required to establish the new path may be critical. A dedicated and skillful adversary may thus identify the most critical nodes and disable their single routing paths, by compromising a small fraction of nodes.

Multipath routing protocols can be resilient to DoS attacks and may protect network availability from faulty or malicious nodes [1]. Indeed, if there exist k node-disjoint paths between two end nodes, the adversary should compromise at least k nodes - and more particularly at least one node in each path - in order to control their communication. A secure multipath routing protocol must be node-disjoint. In order for a multipath routing protocol to be able to guarantee at a certain level the availability of the communication against DoS attacks of a bounded number k of collaborating malicious nodes, it should employ $k+1$ node-disjoint routing paths between two communicating nodes. Otherwise, a malicious node would be allowed to participate and consequently control more than one path. Thus, a single malicious node may manipulate the routing protocol and in this way it may compromise all the available routes between two end nodes.

In order to achieve resilience to DoS attacks, a multipath routing protocol should be properly enhanced with cryptographic means, which will guarantee the integrity of a routing path and the authenticity of the participating nodes. However, the cryptographic protection in the route discovery of the secure multipath routing protocols will naturally increase the control overhead. Until now, the efficiency of the secure multipath routing protocols for ad hoc networks has not been thoroughly evaluated.

In order to reduce the control overhead, in several multipath routing protocols *e.g.* [8], each intermediate node processes each instance of the route request query only the first time it receives it and drops any duplicates. This may lead to discovering less node-disjoint paths from the existing total set of paths between a given source and destination as the second instance of the request is being dropped even if it has propagated through a different neighborhood. This situation is known as the racing phenomenon.

If a protocol requires only end-to end authentication and intermediate nodes participating in a routing path are not authenticated (as for example in SRP), then the protocol is subject to impersonation sybil attacks [19], under which a malicious node may present multiple identities.

3   Description of the Compared Multipath Routing Protocols

In this section we will briefly describe the compared routing protocols.

SRP [8] is a routing protocol that manages to find multiple node-disjoint paths. It uses symmetric cryptography in an end-to-end manner, to protect the integrity of the route discovery. In SRP the route request message contains unique identifiers, assigned by the source, in order to avoid replay attacks. Each intermediate node will process only the first instance of a route request that it receives and it will drop any other recently heard. When an intermediate node receives the request first checks if it has heard it recently and if not appends itself in the routing path and forwards it, otherwise it drops it. When the target node receives a route request query, the node checks the authenticity of the request by using a symmetric encryption key - a security association - which the two end nodes are supposed to share prior to the request. The route reply query will also be protected with the same security association, in order to protect the integrity of the routing paths. Thus, it is very efficient and it protects from several attacks of malicious nodes. However, the route request propagation is inherently weak to the racing phenomenon, which may prevent the discovery of existing node-disjoint paths. Moreover, the intermediate nodes are not authenticated, making the protocol vulnerable to impersonation and sybil attacks [19]. Thus, a malicious node may participate with fake identities to several paths, rendering the multipath routing insecure.

The secure multipath routing protocol Multipath [2] is based on the Ford-Fulkerson MaxFlow algorithm. In this protocol, when an intermediate node receives a request first checks if a maximum hop distance has been reached. If not appends its neighborhood information along with a signature and forwards the packet, otherwise it drops it. When the target node receives the request, it uses the received information in order to estimate the current network connectivity and to construct the complete set of the existing node-disjoint paths. The protocol exhibits high security characteristics, as all the participating nodes are authenticated and the integrity of the routing path is protected. It manages to find the complete set of the existing node-disjoint paths. However, the propagation of the route request query is not efficient in terms of computation and space costs. The cumulative neighborhood information that the message carries may become larger than the message length. Furthermore, the use of digital signatures by the intermediate nodes of each route request message costs both in delay and processing power and may not be affordable for typically available equipment.

SecMR [4] is a complete secure multipath routing protocol that exhibits authentication in end-to-end and in link-to-link levels and manages to protect the integrity of the routing paths. It works in two phases. The first phase is the neighboring authentication phase which is repeated in periodic time intervals and ensures the link-to-link authentication. During this phase, nodes in range are mutually authenticated through digital signatures. Each node n_i constructs a set N_i that contains the identifiers of its authenticated neighbors. During the second phase the source produces a signed request, which grants the system with end-to-end authentication, and each intermediate node processes all the receiving requests, ensuring this way that all possible node-disjoint paths will be finally discovered by the destination. When an intermediate node receives a request through a node that belongs to its authenticated list of neighbors, it will first append itself in the routing path. Secondly, it will construct the neigh-

Table 1. Protocol Comparison

Characteristics	Protocols			Vulnerabilities
	SecMR	Multipath	SRP	(derived from the luck of this characteristic)
end-to-end authentication	yes	yes	yes	luck of data integrity
link-to-link authentication	yes	yes	no	impersonation, sybil attacks
complete	yes	yes	no	less discovered paths
how many requests the intermediate node processes	all	all	only the first	racing phenomenon

borhood information and the exclude-nodes information that are also appended to the message. The neighborhood information will contain all its authenticated neighbors that have not yet received the request and the exclude information will contain all the nodes that have received the message sometime in the past. When a destination receives the request it will check its authenticity by checking its signature, it will construct the node-disjoint paths and will produce a signed reply message, thus protecting the integrity of the used path. Table 1 briefly presents the comparison issues that were discussed in this section.

4   Performance Evaluation

Our study involves a comparison of the route request query between SRP, Multipath [2] and SecMR [4] protocols. We implemented the simulator within the NS-2 library. Our simulation modeled a network of 50 hosts placed randomly within a $1500 \times 1000m^2$ area. Each node has a radio propagation range of 250 meters and channel capacity was 2 Mb/s.

The nodes in the simulation move according to the 'random way point' model. At the start of the simulation, each node waits for a pause time, then randomly selects and moves towards a destination with a speed uniformly lying between zero and the maximum speed. On reaching this destination it pauses again and repeats the above procedure till the end of the simulation. The minimum and maximum speed is set to 0 and 20 m/s, respectively and pause times 0,5,10,20,30 and 40 sec. A pause time of 0 sec corresponds to the continuous motion of the node and a pause time of 40 sec corresponds to the time that the node is stationary.

Ten traffic generators were developed to simulate constant bit rate (CBR) sources. Each source generates data packet continuously until the end of the simulation run. The sources and the destinations are randomly selected with uniform probabilities. The size of the data payload was 512 bytes. Each run is executed for 350 sec of simulation time. We used the IEEE 802.11 Distributed Coordination Function (DCF) as the medium access control protocol. The destination of the traffic waits, if necessary, for 5 seconds until it assumes that all possible paths have been found, selects the node-disjoint ones and generate Reply messages concerning these paths. We generated various traffic scenarios by using the interarrival data packet time. For each traffic scenario, ten different movement patterns were used.

A free space propagation model with a threshold cutoff was used in our experiments. In the radio model, we assumed the ability of a radio to lock onto a sufficiently strong signal in the presence of interfering signals, *i.e.*, radio capture.

In order to compare the performance of the three routing protocols we evaluated them with respect to the following metrics.

Average end-to-end delay or mean overall packet latency: It is the average delay a packet suffers from the time it leaves the sender application and the time it arrives at the receiver.

Destination location time: Is the average time taken for the first instance of a route request to reach the target node (destination).

Request Propagation time: It corresponds to the average total time that a route request message takes to propagate through the entire network. Thus it is an important metric as it can describe the burden that the request process puts into the entire network. In comparison with the *Destination location time* metric, the request propagation time illustrates the time that a request zombies in the network and it is processed by nodes that are not going to participate in the path for a given source and destination.

Drop percentage: The percentage of the packets that are dropped due to various reasons.

Routing throughput: The throughput of the routing control packets in the entire the network, averaged by the total number of nodes.

Figure 1 shows the average delay of the received data packets per data interarrival time and pause time 20 secs. We can observe from the results that both SRP and SecMR outperform Multipath even when the interarrival time is small, which depicts high traffic conditions. In both SRP and SecMR the number of generated messages during the route discovery process are kept in sufficiently low levels while the ones of Multipath tend to flood the network. This happens because in Multipath, each intermediate node forwards all the route requests that reaches it for a given source, destination and sequence number, while SRP forwards only the first and SecMR performs a selective forward with the use of the exclude list. This flooding of the network results in higher delay in the data packet delivery. Figure 2, which presents the average delay of the received data packets to a network that transmits 100 data packets per second per with pause time, strengthens the above observations. Indeed, as shown, the SecMR and SRP protocols handle high mobility conditions better, although with a larger pause time the behavior of Multipath tends to converge to the performance of the other two protocols.

Figure 3 presents the dropping percentage of the data packets in relation to interarrival times and a pause time of 20 secs. All three protocols exhibit comparable performance, but SecMR and SRP manage to drop less packets, especially as the interarrival time is getting larger. The observed high drop-

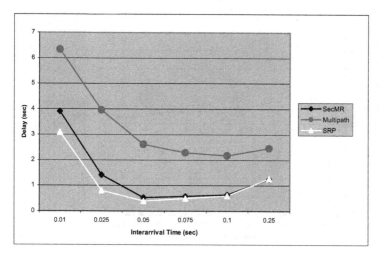

Fig. 1. Average end-to-end data packet delay per interarrival time

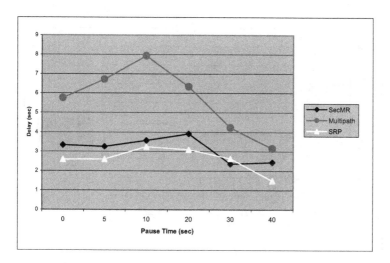

Fig. 2. Average end-to-end data packet delay per pause time

ping ratio, that all three protocols present is mainly due to the configuration of the simulation, namely the expiration time of the paths in the routing tables. Nevertheless, the performance pattern reveals the better performance of SecMR and SRP. Figure 4 shows the dropping percentage of the data packets as this evolves in comparison to various pause times with a data interarrival of 0.25 secs. As figure 4 shows the protocols manage to preserve their dropping pattern under different mobility conditions.

The number of packets that are correctly received by the destination node per interarrival time and a pause time of 20 secs are shown in figure 5. The performance of all three protocols tends to converge as the interarrival time is

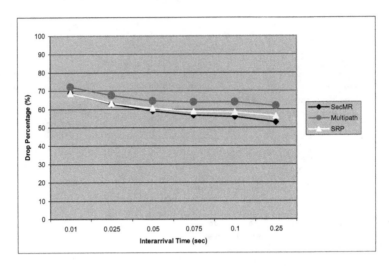

Fig. 3. Data packet drop percentage per interarrival time

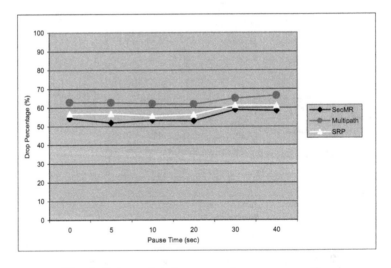

Fig. 4. Data packet drop percentage per pause time

getting larger, that is because in this case the network is facing looser traffic conditions. As one can see SRP and SecMR manage to serve more packets in comparison to Multipath. That is mainly due to the fact that data packets in Multipath encounter higher delay during their propagation and higher dropping rate. All three protocols manage to maintain their behavior with regard to the message delivery ratio under various movability patterns, as shown in figure 6, which represent a data packet interarrival time of 0.01 secs.

Figure 7 presents the average total time that route request messages propagate through the network. In Multipath, as nodes are getting more and more

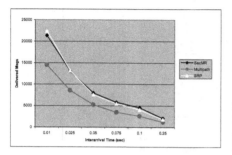
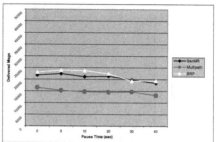

Fig. 5. Number of delivered data packet per interarrival time

Fig. 6. Number of delivered data packet per pause time

stationary the total propagation time tends to get smaller until it reaches a minimum threshold. That does not seem to be the case for SecMR where the use of the *ExcludeList* prevents the request's reception by nodes that have heard it sometime in the past. In the case of SRP the protocol benefits come from the fact that each route request is forwarded only once by each intermediate node.

Figure 8 presents the average time it takes for a request message to reach its destination for the first time. If it is seen in comparison to figure 7, it is obvious that in Multipath the request message zombies into the network for more time than in the other protocols, which causes a degradation to the network's performance. In the secure Multipath, a route request travels for a longer time than in the other two protocols, as the request is being forwarded to all nodes in range, many of which will not be included into one of the discovered paths. The route request of the SRP propagates the request towards the destination faster than the other protocols, since it rejects any variant of a specific request. The route request of the SecMR has slightly longer living times than SRP. This is reasonable as it attempts to ensure the discovery of the complete set of existing node-disjoint paths. Furthermore, the SecMR makes sure that all its neighboring nodes have contributed to the route discovery, either by participating to the *RouteList* (*i.e.* to a routing path) or by avoiding to re-process the same thread

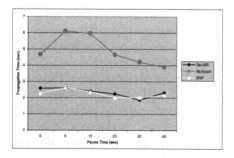
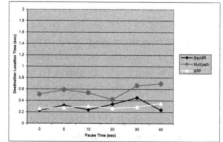

Fig. 7. Request Propagation Time per pause time

Fig. 8. Destination location time per pause time

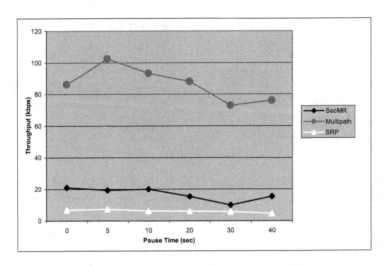

Fig. 9. Routing throughput per pause time

of the query (*i.e.* by participating into the *ExcludeList* of the query thread). The above is, also, illustrated in figure 9, which presents the total throughput derived from control messages averaged by the total number of nodes, for different pause times and with an interarrival time equal to 0.01 secs. SRP produces less control overhead than the other two protocols, thus the reduced control throughput shown in figure 9. The control throughput of SecMR is slightly increased due to the selective forwarding of request messages that it performs. Multipath has the worst performance in comparison with the other two protocols, something that is due to the number of forwards that it performs.

5   Discussion and Future Work

The area of ad hoc networking has received increased attention among researchers in recent years, as the evolution of wireless networking and mobile computing hardware made possible the service of various applications by this type of networks. Security in such environments is a critical issue. Over the past years a variety of new routing protocols have been proposed targeted specifically at the area of secure ad hoc networking, but little performance information and extensive comparisons between these protocols is available.

With this work we intended to examine the routing performance of three secure multipath routing protocols, namely SecMR [4], SRP [8] and Multipath [2], through various traffic conditions and under different mobility patterns. First we briefly examined their security characteristics and studied the security issues that these protocols address. Secondly, the protocols' performance was evaluated under different traffic conditions and mobility patterns.

The simulation results provide significant evidence about the efficiency of the examined secure multipath routing protocols. Our study showed that SRP performs better than the other two protocols, SecMR follows in short distance, while Multipath seems to be the heavier.

Considering the security characteristics (as analyzed in [4]) one can say that, Multipath achieves to provide maximum resilience against DoS attacks of collaborating malicious nodes. It provides completeness in the route discovery process and it explicitly authenticates all the intermediate nodes in each routing path. These features make it appropriate for security critical ad hoc network applications, but its applicability can only be considered in networks with relatively low density. In such environments the risk that of the request information will become larger than the message's length is minimized. Furthermore, congestion and long delays will be avoided.

The SecMR protocol also achieves completeness and provides implicit authentication of the intermediate nodes, since node authentication is performed once for a discrete time period. These features seem to make it appropriate for networks that require high security protection and present medium mobility as well as a rather high node density. In such situations, the SecMR protocol has comparable efficiency with the SRP, while it offers an increased security level.

Finally, the SRP protocol does not provide the complete set of node-disjoint paths, and it provides only end-to-end authentication. Its better routing performance makes it a suitable choice for several network configurations with increased node density. This is caused by the fact that the route request propagation avoids discovery of all the possible routes that each node could participate and in this way it converges faster. This however leads to a non-complete route discovery [4] and reduces the security resilience of the protocol to distributed DoS attacks. Thus, SRP seems suitable for applications with medium security risks.

Regarding possible extensions of our work, we consider examining the behavior of the secure multipath routing protocols in various insecure network configurations.

References

1. M. Burmester and Y. Desmedt, *Secure communication in an unknown network using certificates*, Advances in Cryptology - Asiacrypt '99, Lecture Notes in Computer Science Vol. 1716, Springer, 1999, pp. 274–287.
2. M. Burmester and T. van Le, *Secure multipath communication in mobile ad hoc networks*, Proceedings of the International Conference on Information Technology: Coding and Computing (ITCC 2004) (Las Vegas), IEEE, April 2004.
3. G. Koh, D. Oh, and Heekyoung Woo, *A graph-based approach to compute multiple paths in mobile ad hoc networks*, Lecture Notes in Computer Science Vol.2713, Springer, 2003, pp. 3201–3205.
4. P. Kotzanikolaou, R. Mavropodi, and C. Douligeris, *Secure multipath routing for mobile ad hoc networks*, Proceedings of the WONSS'05 Conference (St. Moritz, Switzerland), IEEE, January 19-21 2005, pp. 89–96.
5. S.-Ju Lee and M. Gerla, *Split multipath routing with maximally disjoint paths in ad hoc networks*, Proceedings of ICC 2001 (Helsinki, Finland), IEEE, June 2001, pp. 3201–3205.

6. M. K. Marina and Samir R. Das, *Ad hoc on-demand multipath distance vector routing*, ACM SIGMOBILE Mobile Computing and Communications Review **6** (2002), no. 3.

7. A. Nasipuri and S.R. Das, *On-demand multipath routing for mobile ad hoc networks*, Proceedings of IEEE INFOCOM99, 1999, pp. 64–70.

8. P. Papadimitratos and Z. Haas, *Secure routing for mobile ad hoc networks*, In Proceedings of the SCS Communication Networks and Distributed Systems Modeling and Simulation Conference (CNDS) (TX, San Antonio), January 2002.

9. A. P. Subramanian, A. J. Anto, J. Vasudevan, and P. Narayanasamy, *Multipath power sensitive routing protocol for mobile ad hoc networks*, Lecture Notes in Computer Science Vol.2928, Springer, 2003, pp. 171–183.

10. A. Tsirigos and Z.J. Haas, *Multipath routing in the presence of frequent topological changes*, IEEE Communications Magazine **39** (2001), no. 11, 132–138.

11. J. Wu, *An extended dynamic source routing scheme in ad hoc wireless networks*, Telecommunication Systems **22** (2003), no. 1-4, 61–75.

12. D. Johnson and D. Maltz, *Dynamic source routing in ad-hoc wireless networks*, Mobile Computing, Kluwer Academic Publishers, 1996, pp. 152–181.

13. Z.J Haas and M. Perlman, *The performance of query control schemes for zone routing protocol*, Proc. of SIGCOMM'98, 1998.

14. C. Perkins, E. Royer, and S. Das, *Ad hoc on-demand distance vector routing*, Proc. of IEEE Workshop on Mobile Computing Systems and Applications, IEEE, February 1999, pp. 90–100.

15. S. Marti, T.J. Giuli, K. Lai, and M. Baker, *Mitigating routing misbehavior in mobile ad hoc networks*, Proc. of the 6th MobiCom Conference, ACM, August 2000.

16. H. Yang, X. Meng, and S. Lu, *Self-organized network-layer security in mobile ad hoc networks*, Proc. of the ACM workshop on Wireless security (Atlanta, GA), ACM, September 2002, pp. 11–20.

17. S. Yi, P. Naldurg, and R. Kravets, *Security-aware ad-hoc routing for wireless networks*, Technical Report, UIUCDCS-R-2001-2241, June 2001.

18. L. Buttyan and J.P. Hubaux, *Enforcing service availability in mobile ad hoc WANs*, Proc. of the 1st MobiHoc Conference (BA, Massachusetts), August 2000.

19. J. R. Douceur, *The sybil attack*, In First International Workshop on Peer-to-Peer Systems (IPTPS '02), American Mathematical Society, March 2002.

Autonomous Network Equipments

Dominique Gaïti[1], Guy Pujolle[2], Mikaël Salaun[3], and Hubert Zimmermann[4]

[1] University of Troyes, UTT, 10010 Troyes, France
Dominique.Gaiti@utt.fr
[2] LIP6, UPMC, 8 rue du Capitaine Scott 75015 Paris, France
Guy.Pujolle@lip6.fr
[3] France Telecom, Lannion, France
Mikael.salaun@francetelecom.fr
[4] Ginkgo-Networks, 8 rue du Capitaine Scott 75015 Paris, France
Hubert.zimmermann@ginkgo-networks.com

Abstract. IP networks are now well established. However, control, management and optimization schemes are provided in a static and basic way. Network control and management schemes using an autonomy based technology offer a new way to master quality of service, security and mobility management. This new paradigm allows a dynamic and intelligent control of the equipment in a local manner, a global network control in a cooperative manner, a more powerful network management, and a better guaranty of all vital functionalities like end to end quality of service and security. In this paper, we provide a way to implement such a paradigm through the use of the agent and multi agent concept. A testbed of an architecture based on autonomous network equipment has been developed. This autonomous architecture is able to optimize the quality of service through the networks.

1 Introduction

The popularity of the Internet has caused the traffic on the Internet to grow drastically every year for the last several years. It has also spurred the emergence of the quality of service (QoS) for Internet Protocol (IP) to support multimedia application like ToIP. To sustain growth, the IP world needs to provide new technologies for guarantying quality of service. Integrated services and differentiated services have been normalized to support multimedia applications. The routers in the IP networks play a critical role in providing these services. The demand of QOS on private enterprise networks has also been growing rapidly. These networks face significant bandwidth challenges as new application types, especially desktop applications. Moreover, voice, video, and data traffic need to be delivered on the network infrastructure. This growth in IP traffic is beginning to stress the traditional software and hardware-based design of current-day routers and as a result has created new challenges for router design.

To achieve high-throughput and quality of service, high-performance software and hardware together with large memories were required. Fortunately, many changes in technology (both networking and silicon) have changed the landscape for implementing

I. Stavrakakis and M. Smirnov (Eds.): WAC 2005, LNCS 3854, pp. 177–185, 2006.

high-speed network equipment. However, scalability problems were discovered with InterServ technologies and statistical problems with DiffServ. Moreover, these technologies are rather complicated to size and we assist to important configuration problems that need specialized engineers.

This paper proposes a new paradigm for providing a smart networking technique allowing a real time network configuration. Indeed, we propose to introduce an autonomy based technology within network equipments to configure themselves depending on the observed state of the network.

The rest of the paper is organized as follows. First, we introduce the autonomous paradigm and the implication on network equipment. Then, we introduce a new autonomy based architecture to support the deployment of the intelligent network equipment. Finally, we describe the agent architecture and we conclude this work.

2   The Autonomous Environment

As user needs are becoming increasingly various, demanding and customized, IP networks and more generally telecommunication networks have to evolve in order to satisfy these requirements. That is, a network has to integrate more quality of service, mobility, dynamicity, service adaptation, etc. This evolution will make users satisfied, but it will surely create more complexity in the network generating difficulties in the control process.

Since there is no control mechanism which gives optimal performance whatever the network conditions are, we argue that an adaptive and dynamic selection of control mechanisms, taking into account the current traffic situation, is able to optimize the network resources uses and to come up to a more important number of user expectations associated with QoS [0]. To realize such functionalities, it is necessary to be able to configure automatically the network in real time. Therefore, all the network equipment must be able to react to any kind of change in the network. Different techniques could be applied but as the most difficult moment is congestion, the technique has to be autonomic and network equipments have to turn into intelligent network equipments.

Autonomic communication paradigm has been mainly defined through the ACF (Autonomous Communications Forum) [1] and particularly as follows: Autonomic communication is centered on selfware – an innovative approach to perform known and emerging tasks of network control plane, both end-to-end and middle box communication based. Selfware assures the capacity to evolve, however it requires generic network instrumentation. Figure 1 outlines a generic framework of a network element that is enhanced by a selfware mechanism to exchange generic policies with groups of other elements and, through embedding of "policies to functionality" rules that control the behavior of an element. Selfware principles and technologies borrow largely from well established research on distributed systems, fault tolerance among others, from emerging research on non-conventional networking (multihop ad hoc, sensor, peer-to-peer, group communication, etc.), and from similar initiatives, like Autonomic Computing of IBM, XG of DARPA, Harmonious Computing of Hitachi, Resonant Networking of NTT, etc.

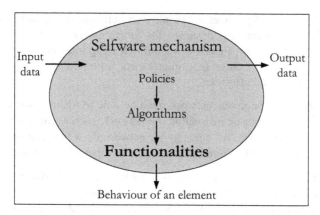

Fig. 1. Generic framework of a network element with a selfware mechanism

A visionary network would be able to (i) configure and re-configure itself, (ii) identify its operational state and take actions to drive itself to a desired stable state and finally (iii) organise the allocation and distribution of its resources. To build such a network, it is necessary to go beyond the improvement of techniques and algorithms by using a new concept, the knowledge plane. This concept was already proposed for managing the Internet. The knowledge plane is able to collect all information available in the network to provide the other elements of the network with services and advice and make the network perform what it is supposed to. There are many objectives to the configuration and reconfiguration of the network, from the optimisation of resources to the use of best available techniques in order to offer the most appropriate service, best adapted to the terminal capabilities.

The network architecture proposed in this paper aims at defining a functional architecture for the interconnection and interoperability of the different autonomous elements (i.e. network equipments as routers, firewall, middle box, etc.) interconnected to form a multiservice network. The architecture has to take into account different aspects for autonomy:

Self-configuration: the autonomic network elements must be able to configure themselves once into the network domain. Self-configuration includes such aspects as IP address, security, QoS among others. Self-configuration should also deal with the technology handover (e.g. going from Wi-Fi to UMTS) and with the parameterisation of each technology to obtain the optimal resource usage and interaction.

Self-management: the autonomic network must be able to self-manage in order to ensure a stable operational state. Whenever a new service must be deployed or a new terminal comes into the network, the self-management functions must drive the network to a stable operational state. This state would be calculated to be optimal with respect to the current operational conditions and the requirements of all available services within the available resources.

Self-diagnostics: the network as a whole must be able to identify its operational state and take action to drive itself to a desired stable state. The network must be able to identify the users accessing the service domain and recognise their profiles including the rights and associated parameters. Finally, an autonomous network consisting of heterogeneous home appliances designed for functions ranging from high complex decoding of video and audio signals to vacuum cleaning must be able to manage the interaction of their interoperation (e.g. interference from one appliance to the others) as well as precedence and priorities.

Self-protection: an autonomous network must be able to identify security threats to the content being carried or treated within the network, such as intrusions or denial of service attacks among others. An autonomous network must take appropriate action to protect itself against such threats and must ensure a transparent experience for the user.

Self-organisation: the autonomous network must be self-organised as regards resource allocation and distribution. Resources should be automatically allocated where necessary or appropriate for the current operational status and service configuration. In addition, taking into account the computational resources available in the network and the different computational grids that can be dynamically formed, the autonomic network must be able to self-organise in an optimal and secure way.

3   The 4-Plane Architecture

The 4-plane architecture approach [1] our proposal is relying on is described in Figure 2. Our proposal does not aim at proposing new algorithms or new schemes in the control plane but rather at selecting the best algorithms and the best values of the parameters at any time to reach the objective (network security, QoS, mobility management, resource optimization, etc.).

This approach will allow reconfiguring the different network elements (routers, switches, mobile elements, firewall, set-up-box, and middle-box) in quasi-real time. The goal of this approach is to secure the network, optimize the performance and control the mobility within the network. This driving process runs in real time and reconfiguration can occur several times per second if necessary. This compares to the configuration schemes used today where networks are configured only at set-up time, the configuration being decided on an average behavior of the network.

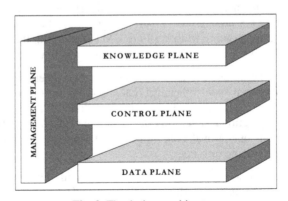

Fig. 2. The 4-plane architecture

The idea lying behind the knowledge plane is to locate our agents and the knowledge they need to act and to help the reactive agents to make the best decisions, take the most appropriate actions in view of attaining the goals set forth. Different types of implementation of the knowledge plane can be provided:

- The knowledge plane composed of meta agents with cognitive intelligence,
- The knowledge plane composed of a Policy Decision Point (PDP) or a set of Local PDPs [2],
- The knowledge plane composed of a supervisor,

A mixed of these different schemes can also offer a solution.

4 Adaptive and Autonomous System

Concerning the implementation of the autonomous system described in the previous section, a multi-agent approach could be a solution. In fact, agents own some features like autonomy, proactivity, cooperation, etc. predisposing them to operate actively in a dynamic environment like IP networks. Agents, by consulting their local knowledge and by taking into consideration the limited available information they possess about their neighbors, select the most relevant management mechanisms to the current situation.

A multi-agent system is composed of a set of agents which solve problems that are beyond their individual capabilities [3]. Multi-agent systems have proven their reliability when being used in numerous areas like: (1) the road traffic control ([4], [5]); (2) biologic phenomena simulation like the study of eco-systems [6] or the study of ant-colonies [7], for example; (3) social phenomena simulation like the study of consumer behaviors in a competitive market [8]; (4) industrial applications like the control of electrical power distribution systems, the negotiation of brands, etc. By its nature, multi-agent approach is well suited to control distributed systems. IP networks are good examples of such distributed systems. This explains partly the considerable contribution of agent technology when introduced in this area. The aim was mainly to solve a particular problem or a set of problems in networks like: the discovery of topology in a dynamic network by mobile agents ([9], [10]), the optimization of routing process in a constellation of satellites [11], the fault location by ant agents [12], and even the maximization of channel assignment in a cellular network [13].

Our approach consists in integrating agents to build an autonomous environment. These agents optimize the network QoS parameters (delay, jitter, loss percentage of a class of traffic, etc.), by adapting the activated control mechanisms in order to better fit the traffic nature and volume, and the user profiles. The agents share a global goal of the network through the knowledge plane. Agents may be reactive, cognitive or hybrid [3], [6], [14]. Reactive agents are suitable for situations where we need less treatment and faster actions. Cognitive agents, on the other side, allow making decisions and planning based on deliberations taking into account the knowledge of the agent about itself and the others. A hybrid agent is composed of several concurrent layers. In INTERRAP [15], for example, three layers are present: a reactive layer, a local planning layer, and a cooperative layer.

The approach we propose is different [16], [17], [18], [19]. In fact, every node has one cognitive agent that supervises, monitors, and manages a set of reactive agents.

Each reactive agent has a specific functioning realizing a given task (queue control, scheduling, dropping, metering, etc.) and aiming to optimize some QoS parameters. The cognitive agent (we call it Master Agent) is responsible for the control mechanisms selection of the different reactive agents, regarding the current situation and the occurring events. By using such an architecture, we aim to take advantage of both the reactive and cognitive approaches and avoid shortcomings of the hybrid approach (coordination between the different layers, for instance).

To get the agent-based autonomous approach, we propose to select the appropriate control mechanisms among:

- adaptive: the agent adapts its actions according to the incoming events and to its vision of the current system state. The approach we propose is adaptive as the agent adapts the current control mechanisms and the actions undertaken when a certain event occurs. The actions the control mechanism executes may become no longer valid and must therefore be replaced by other actions. These new actions are, indeed, more suitable to the current observed state [20];
- distributed: each agent is responsible for a local control. There is no centralization of the information collected by the different agents, and the decisions the agent performs are in no way based on global parameters. This feature is very important as it avoids having bottlenecks around a central control entity;
- local: the agent executes actions on the elements of the node it belongs to. These actions depend on local parameters. However, the agent can use information sent by its neighbors to adapt the activated control mechanisms;
- scalable: our approach is scalable because it is based on a multi-agent system which scales well with the growing size of the controlled network. In order to adaptively control a new node, one has to integrate an agent (or a group of agents) in this node to perform the control.

Our model relies on two levels:

At level 0, we find the different control mechanisms of the node, which are currently activated. Each control mechanism is characterized by its own parameters, conditions and actions, which can be monitored and modified by the Master Agent. Some of the proposed management mechanisms are inspired from known algorithms but have been agentified in order to optimize the performance and to improve cooperation between agents.

Different agents belong to this level (Scheduler Agent, Queue Control Agent, Admission Controller Agent, Routing Agent, Dropping Agent, Metering Agent, Classifying Agent, etc.). Each of these agents is responsible for a specific task within the node. So each agent responds to a limited set of events and performs actions ignoring the treatments handled by other agents lying on the same node or on the neighborhood. This allows the agents of this level to remain simple and fast. More complex treatments are indeed left to the Master Agent.

At level 1, is lying a Master Agent responsible for monitoring, managing, and controlling the entities of level 0 in addition to the different interactions with the other nodes like cooperation, negotiation, messages processing, etc. This agent owns a model of its local environment (its neighbors) that helps him to take its own decisions. The Master Agent chooses the actions to undertake by consulting the current state of

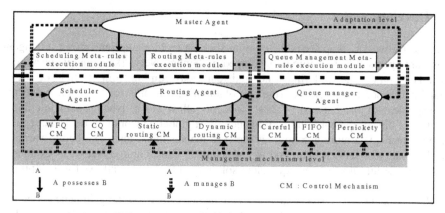

Fig. 3. Two levels of decision within the node

the system (neighbors nodes state, percentage of local loss, percentage of its queue load, etc.) and the meta-rules at its disposal in order to have only the most relevant control mechanisms activated with the appropriate parameters. The node, thanks to the two decision levels, responds to internal events (loss percentage for a class of traffic, load percentage of a queue, etc.) and to external ones (message sent by a neighbor node, reception of a new packet, etc.).

The Master Agent owns a set of meta-rules allowing it to decide on actions to perform relating to the different node tasks like queue management, scheduling, etc. (see Figure 3). These meta-rules permit the selection of the appropriate control mechanisms to activate the best actions to execute. They respond to a set of events and trigger actions affecting the control mechanisms supervised by that Master Agent. Their role is to control a set of mechanisms in order to provide the best functioning of the node and to avoid incoherent decisions within the same node. These meta-rules give the node the means to guarantee that the set of actions executed, at every moment by its agents, are coherent in addition to be the most relevant to the current situation.

The actions of the routers have local consequences in that they modify some aspects of the operations of the router (its control mechanisms) and some parameters of the control mechanisms (queue load, loss percentage, etc.). However, they may influence the decisions of other nodes. In fact, by sending messages bringing new information on the state of the sender node, a Master Agent meta-rule on the receiver node may fire. This can involve a change within the receiver node (the inhibition of an activated control mechanism, or the activation of another one, etc.). This change may have repercussions on other nodes, and so forth until the entire network becomes affected.

This dynamic process aims to adapt the network to new conditions and to take advantage of the agent abilities to alleviate the global system. We argue that these agents will achieve an optimal adaptive control process because of the following two points:

(1) each agent holds different processes (control mechanisms and adaptive selection of these mechanisms) allowing to take the most relevant decision at every moment;

(2) the agents are implicitly cooperative in the sense that they own meta-rules that take into account the state of the neighbors in the process of control mechanisms selection. In fact, when having to decide on control mechanisms to adopt, the node takes into consideration the information received or guessed from other nodes.

5 Development

Ginkgo-Networks company is developing such an architecture integrating intelligent software agents. The technology developed by Ginkgo-Networks is unique because it is linked to a double skill almost non-existent today coming from the field of Artificial Intelligence and networks (the use of intelligent agents for the control and the management of the network). These concepts allow (1) a dynamic and intelligent control of the equipment in a local manner, (2) a global network control in a cooperative manner, (3) a more autonomous network management, and (4) a better warranty of the quality of service in an end to end manner. Thus, Ginkgo-Networks Company provides a solution where no equivalent solution on the market allows for the optimal functioning of the network.

Results of the first testbebs are very convincing. However, the gain depends on the integration of the agents inside the equipment or outside the equipment. We showed in our testbed that if the agents are inside the equipment, the optimal performances are obtained when configuring between every second and hundreds of millisecond. On the contrary, when the agents are implemented outside the equipment (in our testbed outside Cisco routers) the optimum is obtained when reconfiguration take place between one minute and several minutes. The gain in performance with Linux routers and inside agents could be between 10 and 50 %. All these results should appear in a future paper.

6 Conclusion

This paper introduced new implicit communication architecture to better support QoS and new functionalities using the autonomic communication paradigm. A knowledge plane that allows the agents to share a global goal of the overall network introduces this paradigm. Intelligent network equipments are self-configurable using an agent-based control scheme. This architecture and the associated protocols consider not only the policies provided by the business plan but also the constraints of the lower layers of the network. A 4-plane architecture was proposed in the autonomic communication community which help us to provide the selection of control mechanisms to optimize the configuration of the routers and of the protocols. This architecture interacts with the network equipment and protocols in order to configure the network with the selected protocols and parameters. An analysis of our architecture shows that a real time configuration of routers is available and brings an important improvement of the performance. Our proposal has been tested in a simulation environment and gave very good results in terms of delay, and lost packet reduction. Then the agent infrastructure has been implemented in a real environment composed of 9 different routers.

References

1. Castro M., Merghem L., Gaiti D., Mhamed A. – The Basis for an Adaptive IP QoS Management", Special Issue: Internet Technology IV, IEICE Transactions on Communications, vol.E87-B n°3, pp. 564-572, March 2004
2. Proceedings of the first WAC (Workshop on Autonomic Communications), Berlin, October 2004.

3. Verma D. C. – Simplifying Network administration using policy-based management, IEEE Network 16(2), 2002.
4. Ferber J. – Multi-Agent Systems: An Introduction to Distributed Artificial Intelligence, Addison Wesley Longman, 1999.
5. Bazzan A.L.C., Wahle J. and Klügl F. – Agents in Traffic Modelling - From Reactive to Social Behaviour, KI'99, LNAI 1701, pp 303-307, Bonn, Germany, September 1999.
6. Moukas A., Chandrinos K. and Maes P. – Trafficopter A Distributed Collection System for Traffic Information, CIA'98, Paris, France, LNAI 1435 pp 34-43, July 1998.
7. Doran J. – Agent-Based Modelling of EcoSystems for Sustainable Resource Management, 3rd EASSS'01, Prague, Czech Republic, LNAI 2086, pp 383-403, July 2001.
8. Drogoul A., Corbara B. ad Fresneau D. – "MANTA: New experimental results on the emergence of (artificial) ant societies" in Artificial Societies: the computer simulation of social life, Nigel Gilbert & R. Conte (Eds), UCL Press, London, 1995.
9. Bensaid L., Drogoul A., and Bouron T. – Agent-Based Interaction Analysis of Consumer Behavior, AAMAS'2002, Bologna, Italy, July 2002.
10. Minar N., Kramer K.H. and Maes P. – Cooperating Mobile Agents for Dynamic Network Routing. in "Software Agents for Future Communication Systems", Chapter 12, *Springer Verlag*, pp 287-304, 1999.
11. Roychoudhuri R., et al. – Topology discovery in ad hoc Wireless Networks Using Mobile Agents. MATA'2000, Paris, France, LNAI 1931, pp 1-15. September 2000.
12. Sigel E., et al. – Application of Ant Colony Optimization to Adaptive Routing in LEO Telecommunications Satellite Network, Annals of Telecommunications, vol.57, no.5-6, pp 520-539, May-June 2002.
13. White T. et al. – Distributed Fault Location in Networks using Learning Mobile Agents, PRIMA'99, Kyoto, Japan. LNAI 1733, pp 182-196. December 1999.
14. Bodanese E.L. and Cuthbert L.G. – A Multi-Agent Channel Allocation Scheme for Cellular Mobile Networks, ICMAS'2000, USA, IEEE Computer Society press, pp 63-70, July 2000.
15. Wooldridge M. – Intelligent Agents. In « Multiagent Systems : a Modern Approach to Distributed Artificial Intelligence », Weiss G. Press, pp 27-77, 1999.
16. Müller J.P and Pischel M. – Modelling Reactive Behaviour in Vertically Layered Agent Architecture. ECAI'94, Amsterdam, Netherlands, John Wiley & Sons, pp 709-713, 1994.
17. Merghem L., Gaïti D. and Pujolle G. – On Using Agents in End to End Adaptive Monitoring, E2EMon Workshop, in conjunction with MMNS'2003, Belfast, Northern Ireland, LNCS 2839, pp 422-435, September 2003.
18. Gaïti D., and Pujolle G. – Performance management issues in ATM networks: traffic and congestion control, IEEE/ACM Transactions on Networking, 4(2), 1996.
19. Gaïti D. and Merghem L. – Network modeling and simulation: a behavioral approach, Smartnet conference, Kluwer Academic Publishers, pp. 19-36, Finland, April 2002.
20. Merghem L. and Gaïti D. – Behavioural Multi-agent simulation of an Active Telecommunication Network, STAIRS 2002, France. IOS Press, pp 217-226, July 2002.
21. Pujolle G., Chaouchi H., Gaïti D. – Beyond TCP/IP : A Context Aware Architecture, Kluwer Publisher, Net-Con 2004, Palma, Spain, 2004.

Towards Self-optimizing Protocol Stack for Autonomic Communication: Initial Experience

Xiaoyuan Gu[1], Xiaoming Fu[2], Hannes Tschofenig[3], and Lars Wolf[1]

[1] Institute of Operating Systems & Computer Networks,
Technische Universität Braunschweig,
Mühlenpfordtstr. 23, 38106 Braunschweig, Germany
{xiaogu, wolf}@ibr.cs.tu-bs.de
[2] Institute for Informatics, Universität Göttingen,
Lotzestr. 16-18, 37083 Göttingen, Germany
fu@cs.uni-goettingen.de
[3] Siemens AG, Otto-Hahn-Ring 6, 81739 Munich, Germany
Hannes.Tschofenig@siemens.com

Abstract. The Internet is facing ever-increasing complexity in the construction, configuration and management of heterogeneous networks. New communication paradigms are undermining its original design principles. The mobile Internet demands a level of optimum that is hard to achieve with a strictly-layered protocol stack. Questioning if layering is still an adequate foundation for autonomic protocol stack design, we study the state-of-the-art from both the layered camp and its counterpart. We then outline our vision on protocol stack design for autonomic communication with the POEM model and its internals. A novel cross-layer design approach that combines the advantages of layering and the benefits of holistic and systematic cross-layer optimization is at the core of this work. With inspirations from the natural ecosystem, we are working on the role-based Composable Functional System for self-optimization that features proactive monitoring and control. By doing so step-by-step, we envisage reaching the goal of self-tuning autonomic network with high level of autonomy and efficiency, with minimum human management complexity and user intervention.

1 Introduction

What is the Internet? Is it a technology, an industry, a communication medium, or a kind of society? The Internet is all of these and none of these. It is an ecological system - the Internet Ecosystem, and like all the ecosystems it grows, spawns, may be attacked, builds up and declines. Yet, it is extremely complex. Complexity sources from its infrastructure, network management, heterogeneity in devices and access schemes, abundant services and applications. Complexity is amplified by the speed at which the Internet evolves both technologically and in population. With the worldwide wireless buildout, isolations between different communication systems are diminishing. The trend of everything over IP and IP

I. Stavrakakis and M. Smirnov (Eds.): WAC 2005, LNCS 3854, pp. 186–201, 2006.

connects everything is pushing all kinds of networks, wired or wireless, towards integration, composition and interworking.

While the users are benefiting from emerging technologies and convenience, the operators suffer from looming complexity in the construction, configuration and management of such networks. The traditional way of manual planning, configuration, trouble-shooting, policy making and optimization will be exorbitantly expensive or even dominate operational cost, as opposite to hardware/software improvements that continuously help to reduce capital expense. Increasing size of the network infrastructure and shortage of skilled labor for the management of complex systems further convolutes this crisis. In one word, the extent of complexity may eventually exceed the capability of human being, and undermine reliability and end-user trust of the system.

Managing complexity is not the only concern of today's Internet. Rethinking of its design principles represents another urgent agenda. Dated back to the 70's, the early Internet was designed with strict layering and an end-to-end model for its architecture, which was not able to foresee today's pervasive middlebox communications. middleboxes like firewalls, NAT boxes, proxies, explicit/implicit caches basically break the original end-to-end arguments. Other multi-way interactions such as QoS, multicast, overlay routing, and tunneling also contribute to the violations on the layered model. Emergencies of sublayer technologies like TLS at layer 4.5, IPsec at layer 3.5, MPLS at layer 2.5, and wireless networks specific sub-link layers (e.g. RLC, RRC, PDCP) stir up the trouble. To sum it up, the complicated interactions make it difficult to describe using strict layering, and layering often lets some new services fit poorly into the legacy structure.

Apart from the wired network domain, the recent advances in the wireless communications have raised architectural concerns from another perspective. Traffic variability, topology dynamicity, heterogeneity in access technologies, constraints like radio resource, energy in 3G/4G mobile networks, wireless LANs, Mobile Ad Hoc Networks, Micro Sensor Networks, DVB-H Networks, and QoS in real-time interactive mobile multimedia applications, are putting traditional design methodology on protocol stack under examination. A common understanding here is that traditional layering is the source of most performance related problems, and shared information among the protocols layers is critical for performance optimization in wireless networks and the Mobile Internet. With the world-wide push of the wireless communications towards an All-IP infrastructure, the issue of a good architecture is ever more important.

However, giving up layering is extremely difficult, as layering is a natural way of dealing with complex systems. The huge success of the Internet is to a great extent due to its layered architecture. By organizing the communication functions into hierarchical and nested levels of abstractions - the protocols layers, modularity and open interfaces are ensured. This simplified the development of networking protocols and applications, and hence the proliferation of the Internet.

So, the obvious question now is: what would be the right way of structuring the communication software - the protocol stack? We argue that in facing the above-mentioned problems, firstly, there is a need to make future networks self-govern, in the sense that it works in an optimal way with endogenous management and control, and with minimum human perception and intervention. Secondly, a trade-off between architecture and performance has to be in place, and likely a solution for this would be a hybrid architecture that combines the layering for the basic functionalities of the protocol stack, and a non-layered approached for performance-oriented control plane. Such paradigm allows managing complexity, will be better compatible with middlebox communication, and will fulfill the performance requirements in the Mobile Internet.

The rest of the paper is structured as follows. We study the state-of-the-art in the research on network architecture and protocol stack design both from the layered camp and its counterpart in Section 2. This is followed by our vision on the architecture for autonomic communication protocol stack in Section 3, as detailed by the Performance-oriented Reference Model, the AutoComm protocol stack design and prototyping, workflow of self-optimization, as well as the determination of critical control points. Finally we conclude our studies and outline the directions for future research in Section 4.

2   Related Work

2.1   Autonomic Computing

Autonomic computing [1] has in the past few years attracted pretty much attention as a novel computing paradigm. Not only being an area of intensive research in academia, Autonomic Computing has also become a strategic goal of prominent IT companies like IBM, Sun, DaimlerChrysler and Fujitsu-Siemens [2]. Basically, it is a concept of self-managed computing systems with minimum human conscious awareness or involvement, derived from the human autonomic nervous system - a sophiscated computing device and autonomic entity. Still in its early stage, to date, most work on autonomic computing can find its source from neurosciences and biology. In [3], the essence of autonomic computing, architectural considerations, engineering and scientific challenges are thoroughly analyzed. Opportunities and possible research directions of autonomic computing in the system engineering field are well explained in [4]. A bottom-up approach in system design for effective emergency control and handling using so called Observer/Controller architectures is proposed in [5]. Self-organization, self-adaptivity, reconfigurability, and emergence of new properties are topics in a variety of research projects in fields like middleware [6,7], database system [8], and software engineering [9,10]. Cisco, together with IBM is proposing a service framework [11] consisting of a set of potential interface specifications for adaptive remote service and support systems, which enables the customers to interact with the ISPs for autonomic detection, diagnosis, and rectification.

2.2   Autonomic Communication

Despite the heat in the computing area, it is until recently that seeking technical usages of principles observed in natural systems in communication arena has been undertaken. The newly founded Autonomic Communication Forum [12] and its initiative [13] are becoming a call to arms for concerted intellectual efforts towards next generation telecommunication. The University of Bologna is building a framework [14] to support the design, implementation and evaluation of peer-to-peer Internet applications using Swarm Intelligence. A number of projects [15] are going on within the scope of bio-inspired (e.g. from bacteria) approaches for autonomous configuration of distributed systems at the University College London. Based on a chemical reaction model, a new approach [16] with the concept of fraglets for self-healing communication protocol stack has been proposed by University of Basel. All of these efforts hinges on a central theme: autonomic communication.

Autonomic Communication (called AutoComm here after) [17] treats the Internet as an ecosystem - the Internet Ecosystem. By definition, AutoComm represents the study of the inter-relationship between networks or network elements and their situations from a cross-disciplinary perspective, and a methodology of using context-awareness and distributed policy-based control to achieve efficiency, resilience, immunity and evolvability in large-scale heterogeneous communication infrastructure. AutoComm focuses on populations, not individuals, and it seeks balance and optimization on the dynamics of the relationship. A key element in AutoComm is the situation, or called context, which can be understood as a capture for a multi-faceted, uncertain and varying set of communication purposes, policies, conditions, requirements, states, etc. from regulatory, social and private down to technical and engineering. It is of vital importance for AutoComm to understand how network elements behaviors are learned, influenced and modified, how these affect other elements, groups and networks, and how these can offer purposeful inputs on deciding the design principles of the network architecture and protocol stack. The ultimate contribution of AutoComm R&D will be to enable an evolving network platform for sensing, communicating, decision making, and reacting, with high degree of autonomy to ease human efforts and high level of management efficiency in the Operations Support Systems (OSS) in the Telecom industry.

2.3   Cross-Layer Design

Cross-layer Design shares the same motivation of optimal performance of the Mobile Internet as AutoComm. Mobile and wireless networks have a number of characteristics that differentiate them from their wired counterparts, for which one has to think twice before simply borrowing the recipe of the success of Internet and applying its architecture to mobile and wireless networks.

One obvious shortcoming of the two classical models - OSI Reference Model and TCP/IP Model is the lack of information sharing among the protocol layers [18]. This hampers optimal performance of the networks due to the fact that shared layer information is the prerequisite for performance optimization. *Cross-layer design* represents a violation of over-strict layering and too tightly controlled interactions, by encouraging better communications between the protocol layers with holistic and systematic methodology to improve overall system performance.

To date, most existing cross-layer design approaches to a large extent focus on direct interactions between the protocol layers by involving only two or three layers and dragging shortcuts between protocols [19, 20]. Cross-layer design is no easy task, as the cooperation among multiple protocol layers has to be coordinated without endangering conflicts and loops. A common drawback of the current approaches is missing a holistic approach for cross-layer design (not just interactions). Furthermore, once the layering is broken, the luxury of designing protocols in isolation is lost. Also, unbridled cross-layer interactions can create loops, and from control theory's point of view, they become hazards to the stability of the system. Loosely-controlled interactions can also result in "spaghetti code", which basically stifles further innovation and proliferation on the one hand, and increases the cost for upkeep on the other hand. In severe cases, the overall system will have to be redesigned should some key modules change in the future. These problems are detailed in [21] with live examples as proofs.

2.4   Protocol Heap and Role-Based Architecture

If Cross-layer Design is considered as renovation to the architecture of the current Internet, some of the approaches are heading for revolutions - to change Internet's architecture thoroughly by totally giving up layering. Role Based Architecture (RBA) [22] and its Protocol Heap is a good example. Being an ongoing DARPA funded effort toward a new architecture for next generation Internet, it aims to replace layering by roles that correspond to individual communication building blocks. As can be seen from Fig. 1, an arbitrary collection of sub-headers from conventional protocols headers are used to form role data - the Role Specific Headers (RSHs). They are then structured as heap rather than stack to serve as packet headers. RSHs can be added, modified or deleted along the forwarding path.

Obviously, giving up layering can have better functional modulization, flexibility, extensibility, easier in-band signaling, auditability and portability. But these do not come for free. Radical changes of a well established and highly successful architecture will cause compatibility problems. Also efficiency of processing, possible increased complexity and confusion will be questioned. The work is still in conceptual phase, awaiting realization and resolution of many open issues.

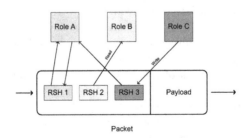

Fig. 1. Role-based Architecture

2.5   Multi-domain Communication Model

Criticizing that protocol stacks are architecturally static and not knowing where the communication is heading for, the Multi-domain Communication Model (MDCM) by Wang et al. [23] proposes to use domains to organize communication building blocks. By concept, domain is a logical construct of the common protocol layers in individual stacks along the communication pathway. Domains are defined by their specific addresses, namespaces and channel properties. Domain specific messages are encapsulated with such definition with correspondence to the protocol header. Hereby, communication can be understood as a process of recursive domain traverse and selections from end to end. Moreover, different from conventional stack approach, MDCM allows dynamic determination of the relationship between the protocol layers using pluggable functions and algorithms. The MDCM builds upon the existing stacks and integrates the next-domain(layer) determination, forwarding and resolution functions into a unified recursive model. Fig. 2 gives an example of using two domains - the IP domain and Ethernet domain to interpret the communication procedure that involves a name resolution with ARP in the LAN.

Although an architecture based on this model allows more relaxed relationship and dynamic binding between the protocol layers, this is more or less a different kind of reasoning of the packet forwarding function of the stack, from a top-down view instead of a bottom-up view which is common in conventional models. What are obviously missing are the new capabilities to enable programmability, self-organization, context-awareness, high degree of autonomy and minimum human intervention, and to deal with prevalent middlebox communication.

Fig. 2. Multi-domain Communication Model

2.6   Region-Based Interworking Architecture

Compared to RBAs approach of functional-oriented granularity, the region-based work tries to divide the Internet using explicit architectural components with the concept of regions. A region in such context is a partition of the network with consistent state, knowledge and control. A collection of interconnected regions represents a connected set of heterogeneous networks. At region boundaries, special gateway or way pointing entities are adopted to facilitate identity mapping, routing information exchange, and message formats representation. Catenet [24] is such a scheme that pioneered the architecture for the Internet with descriptions and criteria in the late 70s. Wroclawski [25] defined the waypoints in his Metanet Model for the description of the transitions from one region to another. This is enhanced by the Regions Project [26] that provides a more generic mechanism for grouping, partitioning, and formalizing boundaries around the groups and partitions. Plutarch [27] is closer to the domain, but it relies on explicit state maintenance along the paths, using a principle similar to the ATMs virtual circuits. Realm-Specific IP (RSIP) [28] and 4+4 [29] went a step further by using different mechanisms to traverse heterogeneous regions.

Regions capture the partitions of homogeneity in the larger-scale heterogeneous communication infrastructure, and focus on issues like interoperability and bridging between heterogeneous networks while leaving the details of the boundary crossing embedded in the waypoints. Unfortunately, most of these work concentrated only on the forwarding function of the communication system, which is only a partial solution to the network architecture as a whole.

2.7   Non-architectural Approaches

Beside architectural approaches mentioned above, many self-optimization schemes have been proposed in recent years. Dated back to 1996, another DARPA project by Tung etc. [30] introduced how to design self-organizing agents that representing finite state automata, to work together collaboratively for maximum optimization in a distributed system. Gausemeier [31] described in a self-optimizing autonomous mechatronic system that consists of intelligent agents, sensors, actuator etc. from four perspectives: target, structure, behavior and parameters. In [32] a proactive online control technique for self-optimization in information system was proposed. The actions that govern system operations are based on optimization of forecasted system behaviors, described using a mathematic model for the specified QoS criteria over a limited look-ahead prediction horizon. Krishnamachari gave a very good overview in [33] on self-optimization in communication with the environment (e.g. sensor networks). Two important views were given. Firstly, the performance of protocol stack must be analyzed with respect to a combination of environment effects, application specifications and protocol parameters. Secondly, protocols must be designed to be self-optimizing, improving autonomously over time by incorporating sensor observations. In [34], a model using so called overall business metric (OBM) was introduced for self-optimizing resources of an IT infrastructure and keeping the infrastructure aligned with business objectives.

3   Our Approach

We consider self-optimization an endogenous process of consistently adjusting the target performance vectors on situational changes, and autonomously adapting the structure, behavior and parameters of a networked ecosystem towards optimal communication efficiency and evolvability. Such a process is a composable/composite function (CF), as can be exemplified by roles like general QoS, resource management, energy efficiency, routing, economic balance etc. Self-optimization should also involve translating business policies into technical counterparts, classifying system policies and map them to the optimization roles, enhancing those policies through learning, context-awareness and conflict resolution, as well as the self-assessment of overall performance using metrics cover both technical and business domains.

One of the enablers in AutoComm will be the innovative approach of the organization of the communication software itself - the cross-layer optimized and situation-aware protocol stack. A self-optimizing AC protocol stack in this context has to face the following challenges:

– Architectural and instrumental considerations with interfacing and compatibility to the current Internet.
– Identification and representation of individual optimization functions and their metrics.
– Dynamic composition and decomposition of self-optimization with functional roles.
– Optimization data processing regulation and execution scheduling.
– Context awareness in self-optimization.
– Distributed and proactive policy-based control in self-optimization.

In answering the challenges, we first give our vision of architectural considerations on interfacing and compatibility to the current Internet with the POEM reference model. We then address the functional considerations of self-optimization with the COP protocol.

3.1 Innovative Approach of the Organization of the Communication Software

We have been working on the Performance-Oriented Reference Model (POEM) (see Fig. 3) that incorporates AutoComm flavors. Conceptually introduced in [35], POEM has no intention to radically change the current Internet architecture by entirely giving up layering. Neither does it follow the protocol heap concept. It is a novel cross-layer design approach that combines the advantages of layering and the benefits of holistic and systematic cross-layer interactions. The basic design criterion is self-optimization is a control plane issue, where the

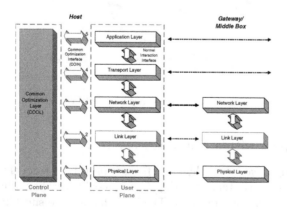

Fig. 3. Performance-oriented Reference Model

normal functions of the protocol stack should not be compromised, and on-top of that to put add-on benefits of controlled cross-layer optimization. As illustrated in Fig. 3, POEM is composed of two conceptual planes: the user plane for normal data flows just like without cross-layer optimization, and the control plane for optimization interaction flows between two protocol layers, between a protocol layer and optimization role specific data, as well as between-roles. The interactions are all done through the defined Common Optimization Interface (COIN). The logical Common OptimizatiOn Layer (COOL) is responsible for offering Self-Optimization Service (SOS), as implemented by its Common Optimization Protocol (COP).

3.2   Cross-Layer Optimized and Situation-Aware Protocol Stack

In an AutoComm system with laws and rules that guiding its efficiency and evolvability, structural and behavioral things are best ways to express the static and dynamic features of self-optimization in the form of protocol. The COP is designated for this task. The main targets of COP are to realize context awareness in community communication, and to perform distributed policy-based control for role-based optimization composite function. Like any protocol, COP has its protocol data unit (PDU). First of all, we propose to organize the ROle-Based INformation (ROBIN) that contains role-based functional entities for stack-wide and node-wide optimization as a heap. Secondly, the conventional protocol stack is structured as a stack, which is left intact due to the reasons mentioned earlier. We then use a frame stack to control the access to the heap and the stack as depicted in Fig. 4. As can be easily understood, the frame stack and the heap are actually corresponding to the header of a COP frame. The necessary stack data of the conventional protocols headers plus the payload of a packet form the payload of a COP frame.

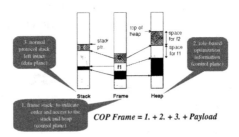

Fig. 4. Data Structure of COP Frame

3.3   Prototyping with a Natural Ecosystem

Interesting enough, we found there exists such prototype from the nature. Consider a simplified ecosystem (see Fig. 5) formed by the lion, the giraffes, the trees, as well as bacteria and fungi. The soil, the air, the sunshine, the water - all the inorganics are the data plane. The plants, animals, microorganisms - all

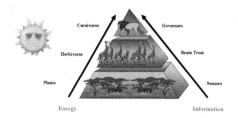

Fig. 5. Inspirations from the Natural Ecosystem

the organisms are our control plane. The trees are sensors that use their roots and leaves to transform sunshine, water, nutrients whatever through photosynthesis into energy to feed the upper-hierarchy animals-the giraffes. The giraffes unfortunately grow up to become the meals of the lion. Noticeable, herbivores do have certain intelligence, and they are able to digest and absorb the food and convert them into flesh to serve the lions (although most likely unwillingly). The flow of energy from the plants to the herbivores and then to carnivores, is just like the way information transverses in the protocol stack, with similar entity mapping as well. On the reverse direction, all old leaves of the trees, dejection (wastes) and dead bodies of the animals are used by the bacteria and fungi - the actuators, who decompose and return some of the elements (the feedbacks) back to the earth to influence its structure. Things work out self-organized and self-optimized. If there are insufficient trees, some of the giraffes will leave or die - the balance is kept.

We have observed at least these from our great nature: Fist of all, it is the rule of "Natural selection" that governs the optimum operations of such ecosystem. Second, the organisms compete to survive, learn to improve, adapt to situate, evolve to prosper, or if they fail to do so, they die or extinct. Third, in doing so, they take the initiative, act proactively rather than reactively, and they often make good use of their environment - the situation, to help to adjust their behaviors. Forth, the intelligence of the organism increases while going up the food chain, the same for the density of the energy contained in the food as more and more processing is involved. Fifth, the consumption of the energy mimics a "pull" mode rather than "push". All of these have motivated us to put more efforts on the inter-disciplinary studies of the natural principles, to extract inspirations and use them to form the foundation for the research in AutoComm protocol stack.

3.4   Matching to Self-optimization in AutoComm

To apply the above paradigm to self-optimization in AutoComm, at layers and sub-layers of normal protocol stack that are relevant to optimization, critical control points (CCP) are set and sensors are correlated for the aggregation of stack-wide context. Sensors are also spread to sense the network elements environment to help to generate and update node-wide and network-wide context. These sub-contexts are then used to form the Common Optimization COntext (COCO). COCO is the basis for carrying out prediction, analysis, learning, conflict resolution, decision and action that are part of policy-based control. This

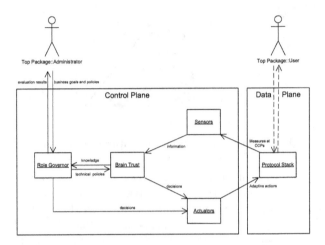

Fig. 6. Interactions Among the Self-optimization Entities

is the task of the Brain Trust as illustrated in Fig. 6. The overall behaviors of a self-optimization function are coordinated by a governor, who is responsible for a number of tasks like translating business policies into technical counterparts, classifying system policies and map them to the optimization role, producing optimization performance metrics (see Fig. 8) that cover both technical and business domains, as well as the self-assessment of overall performance based on the metric(s).

3.5   Self-optimization with Role-Based Composable Function System

We consider a self-optimizing AutoComm System a Composable Functional System (CFS), in which individual optimization functions, the components of such system, can be composed and reconfigured according to needs. This envisions flexibility, extensibility, and evolvability - design for yet unknown. As depicted in Fig. 7, identifying the application domain represents the starting point of the

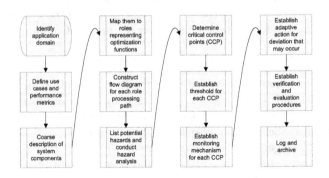

Fig. 7. Sequence Diagram of Role-based Self-optimization

workflow of such system. For example, a domain can be either network management, or network planning, or QoS provisioning, or multimedia service composition - you name it. Use case analysis and the formation of the performance evaluation metrics add to the initial step with greater details. Here the use cases are adopted to capture the intended behavior of the CFS, without having to specify its internal implementation. Requirements are captured, illustrated and implied to help system's end users and domain experts to reach a common understanding. Furthermore, use cases serve to validate the system's architecture and to verify the system as it evolves. To give a few examples: delay optimization, jitter optimization, loss rate optimization, bandwidth consumption optimization, energy consumption optimization, radio resource optimization, processing overhead optimization, storage capacity optimization, financial cost optimization and so on. Each of the use cases can be further divided into sub-use cases, depends on the level of granularity.

Associated with the use cases are the performance metrics for individual functional components, established by mapping the business objectives and policies (e.g. Service Level Agreements), to technical qualitative and quantitative measures (e.g. QoS parameters). This is then followed by the construction of the CFS with coarse description of the system functional elements. Here the static things - the entities and their relationship, and the dynamic things - the activities and interactions among the entities (such as that depicted in Fig. 6 above) are at the core of the work. The use cases are then mapped to the roles that each represents a specific aspect of the composite optimization function. All the related processing routines (can be either for the end-system only, or end-to-end across the network) for a role are described with its own flow diagram afterwards. Potential hazards (factors that will have negative impacts) to the performance are enumerated and analyzed. Critical Control Points (CCP) are determined for

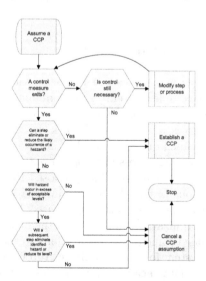

Fig. 8. Flow Chart on Critical Control Point Determination

these factors, such as depicted in Fig. 8. For each effective CCP, threshold(s) is set to provide the reference basis for monitoring and control. The mechanisms of proactive monitoring and control will then be in place. A system like this will feature information gathering and aggregation, context-awareness, learning and knowledge development, distributed policy-based control, consultation and decision making, correction and adaptation etc., depends on the level of autonomy/intelligence and the level of user interaction/intervention desired. Being a self-organized and self-govern system, verification and evaluation (e.g. fitness assessment) have to be conducted to ensure the correct functioning of the system. Naturally, the work along the chain will be noted, if so desired.

3.6   Determination of Critical Control Points

As distributed and policy-based monitoring and control is a most essential part of the model, and this hinges on accurately setting and effectively working critical control points, we explain the logical steps involved in the determination of a CCP in a more detailed way. As can be seen from Fig. 8, at the very beginning, a CCP is assumed. Should there be no control measure or if control is not any more necessary, the assumption of a CCP is dropped. Otherwise, if control is desired even if no measure being present, a control step is re-examined to find an appropriate measure. For each existing control measure, once it is confirmed that a control step can eliminate or at least reduce the possibility of the occurrence of a performance hazard, a CCP is established. Even if the current control step can not eliminate or reduce the likelihood of the occurrence of a performance hazard, and no subsequent control step is able to do so either, should excess of a metric will lead to a performance hazard, a CCP has to be established as well. Only if excess of a control threshold will not be a performance hazard, or there exists a capable subsequent control step down the path, will a CCP assumption be dropped in such context.

4   Conclusions and Future Work

We have reviewed the recent advances of protocol stack design in facing the challenges in managing complexity, emerging communication paradigms, and new performance requirements of the Mobile Internet. We pointed out that one promising direction to go is novel protocol stack design for Autonomic Communication. We then outlined our vision on how innovative organization of the communication software and the cross-layer optimized and context-aware protocol stack can help to realize such goal. The proposed POEM model places the cross-layer control functions beside the normal inter-layer interactions, so that ordinary features of the protocols are not compromised, but with the add-on benefits of well controlled optimization. Rooted from the observations of the nature ecosystem, we have applied some of the inspirations to the design of role-based composable functional system for self-optimizing AutoComm stack.

In addition to the metadata encapsulation and the entity-relationship we have coarsely described, a lot of issues are still open. We plan to perform in-depth investigation on the dynamic composition and decomposition of self-optimization CFS. Optimization data processing regulation and execution scheduling for conflict resolution and loop prevention is a must. Context awareness for effective communicating situation changes is definitely part of the design target, where Directed Diffusion, ACQUIRE and Reinforced Querying algorithms, linear and non-linear optimization methods might help. In distributed and proactive policy-based control, adaptive control theory and the principles distilled form natural ecosystems can be enlightening as well. As we have given only the procedure of CCP determination, other steps in the whole workflow of role-based CFS will be dealt with to complete the design.

The formal system modeling and specifications for POEM and simulation-based investigation of the performance gains are currently ongoing. We expect that the proposed reference model as well as the AC protocol stack design guidelines presented in this paper provide well-defined methodology at a critical time when new network technologies are on the cusps of mass proliferation. By doing so step-by-step, we envisage reaching the goal of self-tuning autonomic network with high level of autonomy and efficiency, with minimum human management complexity and user intervention.

Acknowledgment

This work has been partly supported by the European Union under the E-Next Project FP6-506869. In addition, we would like to thank OPNET Inc. for providing OPNET Modeler for the modeling of POEM.

References

1. R. Murch, *Autonomic Computing.* Upper Saddle River, New Jersey: Prentice Hall, Mar. 2004, iSBN:013144025X.
2. IBM, "Autonomic Computing Initiative," *IBM Press*, 2003, http://www.autonomic-computing.org.
3. J. Kephart and D. Chess, "The Vision of Autonomic Computing," *IEEE Computer*, vol. 36, no. 1, pp. 41–50, January 2003.
4. H. Schmeck, "Autonomic computing - vision and challenge for system design," in *Proceedings of the International Conference on Parallel Computing in Electrical Engineering (PARELEC'04)*, Dresden, Germany, September 2004, pp. 3–3.
5. C. Mller-Schloer, "Autonomic computing: on the feasibility of controlled emergence," in *Proceedings of the 2nd IEEE/ACM/IFIP international conference on hardware/software co-design and system synthesis (CODES+ISSS 2004)*, Stockholm, Sweden, September 2004, pp. 2–5.
6. "The Gryphon Project," http://www.research.ibm.com/gryphon.
7. "The SMART Project," http://www.research.ibm.com/autonomic/research/projects.html.
8. "The LEO Project," http://www.research.ibm.com/autonomic/research/projects.html.
9. "The Recovery Oriented Computing (ROC) Project," http://www.roc.cs.berkley.edu.

10. "The Software Rejuvenation Project," http://www.software-rejuvenation.com.
11. Cisco and IBM, "Adaptive Services Framework (ASF) White Paper," http://www.cisco.com/application/pdf/en/us/guest/partners/partners/c644%/ccmigration_09186a0080202dc7.pdf.
12. "Autonomic Communication Forum," http://www.autonomic-communication-forum.org.
13. "Autonomic Communication Initiative," http://www.autonomic-communication.org.
14. "The Anthill Project," http://www.cs.unibo.it/projects/anthill.
15. I. Marshall, "UCL Bio-inspired Approaches to autonomous configuration of distributed systems," http://www-dse.doc.ic.ac.uk/projects.html.
16. C. Tschudin and L. Yamamoto, "A Metabolic Approach to Protocol Resilience," in *Proceedings of the 1st IFIP International Workshop on Autonomic Communication (WAC 2004)*, Berlin, Germany, October 2004, pp. 2–5.
17. M. Smirnov, "Autonomic Communication: Research Agenda for a New Communication Paradigm," *Fraunhofer FOKUS White Paper*, November 2004.
18. C. M., M. G., T. G., and G. S., "Cross-layering in Mobile Ad Hoc Network Design," *IEEE Computer Magazine*, vol. 37, no. 2, pp. 48–51, February 2004.
19. Z. J. Haas, "Design methodologies for adaptive and multimedia networks," *IEEE Communications Magazine*, vol. 39, no. 11, pp. 106–107, November 2001.
20. S. Shakkottai, T. S. Rappaport, and P. C. Karlsson, "Cross-layer Design for Wireless Networks," *IEEE Communications Magazine*, vol. 41, no. 10, pp. 74–80, October 2003.
21. V. Kawadia and P. R. Kumar, "A cautionary perspective on cross layer design," *IEEE Wireless Communication Magazine*, vol. 12, no. 1, pp. 3–11, February 2005.
22. R. Braden, T. Faber, and M. Handley, "From Protocol Stack to Protocol Heap - Role-Based Architecture," *ACM SIGCOMM Computer Communication Review*, vol. 33, no. 1, pp. 17–22, January 2003.
23. Wang, Y and Touch, J. and Silvester J., "A Unified Model for End Point Resolution and Domain Conversion for Multi-Hop, Multi-Layer Communication," ISI, Tech. Rep. ISI-TR-590, 2004,
http://www.isi.edu/~yushunwa/papers/isi-tr-2004-590.pdf.
24. V. Cerf, "The catenet model for internetworking," *Internet Experiment Notes IEN48*, July 1978.
25. J. Wroclawski, "The Metanet, Research Challenges for the Next Generation Internet," in *Proceedings of Workshop on Research Directions for the Next Generation Internet*, May 1997.
26. K. Sollins, "Designing for Scale and Differentiation," in *Proceedings of the Workshop on Future Directions in Network Architecture(FNDA) at ACM SIGCOMM*, Karlsruhe, Germany, August 2003, pp. 267–276.
27. J. Crowcroft, S. Hand, R. Mortier, T. Roscoe, and A. Warfield, "Plutarch: An Argument for Network Pluralism," in *Proceedings of the Workshop on Future Directions in Network Architecture (FNDA) at ACM SIGCOMM*, Karlsruhe, Germany, August 2003.
28. M. Borella, J. Lo, D. Grabelsky, , and G. Montenegro, "Realm Specific IP: Framework," RFC 3102, IETF, Oct. 2001. [Online]. Available:
http://www.ietf.org/rfc/rfc3102.txt
29. Z. Turanyi, A. Valko, and A. Campbell, "4+4: An architecture for evolving the internet address space back toward transparency," *ACM Computer Communication Review*, vol. 33, no. 5, pp. 43–54, October 2003.

30. B. Tung and L. Kleinrock, "Using finite state automata to produce self-optimization and self-control," *IEEE Transaction on Parallel and Distributed Systems*, vol. 7, no. 4, pp. 439–448, 1996.
31. J. Gausemeier, "From Mechatronics to Self-Optimization," in *Proceedings of International Congress on FEM Technology*, Lake Constance, Germany, October 2002.
32. N. Kandasamy, S. Abdelwahed, and J. P. Hayes, "Self-Optimization in Computer Systems via Online Control: Application to Power Management," in *Proceedings of International Conference on Autonomic Computing (ICAC'04)*, New York, USA, May 2004, pp. 54–61.
33. B. Krishnamachari, "Self Optimization in Wireless Sensor Networks," in *Invited talk at the NSF-RPI Workshop on Pervasive Computing and Networking*, Troy, NY, April 2004.
34. S. Aiber, D. Gilat, A. Landau, N. Razinkov, A. Sela, and S. Wasserkrug, "Autonomic Self-Optimization According to Business Objectives," in *Proceedings of International Conference on Autonomic Computing (ICAC'04)*, New York City, USA, May 2004, pp. 206–213.
35. K. Farkas, O. Wellnitz, M. Dick, X. Gu, M. Busse, W. Effelsberg, Y. Rebahi, D. Sisalem, D. Grigoras, K. Stefanidis, , and D. N. Serpanos, "Real-time Service Provisioning for Mobile and Wireless Networks," Submitted to Elsvier Computer Communication Journal, Oct. 2004.

Towards Service Awareness and Autonomic Features in a SIP-Enabled Network

Giuseppe Valetto, Laurent Walter Goix, and Guillaume Delaire

Telecom Italia Lab, Via Reiss Romoli 274 -10148 – Torino, Italy
{giuseppe.valetto, walter.goix,
guillaume.delaire}@tilab.com

Abstract. Next-generation communication infrastructures can become autonomic only if they can leverage some form of awareness about themselves and the services they deliver. Such awareness can be reached by disseminating across the network a proper amount of multi-faceted knowledge. We have started to identify a set of basic capabilities that provide a degree of service-awareness and enable various autonomic behaviors, oriented towards sustaining communication services within a SIP-enabled network. We have designed network features that support those capabilities, in particular focusing on awareness features relevant to service deployment, monitoring and exposition, and we have built those features natively into the control plane, relying upon the SIP Event Framework specifications. We have also defined a set of scenarios that exploit the service awareness introduced in the network for various autonomic purposes.

1 Introduction

In the area of Information Technology (IT), much attention has been recently given to new ways to respond to the rapid growth in complexity and scale of today's distributed software systems and services. That trend has generated a wealth of new research, often referred to as autonomic computing [2], aiming at providing distributed software ensembles with a set of *adaptive* capabilities geared towards its *self-configuration, self-healing, self-optimization and self-protection* [1]. Also the field of communication technology is experiencing these days a tumultuous growth, similar to that observed in the IT field, and is thus in need of equivalent *self-** provisions; this has recently led to research on *autonomic communication* systems.

All autonomic systems must be capable of automated and dynamic adaptation, and, to that end, one key enabling factor is *awareness*. Both *self-awareness* and *environment-awareness* are necessary, leading to proactive adaptation - based on evaluation of the global state of the system, as well as reactive adaptation – in response to conditions occurring within the execution environment. Next-generation networks can therefore become autonomic in a significant way only if they become aware of (i.e., can tap on and leverage) a wealth of information about themselves and the role they play in delivering communication-intensive services.

I. Stavrakakis and M. Smirnov (Eds.): WAC 2005, LNCS 3854, pp. 202–213, 2006.
© IFIP International Federation for Information Processing 2006

The role of knowledge and awareness is well recognized and accepted in Autonomic Communication; however, it is often summarized with the all-encompassing term of *context awareness* (see for example [21]), which tends to blend together a number of very different aspects. The concept of awareness is instead multi-dimensional by nature: we maintain that – while a common framework for leveraging awareness dimensions is necessary, as advocated for example in [15] – the characteristics of each of those domains should be studied separately, each on the basis of its different contribution.

We are particularly interested in *service awareness*, defined as the ability by the network to understand the services it provides, in terms of their nature, their state and their characteristics. That issue seems quite tightly and directly linked to enabling adaptive mechanisms that relate to the nature of the communication services carried by that network. In fact, communication infrastructures have gradually been moving away from the end-to-end, or "pipe" metaphor: increasingly – as exemplified by active networks [10] – the communication infrastructure is an actor that takes an active interest also in knowing (and, in turn, shaping) how the communication takes place and what service is carried out. However, to fully enable the idea of networks that can adjust themselves to the conditions and demands imposed upon them by services, the level of service awareness in the networks must be raised.

Service awareness is about leveraging knowledge that can be derived – in the terms used in [20] – both from the data and the control plane. We are working on introducing forms of service awareness natively within network protocols. The work we present focuses on endowing an infrastructure for advanced telecommunications services – based on the SIP signaling protocol [3] – with facilities for communicating information related to service monitoring, deployment and advertising (i.e. exposition). In this paper, we describe how we have developed and experimented with those facilities on top of SIP, the kind of information they provide, and how the level of service awareness made available enables a number of service-oriented autonomic scenarios.

2 Background

2.1 SIP

SIP (Session Initiation Protocol) [3] is an application-layer control (signaling) protocol for creating, modifying, and terminating multimedia and multiparty sessions, e.g. Internet telephone calls over an IP network.

SIP entities (typically, users or services) are accessible through a SIP address, or SIP URI, whose format is *sip:username@domain*.

A typical SIP network is made of *proxy servers* or *proxies*, to route requests to the user's current location, which can also authenticate and authorize users for services and implement call-routing policies. Typically, services are provided by SIP *Application Servers* (AS), which run software that implements the application logic, and are accessed by users through the proxies.

2.2 The SIP Event Framework

The SIP Event Framework [4] is a mechanism for asynchronous event notification over SIP; it comprises the following logic entities:

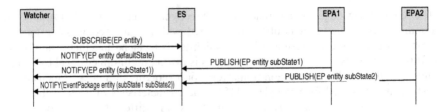

Fig. 1. Generic call flow of the SIP Event Framework

- **Watcher:** an entity interested to receive information: it declares its interest via an explicit subscription created with a SIP SUBSCRIBE message.
- **Event Server** (ES): notifies Watchers with the requested information, through a SIP NOTIFY message. How the ES produces or retrieves the information to be notified is out of the scope of the framework itself.
- **Event Publication Agent** (EPA) - or Event User Agent (EUA): provides the ES with information through the SIP PUBLISH [5] method. Multiple EPAs can send PUBLISH requests to the ES on behalf of the same "entity" and with respect to the same subscription. In that situation, the ES must aggregate the information prior to notify Watchers.

The above-described mechanism is general enough to allow a variety of notification services, named "Event Packages" (EPs). A generic event notification scenario using the SIP Event Framework is described in Fig. 1.

The ES is the central point of the SIP Event Framework, which gathers and propagates information from and to SIP entities, carrying out the manipulation of data between the information published by the EP and the information notified to watchers. Such manipulation includes for example data filtering, aggregation or transformation. It is also in charge of protecting sensitive information, by managing subscription access based on the identity of the watcher and its related privileges.

Each EP defines its own XML format to describe specific events. It also defines any composition and transformation rules that must be applied by the ES to aggregate information PUBLISHed from several EPAs, before proceeding to NOTIFY them in a single event carrying an aggregated XML document. That aggregation feature is a peculiarity of the SIP Event Framework with respect to most event-based communication buses. Before sending this aggregated document to the Watchers, the ES can further apply filters defined by local policy or by Watchers when subscribing.

3   Introducing Service Awareness in a SIP Network

We have developed some service awareness capabilities in a SIP-enabled network, as a set of Event Packages, for the exchange of non-functional information related to the state of services as well as network elements. Our interest in SIP comes from the possibility to enable an **existing control layer** that is being currently deployed in advanced service-based networks, to evolve gracefully to include autonomic behaviors, without the introduction of ad-hoc additional protocols. That provides a good framework for practical experimentation and evaluation of service awareness features in next-generation networks.

Our work also allows encompassing SIP-based **terminals** in a service-aware communication environment, which is advantageous in a telecom context, given the large number and wide variety of terminals operating on the service network.

By choosing SIP, we have also been able to leverage the characteristics of its well-specified Event Framework, which seems particularly indicated for the native integration of situation (and – specifically- service) awareness within the control plane of Next Generation Networks. The relative simplicity of defining Event Packages for the SIP Event Framework supports the process of **extending the awareness information** with additional dimensions. It thus permits to promptly provide more sophisticated levels of service awareness, which could be used for autonomic concerns[1].

For starters, we have focused on Event Packages dealing with *service monitoring, deployment and advertising*, whose semantics enable to communicate about and to control what happens during the post-development lifecycle of a service, starting from the moment in which it is ready to be rolled out on the network. Hereby we motivate our choice, highlighting the importance of monitoring, deployment and advertising information for autonomic scenarios.

The **monitoring Event Package** provides a generic capability to collect information about and inspect the state of entities within the network, as they evolve over time. That is a basic feature for any autonomic system, since it enables the necessary levels of self-awareness needed for introspection and diagnosis. The monitoring package, as we designed it, supports the collection of different kinds of data, and can be tailored towards different network elements and services. The subscription mechanism, furthermore, allows filtering of events in various ways, and the construction of multiple monitoring views that respond to different reporting needs and can be directed to diverse recipients.

While monitoring constitutes on its own only a pre-requisite for the development of autonomic behavior, the **deployment Event Package** addresses directly some concerns related to the four major self-* autonomic areas. Deployment events provide notification means and commands for the installation, configuration, activation, deactivation and retirement of service and service components on SIP Application Servers, as well as the arming and disarming of SIP proxy triggers used to route requests to those services. As such, deployment events provide a set of primitives that can be leveraged, in the first place, for the automated and controlled (re-) configuration of services on top of the nodes of a SIP network. In particular, a combination of deployment actions enable the configuration of arbitrarily complex composed services, by moving, adding or substituting components on a topology of available network elements. Deployment primitives remain coarse-grained and as such do not cover the whole (re-)configuration spectrum. However, especially when used in conjunction with the monitoring Event Package, they can also be effective

---

[1] Notice how SIP already includes protocol-level means to provide a level of person awareness to the network, called SIP Presence [6]. Presence information in a SIP network is aimed at representing the state of users and their terminals and devices; the basic presence information can also be extended to capture more details, as required by applications.

towards other autonomic concerns: for example, self-healing and self-protection- as exemplified by the Willow survivability architecture [17], which leverages the Software Dock deployment engine [18]; or self-optimization, for example to increase the instances of critical services or components present on the network, in the face of bottlenecks and request surges, or to modify the routing scheme, in order to streamline communications.

Finally, the **advertising Event Package** deals with service exposition, i.e., provides interested watchers, such as (but not limited to) user agents and terminals, with notifications on available services and their characteristics. The advertising Event Package is used to communicate information such as the access modality and channel to the service, and the entry point to the service network (proxy). Advertising events, therefore, are instrumental to push service awareness towards the edge of the network. Service exposition, as we implemented it through the advertising Event package, can be tightly coupled with deployment; that helps in making new services immediately and automatically available / unavailable to all user agents (or to selected subsets) as they are successfully activated / deactivated, as a logical extension of configuration or healing scenarios. Such feature may also affect optimization, as seen from the network edge, for example for dynamically planning and dimensioning the ratio between user agents and proxies. Finally, advertising can also be used within the network itself to adapt on the fly the view of available services also to other services, which has clear beneficial implications for autonomic communication scenarios that intend to deal with spontaneous service aggregation and composition.

The rest of this Section describes in detail the Event Packages introduced above.

3.1 Monitoring

The goal of the monitoring Event Package is to provide information about the state of a network element or a service over time to interested monitoring parties, or *watchers*. Network elements typically publish information about their own state, and separate information on behalf of each service (or trigger, in case of a proxy) they host.

In the simplest case, the monitoring information consists in on/off availability data. To detect the unavailability of network or service entities, keep-alive mechanisms are used, requiring monitored elements to frequently refresh their availability information. That makes available a "heartbeat" that enables to notify watchers in near real-time whenever a refresh event is not received in time.

Enrichment of the basic heartbeat data is made possible by the extensibility of the XML schema for monitoring events. Our architecture envisions that a single network element may incorporate many monitoring components, each of them publishing different information, using different XML namespaces. The notification of arbitrarily rich monitoring data produced within a service component and issued by the hosting node is therefore easily enabled, and entirely application-dependent.

On the other side, watchers can subscribe to raw data or – taking advantage of the filtering and data composition mechanisms provided by the Event Framework – to some aggregate data, based on several criteria, such as network element identifier, or type (e.g. AS or proxy), service identifier, etc. Some examples of subscriptions with different purposes are given below:

- *Auto-discovery*: all known information about all network elements
- *Network Element monitoring*: all known information about a single network element, such as an AS, including the state of the node and all its hosted services
- *Service monitoring*: all known information about a service, across all network elements that host it
- *Load distribution*: load information compiled for all elements of a certain type (AS, proxy, service)

3.2 Deployment

The deployment Event Package is substantially more complex that the monitoring one, since its goal provide a complete view of what happens on a SIP network during the service deployment, retirement and upgrade process.

Deployment events circulate in the network on the basis of an XML deployment document. That document is intended to prescribe a workflow of deployment steps (or *actions*). When the document is created, a corresponding event is routed to network elements involved in some step of the workflow, i.e., the *deployment targets*. Each target checks whether the action(s) prescribed for that target can be performed immediately, or depends on actions that must yet take place upon other targets.

In the former case, the target sends feedback through a PUBLISH request, indicating that it has finished processing the action, to indicate either success or failure. In the latter case, the network element will perform its action only when it receives the notifications indicating that those other actions are completed.

This way, once the process is under way, notification events that are issued and propagated from network elements involved in the workflow collectively enable to maintain a coherent state trace of the deployment procedure across all targets.

XML extensibility is also leveraged in the Deployment EP, for instance to provide network elements with data relevant to the deployment of specific services, such as their initial configuration, etc.

At startup, we envision that a network element subscribes to all deployment processes. If it is an intended target in some already running deployment processes, the Deployment Server returns all relevant events aggregated in the body of a NOTIFY. This mechanism enables network elements to remain up to date about deployment processes that may have been triggered on the network at any time in which they were not connected, which is helpful for automated recovery and leads to self-stabilization.

Fig. 2. exemplifies a typical automated deployment scenario, in which a new service must installed and activated. The whole deployment process is activated through a single PUBLISH request from some deployer entity, which may initiate the process autonomously, or due to an external stimulus, such as an event containing monitoring information. This scenario envisions four automated steps: (1) a service is deployed on a SIP AS, (2) a corresponding trigger is installed on a SIP proxy at the same time, (3) once the trigger is installed, the service may be invoked, so the service is activated on the AS, (4) once the service is active, the proxy can activate its trigger to route incoming requests to the AS. In this scenario, the deployer entity in the network subscribes to the deployment process itself, in order to watch over its progress.

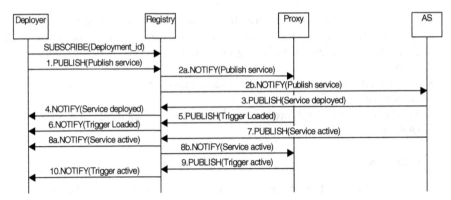

Fig. 2. A deployment scenario - installing and activating a service

3.3 Advertising

The advertising Event Package communicates rich information about available services: we have defined it on the basis of the Presence Event Package [6]. Our package defines an extension to the Presence XML schema [7], which enables to notify a watcher (e.g. a user device) with service availability, how and where to access the service, and optionally any additional information useful for service invocation. It has been designed also to allow subscribing to list of services, according to categorization mechanisms that can be user-, application- and domain-specific. In some cases, the number of services within each list may vary over time.

Typically, when starting up, user terminals subscribe to one or more list of services using the standard presence subscription mechanism defined in [6], and get notified about services available on the SIP network, together with the corresponding additional advertising information. Whenever a new service is available, or becomes unavailable, a PUBLISH request containing new service information is issued to the advertising ES, which triggers notification of this change to all interested watchers. For scalability reasons, a hierarchical architecture has been designed that geographically distributes terminal subscriptions over outbound advertising ES, which in turn subscribe to service lists on a central master server.

Notice that, advantageously in an autonomic scenario, advertising can be linked to monitoring or deployment information, to dynamically reconfigure user terminals in response to new services as well as changes in availability of existing services.

4 Building Autonomic Scenarios Using Service Awareness

In this Section, we outline how some typical autonomic scenarios that are often used to exemplify common self-* concerns are enabled in a SIP network by the service awareness features we have introduced, in particular using the above-defined Event Packages. Our purpose is to remark how even relatively straightforward forms of service awareness provide a sufficient level of support to a variety of basic autonomic applications.

4.1 Self-configuration

The first, obvious configuration scenario regards the *adaptive deployment* of a service on a certain portion of the network, that is, on a certain number of AS and proxy elements, each of which must be individually equipped with the appropriate triggers and software features. The foundation of this scenario is of course the automated deployment facilities provided by our deployment Event Package.

Automated deployment immediately becomes adaptive when some controlling entity, e.g., the original deployer entity, as seen in Fig. 2, can modify on the fly our deployment document, and issue it as a subsequent event that updates the deployment workflow. That controlling entity can act for various reasons: on the basis of a particular state reached in the workflow, or in response to external stimuli (e.g. incoming monitoring information), or to modify proactively the deployment process, based on some policy decision. Notice how the SIP Event Framework naturally supports adaptive deployment, through its keep-alive, aggregation and trasformation features.

Another configuration scenario regards the self-adaptive *static load balancing* of a service[2], which relies principally on monitoring events. In this scenario, a high-throughput proxy acts as a load balancer to a cluster of AS nodes, all providing the same service. The proxy can subscribe to monitoring events published by the AS in the cluster and the running service instances hosted by that cluster. That way, it is made continuously aware of the runtime properties of services (number of requests being processed, number of error generated, etc.), as well as the state of the elements in the cluster (CPU load, network load, length of request queue, etc.). Further monitoring information, expressed in whatever measurement unit is appropriate for the service at hand, can also be considered. The proxy has installed specialized logic and algorithms, which, by subscribing to and processing that monitoring information, decides on the fly how to distribute incoming requests. Additionally, another monitoring entity can be added to this scheme, in charge to decide when to equip the proxy with a new logic, in order to modify on the fly the distribution profile it uses, and optimize it with respect to higher-degree circumstances.

4.2 Self-healing

The combination of monitoring and deployment events can address various *fault recovery* scenarios. Monitoring events can be issued by AS elements to signal different kinds of application-level faults, such as failures of the deployed service software. Also network-level faults can be similarly signaled by the affected AS and proxy elements. A deployer entity with specialized logic on board that subscribes to those events can react by issuing appropriate re-deployment events as a counter-measure. For application-level faults, such as software crashes, it can decide to re-activate the crashed component on the same AS. For network-level faults, it can re-configure the service by deploying onto different network elements new instances of the unreachable service, and/or change the routing of proxies, to overcome network partitioning.

---

[2] Static load balancing, as opposed to dynamic load balancing (described in Sec 4.3.), involves the over-provisioning of the service, with enough active instances activated on a number of AS nodes, to ensure they can collectively handle any reasonably foreseeable request load.

The same approach can also be used for *fault avoidance*, assuming sufficiently rich monitoring information, coupled with sophisticated analysis and proactive logic on board of the deployment service acting as a subscriber. In that case, the accumulated information about the service state can reveal conditions that might lead to a fault, and the proactive interventions can consist of an appropriate set of deployment events. For instance, in case the memory occupation by some entity can be monitored, whenever a certain threshold is passed a fault avoidance strategy might be to de-activate and then shut down the offending entity before it crashes hard - possibly taking down a number of important user sessions - and start a fresh copy anew. Notice that this scenario can be made increasingly sophisticated by embracing the idea of controlled, hierarchical reboots, like in Recursive Restartability approaches [19]: if we envision a service as a collection of features running on various AS instances, it may be possible to shutdown and restart a single feature, or a single AS, or subsets of features and AS, in order to get rid of the problem at the finest possible granularity level, without perturbing service components that remain in good shape.

4.3 Self-optimization

An optimization scenario is that of *dynamic load balancing*. On the basis of load monitoring events originating from the AS elements and their resident services, a subscribing deployer entity can react to a request peak for a certain service by *scaling up*, that is, deploying on the fly one or more additional instances of that service onto AS elements that happen to be able to handle additional computational and communication loads, and re-configuring the service trigger of the outbound proxy for that service accordingly. Conversely, when the peak is over, the deployer entity can decide to *scale down*, deactivating or retiring some of the newly deployed service instances, thus releasing some resources. Dynamic load balancing reduces the need for over-provisioning; thus, it enables optimized usage of the resources of the network, without compromising availability and responsiveness.

Another optimization scenario, at the edge of the network, involves advertising in combination with monitoring. It aims at resolving cases in which a proxy is overloaded by the traffic generated by its assigned user agents: to optimize its ability to efficiently route requests, that situation can be monitored, and advertising events can be issued to re-assign some of the user agents involved to less busy proxies.

4.4 Self-protection

Many protection scenarios, which aim at easing the effects of or altogether thwarting attacks to the service network, can be approached in ways that are analogous to those used for self-healing. From the point of view of autonomic communication, the concepts of fault recovery and fault avoidance apply also to self-protection, with the "semantic" difference that the fault in this case is not fortuitous, but maliciously inoculated into the network. Another difference is that faults caused by attacks tend not to remain local, but to actively spread across the network.

Deployment features can be helpful to counter that to a degree, in order to willingly cause the partitioning of the service network, and "quarantine" the portion under attack, thus limiting the effect of that attack. Changing the deployed triggers on

SIP proxies is a very effective way to cause transient network partitioning as needed. To achieve self-protection via partitioning, however, it is necessary to put in place rather sophisticated and efficient (hence resource-intensive) analysis facilities of the monitoring information that circulates in the network, to quickly and proactively recognize anomalous patterns and behaviors that reveal an attack before it spreads too widely. How to deal with that challenge is a large part of the research on intrusion detection, which is itself based on awareness about the behavior of entities residing on the network, and the recognition of "suspicious" or non-motivated variations of that behavior. We are currently investigating how to inject that kind of awareness specifically in a SIP network, in particular as an evolution of our monitoring facilities.

5 Related Work

Several awareness domains are actively being researched. For example, *person-awareness* in networks (e.g. with PANs), as well as services, increasingly tries to tie the individual world of users to their usage of communication, by taking into account intrinsic and explicit personal demands (see for example I-centric communications [13]). Another example is *space-awareness* which is obviously very important in contexts like mobile communications and sensor networks, but more subtly also for situating an ensemble of communication entities within a generic multi-dimensional metric space, which can be used to define and measure inter-relationships among those entities (as proposed in spatial computing [14]). Many other dimensions are sometimes collectively encompassed by the generic term of *context awareness*, and variously exploited, for example in ad-hoc networking (as in [16]). In general, however the systematic investigation of each single awareness dimension, and its relevance for enabling autonomic communication, remains an open research endeavor. In the extreme, that investigation can lead to the introduction of a full-fledged knowledge plane [15] in between the data transport plane and the application plane.

In the context of SIP networks, the Event Framework is receiving significant interest especially as a means for adapting services on the basis of users' context (user awareness). Works like [8] and [9] follow that approach, in particular for the adaptive configuration of User Agents, which we try to support in a general way with our service advertising Event Package. We are not aware, however, of other works – besides ours – which use the SIP Event Framework to embed deployment and monitoring primitives in the network, in order to support adaptation of services.

Similar capabilities are instead found in the field of active networks. NESTOR [11], for example, with the complement of the JSpoon language [12], makes service components aware of and accessible by a self-configuration management layer, and can achieve reactive autonomic behavior similar to that enabled by our monitoring and deployment Event Packages.

6 Conclusions and Future Work

Our work aims at experimenting with service awareness capabilities, as enablers for a variety of autonomic behaviors within a SIP-enabled network. The originality of our contribution is twofold: first of all, since we have chosen to implement those capabilities as Event Packages on top of the SIP Event Framework, they have become native to the SIP control network; second, they are modular and can be easily extended to cover other awareness aspects. Together, those two characteristics provide the foundations of an awareness framework native to SIP, which we are incrementally designing and building within the Telecom Italia SIP-based service control platform. We are currently investigating the usage and extension of that service awareness framework, in particular along the following directions:

- Application of monitoring / deployment at the network edge, i.e., on user agents.
- Extension of deployment events to provide a highly flexible sensor configuration. That would allow new sensors to be deployed and configured on network elements as needed, to dynamically enhance reporting through the monitoring feature. For example, preliminary monitoring events could be used to trigger a deployment of new specialized sensors for a given situation. We are examining dynamic Aspect-Oriented Programming (AOP) techniques to that end.
- Application of policy-based mechanisms, as a natural complement to the service awareness events that circulate in our Event Bus. Policies can be used for a high-level, easily re-configurable implementation of the adaptation logic that is present in the autonomic scenarios proposed in Sec. 4, as well as future scenarios. Policies could be dynamically installed via deployment on watchers for specific EPs.
- Extension of advertising to support spontaneous aggregation behavior. Advertising can disseminate information that is needed by service components to find, negotiate with and interact with other components, for dynamic self-composition.

References

1. A.G. Ganek, T.A. Corbi, *The Dawning of the Autonomic Computing Era,* IBM Systems Journal, 42(1): 5-18, January-March 2003.
2. P. Horn, *Autonomic Computing: IBM's Perspective on The State of Information Technology*, at Agenda 2001 conference, Scottsdale, Az., USA, October 15, 2001, http://www.research.ibm.com/autonomic/manifesto/agenda2001_p1.html
3. Rosenberg et al., SIP: Session Initiation Protocol, RFC3261, June 2002 http://www.ietf.org/rfc/rfc3261.txt
4. B. Roach, Session Initiation Protocol (SIP)-Specific Event Notification, RFC3265, June 2002 http://www.ietf.org/rfc/rfc3265.txt
5. Niemi, Session Initiation Protocol (SIP) Extension for Event State Publication, RFC3903, October 2004 http://www.ietf.org/rfc/rfc3903.txt
6. J. Rosenberg, A Presence Event Package for the Session Initiation Protocol (SIP), RFC3856, August 2004, http://www.ietf.org/rfc/rfc3856.txt
7. Sugano, et al., Presence Information Data Format (PIDF), RFC3863, August 2004, http://www.ietf.org/rfc/rfc3863.txt
8. L. Wei, A Service Oriented SIP Infrastructure for Adaptive and Context-Aware Wireless Services, in Proceedings of the 2nd Intl. Conference on Mobile and Ubiquitous Multimedia, Norrköping, Sweden, December 10-12, 2003.

9. D. Pavel, and D. Trossen, Context Provisioning and SIP Events, in Proceedings of the MobySys 2004 Workshop on Context Awareness, Boaston. Ma., USA, June 6, 2004.
10. D.L Tennenhouse, J.M. Smith, W.D. Sincoskie, D.J. Wetherall, G.J. Minden, A survey of active network research, IEEE Communications Magazine, 35(1): 80-86, 1997.
11. A.V. Konstantinou, Y. Yemini, and D. Florissi, Towards Self-Configuring Networks, in Proceedings of the DARPA Active Networks Conference and Exposition, San Francisco, Ca., USA, May 2002.
12. A.V. Konstantinou, and Y. Yemini, Programming Systems for Autonomy, in Proceedings of the IEEE Autonomic Computing Workshop, Active Middleware Services (AMS 2003), Seattle, Wa., USA, June 2003.
13. S. Arbanowski, P. Ballon, K. David, O. Droegerhorn, H. Eertink, W. Kellerer, H. van Kranenburg, K. Raatikainen, and R. Poplescu-Zeletin, I-centric Communications: Personalization, Ambient Awareness, and Adaptability for Future Mobile Services, IEEE Communication Magazine, 42(9), September 2004.
14. F. Zambonelli, and M. Mamei: Spatial Computing: an Emerging Paradigm for Autonomic Communication and Computing, in Proceedings of the 1st IFIP WG6.6 International Workshop on Autonomic Communication (WAC 2004), Berlin, Germany, October 2004.
15. D.D. Clark, C. Partridge, J.C. Ramming, J.T. Wroclawski: A Knowledge Plane for the Internet, in Proceedings of ACM SIGCOMM 2003, Karlshrue, Germany, August 2003.
16. R. Gold, C. Mascolo: Use of Context-Awareness in Mobile Peer-to-Peer Networks, in Proceedings of the 8th IEEE Workshop on Future Trends of Distributed Computing Systems (FTDCS'01), Bologna, Italy, October 31-November 2, 2001.
17. J.Knight, D. Heimbigner, A. Wolf, A. Carzaniga, J. Hill, and P. Devanbum: The Willow Survivability Architecture, in Proceedings of the 4th Information Survivability Workshop (ISW-2001), Vancouver, B.C.,18-20 March 2002.
18. R.S. Hall, D. Heimbigner, and A.L. Wolf, A Cooperative Approach to Support Software Deployment Using the Software Dock, in 21st International Conference on Software Engineering, May 1999.
19. G. Candea, and A. Fox: Recursive Restartability: Turning the Reboot Sledgehammer into a Scalpel, in Proceedings of the 8th Workshop on Hot Topics in Operating Systems, Schloss Elmau, Germany, May 2001.
20. R. Sterrit, M. Mulvenna, and A. Lawrynowicz: A Role for Contextualized Knowledge in Autonomic Commnication, in Proceedings of the 1st IFIP WG6.6 International Workshop on Autonomic Communication (WAC 2004), Berlin, Germany, October 2004.
21. A.K. Dey: Understanding and Using Context, Personal and Ubiquitous Computing Journal, 5(1):4-7, 2001.

Integration of Decentralized Economic Models for Resource Self-management in Application Layer Networks*

Pablo Chacin, Felix Freitag, Leandro Navarro, Isaac Chao, and Oscar Ardaiz

Computer Architecture Department, Polytechnic University of Catalonia, Spain
{pchacin, felix, leandro, ichao, oardaiz}@ac.upc.es

Abstract. Resource allocation is one of the challenges for self-management of large scale distributed applications running in a dynamic and heterogeneous environment. Considering Application Layer Networks (ALN) as a general term for such applications including computational Grids, Content Distribution Networks and P2P applications, the characteristics of the ALNs and the environment preclude an efficient resource allocation by a central instance. The approach we propose integrates ideas from decentralized economic models into the architecture of a resource allocation middleware, which allows the scalability towards the participant number and the robustness in very dynamic environments. At the same time, the pursuit of the participants for their individual goals should benefit the global optimization of the application. In this work, we describe the components of this middleware architecture and introduce an ongoing prototype.

Keywords: Resource Allocation, Autonomic Systems, Decentralized Economic Models, Middleware Architecture.

1 Introduction

"Autonomic Communication is a paradigm in which the applications and the services are not ported onto a pre-existing network, but where the network itself grows out of the applications and the services that end users wants" [ACCA04].

Under this vision, large scale Application Layers Networks (ALNs), including computational Grid, Peer-to-Peer and Content Distribution Networks, are evolving towards the notion of "Selfware", which achieves local autonomic control and global self-organization applying management policies in a decentralized way. One of these key polices is the assignment of resources to ALN's services.

Within such dynamic and heterogeneous environments, centralized allocation instances are limited in performing an efficient resource allocation task. To operate in such environments, the decision making processes within the application needs to be transferred to decentralized components with autonomic behavior.

We propose a resource allocation middleware architecture which facilitates the application of resource management in a decentralized, autonomous and infrastructure

---

* This work was supported in part by Spanish Government under Contract TIC2002-04258-C03-01, and the European Union under Contract CATNETS EU IST-FP6-003769.

I. Stavrakakis and M. Smirnov (Eds.): WAC 2005, LNCS 3854, pp. 214–225, 2006.

independent way. It offers a generic decentralized negotiation framework, on which specialized negotiation strategies and policies can be dynamically plugged to adapt to specific application domains and market designs.

This middleware's architecture is based on the ideas of the decentralized economic model known in the economic community as "Catallaxy", on which a state of coordinated actions, the "spontaneous order", comes into existence through the bartering and communicating of economic agents with posses only partial knowledge of the market participants and price's evolution.

The rest of this paper is organized as follows. Section 2 presents requirements for resource allocation in ALNs, exploring the characteristics of this kind of distributed applications, the issues related to resource self-management and the applicability of decentralized economic models to address those requirements. Section 3 describes the proposed middleware architecture, presenting its design principles and how the components interact to address resource allocation requests. Section 4 presents the related work. Finally, section 5 present our conclusions and proposes some future work.

2   Resource Self-management in ALNs

Application-layer networks (ALN) such as Grid, Peer-to-Peer (P2P) and Content Distribution Networks (CDN) are envisioned as large-scale distributed applications that allow the provisioning of services using the needed resources from a large, heterogeneous and dynamic resource pool. However, allocating and scheduling the usage of computing resources in ALNs is still an open and challenging problem.

In this section we introduce the characteristics of the targeted ALNs, the specific requirements for resource allocation and the principles of decentralized economic mechanisms that allow an efficient resource allocation in this kind of environments.

2.1   Characteristics of Large-Scale Application Layer Networks

Applications that are targeted have the following common characteristics:

- Dynamic: changing environments and the need for adaptation.
- Large: having such number of elements that locality is required in order to scale
- Partial knowledge: it is not possible to know everything in time. This can be caused by scale issues such as a large number of elements, number of messages, or communication latency.
- Evolutionary: open to changes which cannot be taken into account in the initial set-up.
- Diverse: requests may have different priorities and responses should be accordingly assigned.
- Complex: many parameters must be taken into account to take decisions. Learning mechanisms are necessary to self-adjust or adapt to changes, and optimal solutions are not easily computable.

In order to identify the application classes, we map the parameter space into two dimensions. We consider *Configuration Complexity,* which includes the dynamics of

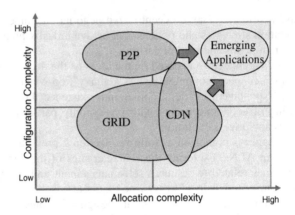

Fig. 1. Target application space

the configuration, lack of global knowledge and evolutionary environment and *Allocation Complexity*, which includes the diversity of requirements and complexity of allocation demands. Figure 1 shows a two-dimensional map with an approximate location of three important application classes.

It can be seen in Figure 1 that that our target application space is situated in the upper right area of the diagram. In our view, none of the three application classes do fully exploit this space, but we expect that distributed applications still to come are aimed to work in this environment. This fact emphasizes the need for a description of a software architecture which integrates decentralized components.

Within such environment, applications with a centralized allocation instances are limited in performing an efficient resource allocation task. To operate in such environments, the decision making processes within the application needs to be transferred to decentralized components with autonomic behavior.

2.2 Resource Allocation and Self-management in ALNs

We expect ALN to be built from basic services that can be dynamically combined to form value-added complex services. These basic services require a set of resources, which need to be co-allocated to provide the necessary computing power.

Therefore, the introduction of new services into this kind of networks, due to the dynamic nature of the environment, precludes any manual or static configuration and demands a self-organization approach, where services should be able to self-configuration, self-optimization and self-healing [WHW+04].

One goal of self-managed network services is to move away from individual system configuration management to policy management. This approach brings a higher level of abstraction to management by introducing a policy from which the configuration is derived, allowing components of the infrastructure to apply these derived configurations to the individual systems across the environment.

In this context, self-managing service's resources involves defining SLA policies for services and resources, mapping required SLA to resources needs, discovering resources that guarantee an adequate QoS, allocating resources ensuring that allocation policies are meet and providing a management interface to monitor an control service life-cycle. Because of the dynamicity of the environment we envision, the service allocation framework must address some specific issues:

- Situateness: services must be aware of its location and the closeness of peer services to collaborate
- Dynamic (re)configuration: usage patters from service users are unpredictable, therefore neither the location nor the number of service instances could be known in advance. New instances must be created and located as needed
- Topology neutrality: services deployed in the ALN could have very different interaction topologies. Some will be structured in a rather hierarchical overlay, like content distribution, while other interact in a closely connected P2P overlay.
- Autonomy: service and the resources it uses will span multiple administrative domains so each of them should be allowed to take decisions autonomously.

We propose a resource allocation middleware architecture based on decentralized economic models, which facilitates the application of resource management polices according to the above requirements (i.e. in a decentralized, autonomous and infrastructure independent way).

This resource allocation middleware has been envisioned as a set of economic agents (representing the Client Applications, Services and Resources of the ALN) that interact between them and with the software components of the underlying ALN, to coordinate, in a decentralized way and using economic criteria, the assignment of resources, as can be seen in the Figure 2.

Direct agent to agent bargaining allows participants to use thee negotiation strategy more suitable to its objectives and current circumstances. Local bilateral bargaining also facilitates the scalability of the system and the quick adaptation to local fluctuations in resource allocation dynamics.

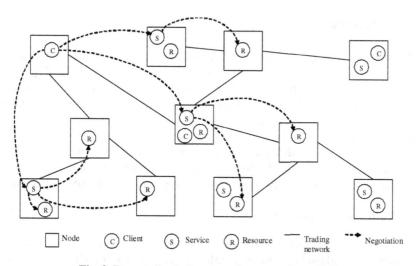

Fig. 2. Decentralized allocation of resources in an ALN

2.3 Decentralized Economic Models for Resource Allocation

The decentralized economic models applied in our work are based on the ideas of the 'free market' economy, the 'Catallaxy' proposed by Friedrich A. von Hayek, as a self-organization approach for information systems [EyPa00]. It is opposed to "plan economy" where a central entity has global knowledge of the system and commands every entity decisions. In Catallaxy, in fact, a central presumption is "constitutional ignorance", assuming that it is impossible have global knowledge.

The Catallaxy concept bases on the explicit assumption of self-interested participants who try to maximize their own utility and choose their actions under incomplete information and bounded rationality. Agents subjectively weigh and choose preferred alternatives, and communicate using commonly accessible markets, where they barter about access to resources held by other participants. The market here is nothing more than a communication bus – it is not a central entity of its own and does not participate in matching participants' requirements using some optimization mechanisms.

The goal of Catallaxy is to arrive at a state of coordinated actions, the "spontaneous order", which comes into existence through the bartering and communicating of the community members with each other and thus, achieving a community goal that no single user has planned for. It promotes ideas that ultimately underpin self-configuring, self-healing, self-organizing and self-protecting computer systems like envisioned in the Adaptive & Autonomic Computing [IBM01] and Autonomic Communication [ACCA04] research initiatives.

The applicability of this approach for resource allocation in the context of ALNs has been evaluated in simulation studies which shown it is particularly well suited to handle highly dynamic environments [Catn03]. We address the task to develop a middleware architecture that helps to embody this concept in diverse applications domains.

3 Architecture

We believe the requirements imposed by the application scenarios analyzed demand an innovative approach for the construction of the resource allocation middleware. The proposed approach is the construction of a framework that offers a set of generic nego-tiation mechanism, on which specialized strategies and policies can be dynamically plugged to adapt to specific application domains or market designs. The middleware should therefore offer a set of high level abstractions and mechanisms to locate and manage resources, locate other trading agents, engage agents in negotiations, learn and adapt to changing conditions. We will first analyze the architectural requirements that need to be addressed to fulfill this vision and then present the proposed architecture.

3.1 Architecture Requirements

The more astringent architectural requirements come from the need for self-organization and adaptability to very different ALN scenarios. These requirements can be summarized as follows:

- The dynamicity of the network prevents an a priori configuration of the peers or the maintenance of centralized configuration services. A peer needs to discover continuously the network characteristics and adapt accordingly.
- The fully decentralized nature of the approach requires the distribution of some critical system functions like security, resource management, topology management, without requiring specialized nodes.
- As all the system function should be implemented in all peers and they have heterogeneous properties and configurations, the P2P system should make little assumptions about the underlying platforms.
- Different ALN architecture will lead to different ways to deploy the middleware components, which cannot make any assumption about the location of other components, to facilitate their (potentially dynamic) redistribution.
- Given the multi-service nature of today's ALNs, one important goal of the architecture is to allow the coexistence of diverse specialized market models on top of a single middleware infrastructure.
- The middleware should allow pluggable policies, strategies and mechanisms, which could be dynamically activated to adapt the system to different environments.

3.2 Proposed Architecture

We propose a layered architecture shown in the figure 3. This layered approach offers the palpable benefic of a clear separation of concerns between the layers, which beside helping in tackling the complexity of the system, also facilitate the construction of a more adaptable system as the upper layers can be progressively specialized (by means of pluggable rules and strategies) into specific application domains.

Agents in the *Economic Algorithms Layer* are responsible for implementing the high level economic behavior contained in the economic algorithms layer (negotiation, learning, adaptation to environment signals, other agent's strategies and its own outcomes). Applications themselves do not participate (and are not actually aware of) the negotiation, but delegate it to the economic agents.

Economic agents rely on a lower level layer, the *P2P Agent Layer*, for the self-organization of the systems and the interaction with the base platform that ultimately

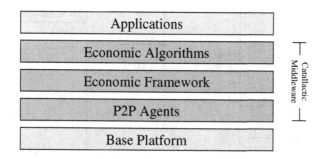

Fig. 3. A layered architecture for resource allocation

manages the resources being traded. This layer offers key functions like the maintenance of the trading network topology following a P2P paradigm, the decentralized resource discovery and the group communication among agents.

In this context the term "P2P" should be interpreted as a general approach for distributed system design, characterized by the ad-hoc nature of the system's topology and the functional symmetry of its components, which can be realized under very different architectures, ranging from unstructured and disperse networks to very hierarchical systems.

Between those two layers, a *Framework Layer* isolates economic agents from technical complexities; much in the same tenor that modern online trading platform allows non expert users to trade stocks. This framework offers basic functions like searching for suitable providers given a resource specification, handle the exchange of messages during the negotiation process, keeping track of the evolution of the negotiation for further adaptation of strategies.

3.3 Dynamic View

To appreciate the interrelationships between the components of the architecture, it is necessary to see how they interact in different scenarios, being the more relevant the initial registry of agents, the distributed object location, which shows how the underlying P2P platform can be used to achieve a high degree of decentralization in this critical function, and the initiation of the bargaining process.

3.3.1 Registering Resources and Agents
Negotiation for resources is carried out by agents that represent the client requesting a resource and the providers that offers that resource. How those agents are actually created is very dependant on the architecture of the systems requesting the resource and offering it. Figure 4 shows a generic situation.

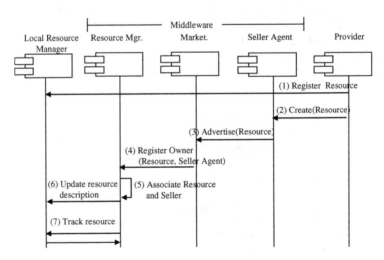

Fig. 4. Registering Agents and resources

The Resource *Provider* application, registers a resource with its local platform specific *Local Resource Manager* (which is part of the execution platform and outside the middleware), and instantiates a *Seller Agent (SA)* to represent it in bargaining for a specific service. The *SA* registers itself to the local *Market* agent, which uses the middleware's *Resource Manager Agent (RMA)* to associate the *SA* with the resource. The *RMA* can, optionally, update the resource's information in the *Local Resource Manager* to reflect, for instance, that the resource is already reserved by the middleware and cannot be offered to other application. Finally, the *RMA* keeps track of the resource state (e.g. availability and usage level) and uses this information to answer queries for resources given a certain characteristics.

3.3.2 Negotiating for Resources

Negotiation process begins when a *Client Application* (*CA*) requests a resource to the *Broker Agent (BA)*, giving some contractual conditions (e.g. available budget) and technical specifications. How this is accomplished depends on the application scenario. The *CA* can invoke directly the *BA* or it can be invoked by a component in the *CA's* platform (a local resource manager, for instance) in response for a request for resources. Also, the conditions and specifications can be explicitly given by the *CA*, be part of the middleware configuration or a result of the *BA* learning during past negotiations.

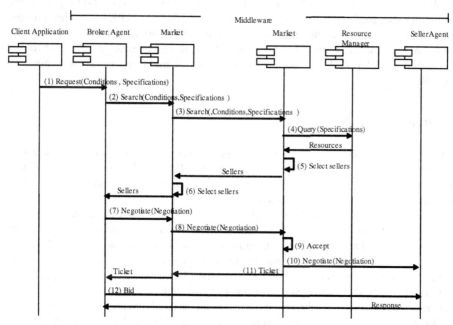

Fig. 5. Negotiating for resources

After receiving the request, the *BA* asks its local *Market Agent* (*MA*) for a list of potential *Seller Agents* (*SA*). The *MA* performs a distributed search among neighbor nodes. On each neighbor node, the local *MA* requests the *Resource Management*

Agent (RMA) a list of resources that match the specifications, and their related *SAs*. Then the *MA* selects the appropriated *SA* according to the contractual conditions and sends the list back to the *MA* that started the search. Finally, the *BA* select the *SA*(s) it want to trade with and starts the negotiation process. The *MA* in both sides (broker and seller) can additionally enforce some trading rules based on the participant's reputations, past experiences and local allocation policies, filtering negotiation requests and responses.

3.4 Ongoing Prototype Implementation

Out of the layers of the architecture, the P2P Agent Layer is currently being implemented. This prototype can be viewed as an early validation of the proposed architecture with a threefold objective. First, test to what extent the middleware can be constructed using already existent toolkits. Second, validate the feasibility to compose the middleware following the proposed separation of concerns in multiple interacting agents. Finally, allow to test that the middleware can handle the required levels of decentralization and scalability. The results of these tests are expected to raise additional architectural requirements to be included in following iterations of the design process.

The implementation of the middleware builds on the use of different middleware toolkits, namely the DIET agent platform [Diet05], JXTA [Jxta05] and the WSRF/OGSA implementation of Globus Toolkit 4 [Glob05]. DIET provides a modular, lightweight and scalable execution platform, JXTA offers a rich P2P networking environment and GT4 provides full support for resource management in different scenarios. A detailed description of the selection of middleware toolkit is given in [Catn05].

4 Related Work

Many market based resource allocation systems have been proposed in the literature [YeBu04]. However, all of them fail to entirely fulfill two key features needed in a resource allocation mechanism for autonomic systems: fully decentralization and openness to evolutionary environments.

The vast majority is based on a sort of bidding or utility maximization process and relay in a facilitator to accomplish the allocation of resources, introducing a high degree of centralization. One example of this approach is the GridBus project [BuVe04], which applies concepts from the utility markets (e.g. power market) for resource allocation in grid applications. GridBus is based on a Service Market Directory, where application services are published, and a Service Broker, which matches the requests from users to the available resources considering the execution const and diverse QoS parameters and looking for the optimization of the system wide utility. Our model, on the contrary, is a fully decentralized direct bargaining between producers and consumers and does not require any centralized market mechanism. This decentralization brings a higher scalability and a better adaptability to local resource requirements and to highly dynamic environments. The drawback is, however a less than optimal allocation of resources [Catn03].

Some few decentralized frameworks have being proposed in the literature, remarkably OCEAN [PHP+03] and Tycoon [LHF04]. OCEAN (Open Computation Exchange and Network) provides an open and portable software infrastructure to automated commercial buying and selling of computing resources over the Internet. Each OCEAN node that wants to buy resources uses a Matching service, which implements an optimized P2P search protocol, to find a set of potential sellers based on the description of the resources being requested. Then, an automatic negotiation process starts with each seller, based on the rules dynamically defined in a XML format. The ability to define negotiation rules is a remarkable characteristic of OCEAN that allows the adaptation of the economic model to diverse applications. The main limitation we found in this rule based approach is the lack of mechanisms for learning and adaptation to evolving environments. We found an agent based approach more suitable to achieve this level of adaptativeness.

Tycoon is a distributed market-based allocation architecture based on a local auctioning for resources on each node. Auctioneers receive fine grained requests of local resources from agents acting on behalf of applications and schedule them using efficient sealed bid auctions in a way that approximates proportional share, allowing high resource utilization rates and the adaptation to changes in demand and/or supply. One interesting feature of Tycoon is that it separates the allocation mechanism from the agents which interprets application and user preferences. This allows the specialization of agent different applications. Tycoon however doesn't offer any framework for the construction of those agents.

A major limitation of Tycoon is that the resource allocation mechanism is already fixed in the system design and no extension or adaptation methods are offered. To overcome this limitation, our proposed framework is capable to plug key components to adapt to specific application domain in environments with heterogeneous or changing resource allocation requirements. Also, we offer a set of high level tools to develop those components, alleviating the implementation burden for new market designs.

5   Concluding Remarks

We expect that the proposed architecture could guide the implementation of future large scale distributed applications which integrate decentralized and autonomic resource allocation components, employing economic mechanisms.

The proposed architecture brings a set of important benefits for the implementation task, namely an appropriated separation of concerns that will facilitate the implementation process, a great deal of flexibility and a strong "agnosticism" regarding the underlying platforms, application domain and economic model, which will make more adaptable to evolving environments.

However, we believe that some critical issues that must still be addressed, which constitutes our proposed research agenda in the field:

- A flexible framework that allows a consistent view and management of resources using a uniform set of abstractions, independently of the how each base platform handles the allocation and monitoring of its resources.

- A generic interface to pass the description of the resource requirements along with the desired conditions (preferences) from application layer to the economic agents and to automatically fill any missing information that can not be provided by the application could be automatically filled. One example of such information is the application's budget to negotiate for resources. This brings some important consideration for the mapping from generic economic parameters (e.g. price) and the underlying technical parameters in the base platform (e.g. CPU workload).
- A set of interaction patters between the P2P Agent Layer and the Economic Algorithms Layer, to allow the adaptation of the trading network and search mechanisms to the results of the economic negotiations and the system's performance.
- Implementation of a fully decentralized accounting and payment service to handle the user budgets and execution costs, to incentive cooperation and prevent the "free riding" of the system.
- Definition of metrics to measure the performance of the system and model to analyze them from both a technical and economic perspectives and their instrumentation in the middleware.

References

[ACCA04]   Autonomic Communication, Report on FET consultation meeting on Communication paradigms for 2020, Brussels, 3-4 March 2004.
           http://www.autonomic-communication.org/publications/doc/AC_report-mar04.pdf
[BuVe04]   R. Buyya, S. Venugopal (2004), "The Gridbus Toolkit for Service Oriented Grid and Utility Computing: An Overview and Status Report, Technical Report, GRIDS-TR-2004-2", University of Melbourne, Australia, April 2004
[Catn03]   CATNET Project (2003), "Catallaxy Simulation Study. Report No. D2", http://research.ac.upc.es/catnet/pubs/D2_Simulation_Study.pdf
[Catn05]   CATNETS Project (2005): "Deliverable D3.1: Selection of middleware toolkits and options for integration of catallactic mechanisms in current middleware used in peer-to-peer and grid implementations, March 2005. http://www.catnets.org
[Diet05]   http://diet-agents.sourceforge.net/Index.html
[ERA+03]   T. Eymann, M. Reinickke, O. Ardaiz, P. Artigas, F. Freitag, L. Navarro (2003), "Self-organizing resource allocation for autonomic network", Proceedings. 14th International Workshop on Database and Expert Systems Applications, Germany, 656- 660
[EyPa00]   T. Eymann and B. Padovan. "The Catallaxy as a new Paradigm for the Design of Information Systems". Proc. of The World Computer Congress 2000 of the International Federation for Information Processing. 2000.
[Glob05]   http://www.globus.org/
[IBM01]   IBM Corp.: Autonomic Computing. Yorktown Heights, NY: IBM 2001,
[Jxta05]   http://www.jxta.org/
[LHF04]   K. Lai, B. A. Huberman, and L. Fine, "Tycoon: A Distributed Market-based Resource Allocation System," HP Lab, Palo Alto, Technical Report cs.DC/0404013, Apr. 2004.

[PHP+03]   P. Padala, C. Harrison, N. Pelfort, E. Jansen, M. P Frank and C. Chokkareddy. OCEAN: The Open Computation Exchange and Arbitration Network, A Market Approach to Meta computing. In proceedings of the International Symposium on Parallel and Distributed Computing (ISPDC'03), Oct 2003

[WHW+04]   S. White, J. Hanson, I. Whalley, D. Chess, J. Kephart (2004), "An Architectural Approach to Autonomic Computing", International Conference on Autonomic Computing

[YeBu04]   C. S. Yeo and R. Buyya, A taxonomy of market-based resource management systems for utility-driven cluster computing, Technical Report, GRIDS-TR-2004-12, University of Melbourne, Australia, December 8, 2004.

Service Discovery and Provision for Autonomic Mobile Computing*

George C. Polyzos, Christopher N. Ververidis, and Elias C. Efstathiou

Mobile Multimedia Laboratory, Department of Computer Science,
Athens University of Economics and Business,
Athens 10434, Greece
{polyzos, chris, efstath}@aueb.gr

Abstract. Service discovery and related service advertisements, redirection and provision decisions are essential processes in networks supporting mobile communications in order for these systems to be self-configurable with zero or minimal administration overhead. More so in mobile networks, i.e. networks where the network infrastructure is moving and the topology is constantly changing. Finally, the servers themselves offering the services might be mobile, wirelessly connected and battery powered and thus power limited and energy constrained, with a finite horizon of operation and service availability. For this reason they will probably have rather selective policies for service advertisement and provision. In this paper we review our previous work on topics in this area and put it under this new perspective, providing our vision for a general autonomic framework for service advertisement, discovery, provision decision, redirection etc.

1 Introduction

The primary goal for Autonomic Systems is to enable systems to manage themselves given high-level objectives from administrators. Self-management of systems involves self-configuration, self-optimization, self-healing and self-protection. In mobile communications and pervasive environments a crucial facet of self-management is *service discovery*. If a flexible service discovery framework is lacking, nodes in a mobile, and in particular in infrastructure-less and constantly changing, environment will be confined to use only their own services and resources (or at most any such services and resources that are pre-configured by system administrators). Service discovery is thus paramount for operation in unknown environments (and for practical reasons in mobile and pervasive environments). Even though many service discovery protocols and architectures have been proposed, particularly for volatile environments such as those of Mobile Ad hoc NETworks (MANETs), none addresses all of the aforementioned aspects of self management and most are very far from being considered autonomic.

Autonomic service discovery is the gateway to autonomic mobile computing, since it enables heterogeneous nodes to interact with each other, learn and adapt to their

---

* Supported in part by IST FET ACCA coordination action.

I. Stavrakakis and M. Smirnov (Eds.): WAC 2005, LNCS 3854, pp. 226–236, 2006.

environment. Without this component no self-configuration, self-optimization, self-healing or self-protection can be performed. For example, imagine a scenario where a node joins a MANET and initially has no knowledge of any resources located around it, neither does it host a well-known protocol for discovering services and resources. Up to now all the proposed approaches to service discovery assume that all nodes run the same discovery protocol, which may be very far from true in such heterogeneous environments, either due to the different capabilities of mobile hosts (e.g., differences of orders of magnitude in available local resources and performance metrics when comparing laptops to PDAs), or to their diversity (type of device and type of use and circumstances, e.g., slow or very fast movement). It is clear that in such a case nodes would not be able to demonstrate even the simplest autonomic behavior. What would change the situation in the above scenario would be a discovery approach allowing nodes to negotiate how to implement the discovery process. In this paper we shall discuss the need for a general framework for autonomic service advertisement, discovery, provision decision (on whether and to whom) in such networks along with our vision for the basic components of such a framework.

In our effort to introduce our vision of a universal autonomic framework (and also show the way towards its realization) we will present three ways for applying the concept of autonomy in service discovery and provision in different architectural contexts and over different network technologies. In order to realize autonomic systems we believe that we need to begin from service discovery, which itself should be inherently autonomic. Thus, first we will discuss how service discovery for MANETs may be enriched with autonomic properties. Then, we will provide an overview of a proposed architecture for global service publishing, discovery and access over a fixed infrastructure, where a critical contribution towards the autonomic goals is the use of ontologies for the description (and thus discovery) of services. As we will see in that section, this is a key idea for supporting really heterogeneous environments. In addition that architecture supports and emphasizes context and scheduling for service selection and provision. Finally, we will focus on service provision decisions. We have applied autonomic principles in the design of a fully decentralized system for service provision (with indirect reciprocity) in Wireless LANs (WLANs) with no central managing authority and even no strong (persistent, verifiable) identities.

2 Autonomic Service Discovery

Significant academic and industrial research has led to the development of a number of protocols, platforms and architectures for service discovery such as JINI [1], Salutation [2], UPnP [3], UDDI [4], Bluetooth's SDP [5] and SLP [6]. Most of these approaches were designed for static networks employing centralized approaches and implying reliable communication and enough bandwidth provided by the underlying networks. Recently newer approaches such as Allia [7], GSD [8], DEAPspace [9], Konark [10] and SANDMAN [11] were developed with pervasive computing environments in mind.

From our point of view a service discovery protocol or framework, especially for the harsh environment of MANETs, in order to be autonomic, should posses the following properties [12] or at least address the following issues:

- Distributed–decentralized: This means that it should not rely on fixed or well known a priori infrastructure (e.g., a centralized service directory), but it should be able to discover services in a cooperative, on-demand, Peer-to-Peer (P2P) way. However, this does not exclude cases where specific powerful mobile nodes take up (temporarily) advanced roles, acting as service registries and allowing other potentially weaker nodes to publish and obtain services through them.

- Ontology-based: An autonomic service discovery framework should not rely on a priori knowledge of how services get to be known, but it should be able to semantically match services (with extensive use of ontologies). For example, two "mobile servers" may provide a similar service for currency conversion. However one of them may announce it as "currency conversion" and the other one as "currency exchange." So when a node is in need of a "currency conversion" service, by consulting the ontology, it should be able to identify both of these services as candidates for invocation.

- Context awareness for self-adaptation: The autonomic service discovery framework should be able to sense the environment by taking into account context information. This way service discovery can be self-adapted or self-configured based on high-level policies like "energy consumption minimization."

- Policy driven–election mechanism: Nodes should be free to choose the way service discovery is performed given their capabilities and goals. The service discovery framework should provide techniques to allow nodes to negotiate how service discovery will be performed. An election (majority) mechanism might be needed to decide on the preferred policy.

- Recovery mechanism: An autonomic service discovery should undoubtedly implement a "self-healing" process for recovering from service failures (e.g., due to a path break) after a node has invoked a service.

Our vision for autonomic service discovery points to a general framework responsible for disseminating the way that service discovery should be performed according to the administrator's goal. Since in MANETs there are no central (domain) administrators, this goal could be "translated" to being the result of the "common" goal of the users, for example "discovered" through an election performed by nodes participating in the MANET. In addition, different parts of the MANET could select different goals based on their local needs and requirements and hence tune the framework and its components to perform a different kind of service discovery in their area.

Our proposal is to split the service discovery into components, which can be tuned by every node according to their capabilities and policies. Hence, constrained nodes may select to use/participate in a more lightweight discovery process (e.g. directory-less and zone-limited), while more powerful and resource rich nodes may select advanced discovery policies. Important issues, such as aggregation and communication of information between neighboring areas with different goals and hence different ways to perform service discovery, are also addressed by the framework. A similar approach was proposed in [13] for autonomic routing.[1]

---

[1] A "... programmable routing framework that creates an adaptable routing service for sensor networks. In this framework, a routing service is divided into several programmable components. Based on this division, a universal routing service is developed that allows the introduction of different services through a set of tunable parameters and programmable components."

We also break service discovery into programmable components and allow tuning depending on application needs and management goals. The basic components are:

- Service advertisement (query vs. announcement, directory vs. directory-less, flooding vs. zone)
- Service selection (location based vs. energy conservation based vs. load based)
- Service recovery (statefull vs. stateless)

For example, a management goal could be: minimize total energy consumption and avoid draining a single server. This could be interpreted by the framework as: avoid using flooding, perform localized service discovery (e.g. up to 2 hops), implement pull techniques (i.e., queries from clients) for discovery and not push (i.e., advertisements from servers), encapsulate battery and current load information into service replies, automatically select closest server with minimal load etc.

Given the above approach, a high level of autonomy can be introduced into the nodes so that they can automatically cope with the increased levels of heterogeneity and volatility, which are present in a MANET environment. However many questions remain open. Among them:

- How do we select tunable components to include in the framework?
- How often can we change network or area-wide policies and goals?
- How can we allow and support different deployments of the service discovery approach in different areas?
- How can we aggregate information from neighboring areas with different goals and hence different ways to perform service discovery?

In the next paragraphs we continue discussing service discovery, but in a different context, that of a fixed infrastructure, highlighting the autonomic properties of a proposed architecture called *Mobishare*. We also discuss service publishing and optimized service access.

3 An Architecture for Mobile Service Publishing, Discovery and Optimized Access over Fixed Infrastructures

The rapid advances in wireless communications technology and mobile computing have enabled small, personal mobile devices that we use in everyday life to become information and service providers by complementing or replacing fixed-location servers connected to the wireline network. In the following we briefly describe the *MobiShare* architecture [14], which allows mobile servers to publish their services and mobile clients to discover and access these services over a distributed, possibly global, autonomic system, i.e., with no human intervention for low-level management. This system is capable of capturing context and uses it to self-optimize its responses when services are requested. Mobile servers publish their services through wireless networks to Administration Servers (ASs) that manage an area. These ASs are responsible for maintaining a list of services available (published) within their area of authority and also to provide a semantic discovery capability to assist the users with locating the services that can fulfill their needs by performing *context-based filtering* of query results.

ASs are aware of and interconnected to other ASs over the (most probably fixed) network. Although they can function autonomously, in certain cases they may cooperate to perform inter-AS service discovery, mobile server positioning (for redirecting clients), and service migrations.

A major concern of an AS is to preserve service integrity throughout the system, since the same service description maybe available to more than one AS. So when publishing the service the user has to (1) declare an initial area where they wish the service to be available (user-defined policy) and (2) specify whether the service is fixed (i.e., an on-site service giving information about a monument) or mobile. In the case of a mobile service (i.e., a picture sharing service on the mobile "phone" of a tourist) the user (or its agent) must decide if the service availability area should be extended to the areas where the device moves (mobility-based policy) or to the areas where it is requested (request-based policy). This advertisement profile of a service is part of the service description and the ASs of the areas that publish the service are aware of it, since it determines whether they should proactively push the description to their neighbors. Every service description is initially stored to the AS to which the mobile server has published it for the first time. This AS is responsible for maintaining a list of all ASs to which the service may be propagated in the future. In this way whenever an update is issued to any AS which hosts the service description, this AS automatically obtains (from the original AS) the list of all ASs that store the service description and contacts them in order to update them too.

In order to minimize administrative and management overhead and provide a convenient level of autonomy when discovering and accessing services, a data-centric and services oriented approach is employed. Service providers publish their services to the infrastructure and service requestors discover them through the infrastructure. There is no human-to-human negotiation for publishing a service and the whole process is automated.

- Since a service is published there is no administrative overhead for managing it, but the system, according to the publication policy which is submitted along with the service, 'self-configures' it, determining the way that the service will be visible to requestors and to the rest of the system.
- For the discovery phase, the system's context sensing modules and semantic matching capabilities (with the help of an ontology), provide the system with 'self-optimization,' managing to keep bandwidth consumption low and user satisfaction high by returning only the most appropriate services to the user. (A non-optimized, not context-based, system would return a large number of candidate services, using excess bandwidth and overwhelming users with redundant and possibly useless information.)
- Also, in the case of a loss of a service provider, an AS may take up the role of executing the service (if at all possible), or alternatively finding another available service provider providing the same or a similar service, hence giving the system the so much anticipated 'self-healing' property.

Providing a convenient level of self configuration, self optimization and self healing the described architecture for service discovery, provision and access over fixed networks, shows the path towards realizing autonomy in such environments. Future research in this area could be in the direction of intelligent mechanisms for providing self protection to the system by automatically identifying and blocking malware-services or possible DoS attacks.

In the next section we continue the presentation of our vision for autonomic computing by presenting a fully decentralized system that provides incentives and promotes cooperation of independent agents for service provision to peers.

4 Autonomic Indirect Reciprocity-Based Peer-to-Peer Service

The proliferation of low-cost networked digital devices enables new provisioning models for digital services. Traditional service models assume logically centralized providers, responding to service requests from authorized consumers. This general model covers a wide variety of network-based services ranging from basic Internet and content access to the provisioning of storage and computing services. This traditional model has several advantages. For example, the existence of a logically centralized authority, i.e. the service provider, which is usually assumed to be trustworthy, allows for simple solutions to common provisioning problems. From the provider's perspective, the main problem is the design of an appropriate charging model that would allow the provider to maximize profits. Traditional economic models can be used; for example, a utility-based mechanism design approach can assist in deriving the relevant parameters. Moreover, the existence of a centralized authority allows for the application of straightforward security solutions that protect the service from disruption and unauthorized usage. Finally, even though in many cases the service provider maintains a distributed network of service access points, this does not alter the basic "client-server" provisioning model. The physical distribution of services is usually adopted to address scalability and availability concerns. The physically distributed service is still under the control of a single authority, and there is usually a clean interface between the service's public and private layers. This physically distributed but logically centralized model encompasses, e.g., cellular operators and their networks of base stations, ISPs and their point-of-presence networks, and content providers and their networks of caches.

Recent research on what is generally referred to as the "peer-to-peer" (P2P) model of service provisioning revisits several of the assumptions above. Proposed P2P systems, such as multi-hop ad hoc and mesh networks, storage and file sharing networks, and grid computing networks, are all examples of a novel service provisioning model in which logically centralized (and profit-maximizing) authorities are absent. In the new model, the peers, in addition to consuming services, also take charge of service provisioning. The P2P model is now becoming possible due to Moore's law, the drop in digital storage costs, the abundance of network bandwidth, and the existence of unlicensed wireless spectrum. Some of the most common P2P advantages cited include increased scalability and fault-tolerance, increased efficiency (in terms of pooling under-exploited resources), lower provisioning costs, and an increased sense of community-building among peers.

A less cited advantage of the P2P model, however, is the fact that it allows for near-zero-cost service configuration. To show how this is possible, in the following analysis we will assume a "pure" P2P system, i.e. a system in which centralized authorities are absent *even during system initialization*. In the traditional model, an important component of the overall cost is the cost incurred when consumers *register* with providers. This can include the cost of a provider-approved device, configured with the appropriate security credentials, or the administrative cost of an online or offline registration procedure that would provide the client with a unique system identity.

We will present a specific near-zero-cost configuration model that is applicable to a wide variety of pure P2P systems. In our previous work [15], we applied this model to a P2P wireless LAN roaming system of our design. The theoretical incentives issues behind this type of P2P model have been analyzed before by Feldman *et al.* [16].

Central to our model is the concept of a zero-cost system identity. Many P2P models assume a unique identity for every peer; they also assume that peers will not create more than one such identity (in order to avoid Sybil attacks[2] [17], i.e., a no-cost change of identity). However, it is unclear how the "one peer, one identity" requirement can be met without a centralized trusted authority. Even though some designs acknowledge that this authority does not have to be online during the lifetime of the system, consumers are required, at the very least, to register with the authority when they first join the system.

In our registration-free model, identities are simple private-public key pairs. We assume that the private key is kept secret by the peer who wishes to adopt the specific identity, and that it is computationally infeasible to derive the private key from the digitally signed statements that are created with it. Obviously, peers can create multiple such identities if they so wish.

A peer's private key is used to sign a *receipt* each time service is consumed by a providing peer. Whenever such a *transaction* takes place (i.e. whenever a providing peer provides a service to a consuming peer), a receipt is generated and stored in the system. The receipt contains the public keys of the providing and consuming peer, a timestamp, and a "weight" which corresponds to the amount of service offered. The receipt is signed by the private key of the consuming peer. Our model assumes that all peers are selfish and rational and that peers will attempt to provide as few services as possible, while consuming as many services as possible. At the same time peers will attempt to hide any evidence of service consumption and forge evidence of service contribution. To increase the chances that the peers do actually sign receipts for the services they consume, the providing peers can request several (intermediate) receipts during service provisioning. This way, the risk of providing service and not obtaining the corresponding receipt can be minimized. The consumer has no choice but to sign whatever the provider requests in order for service provisioning to continue without interruption. If the consumer believes the provider is making unreasonable claims the only choice would be to abort the transaction. This, however, would cost the provider a useful receipt since receipts are to be used as evidence of service contribution.

The main idea behind the service model is to enable *indirect reciprocity*, a cooperation model that has been shown [16] to enable the evolution and stabilization of cooperation in self-organizing communities of selfish peers in the absence of authorities. To allow the evolution of cooperation, however, two additional issues need to be addressed. First, a peer's history must be visible to other peers. Second, collusion-based attacks that enable a peer to appear cooperative without really contributing to the

---

[2] Sybil attack [17]: the creation of multiple identities per entity; a fundamental problem in open and self-organized electronically mediated communities without identity-certifying authorities. Sybil attacks can invalidate any number of system assumptions.

community must be difficult or impossible to launch; the possibility of Sybil attacks (which are allowed in our model) makes the second task harder.

In [15] we presented a novel distributed history implementation that is incentive-compatible and allows peers to view subsets of the receipt graph and apply one of a family of collusion-resistant decision functions, which can guide peers in their provisioning decisions. Two main points that should underlie any distributed history implementation are the following. First, any gossiping algorithm that peers use in order to share their views of the receipt graph must take peer selfishness into account. It is safe to suggest that, irrespective of decision function, peers would want to hide their consumption actions and advertise only their contribution actions. Second, only short-term history can be maintained. This practical limitation (finite storage capacity at the peers), however, works to the system's advantage: short-term history means that contributions and consumption actions that are in the past are erased from system memory. In an environment that relies on reciprocity this means that peers cannot rely on their previous contributions indefinitely, and that continuous cooperation is necessary.

A specific gossiping algorithm works as follows. Each peer maintains a local repository of receipts. The general replacement rule for this repository is "oldest one out," that is, when the repository is full, receipts with newer timestamps replace receipts with the oldest timestamps. This local repository certainly includes the receipts that have been directly "earned" by the peer when cooperating as provider (the consuming peers present these new receipts to the providers directly). However, these receipts represent only a subset of the receipt graph. We observe, however, that requesting peers have an incentive to show their own directly-earned receipts to their prospective providing peers (irrespective of specific decision function, as we mentioned above). If peers share receipts in this manner, after a few iterations each peer becomes aware of many more receipts than the ones it earned directly when acting as a provider.

5 Conclusions

This paper is motivated by the need for introducing autonomy in service discovery and provision architectures for mobile computing environments. With this in mind, we have reviewed new architectures we have recently proposed that are demonstrating some autonomic characteristics. Collectively, these architectures address self-configuration (or zero configuration), policy-based self-configuration and adaptation (nodes automatically adapt to a service discovery policy based on a high-level goal), self-optimization and self-healing. In figure 1 we illustrate the basic facilitators for inducing autonomy in service discovery and provision for different mobile environments as explained in the above paragraphs.

It is evident that context awareness and the use of ontologies are of paramount importance in both MANETs and more centralized architectures for enabling self-configuration, self-adaptation and self-optimization. Election mechanisms for deciding the preferred advertisement, discovery and recovery policies are also mandatory for MANETs so that autonomic service discovery can be performed taking into account the capabilities and needs of the nodes. Finally, indirect reciprocity mechanisms are useful in decentralized environments for allowing autonomic service provision without the need for central coordinating entities.

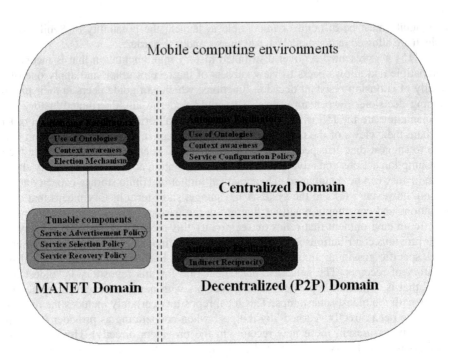

Fig. 1. Mobile Computing and Autonomy Facilitators for Service Discovery and Provision

At first glance there is no commonality between the autonomy facilitators presented across the different mobile environments. However, if we take a closer look, we can see that they can actually be merged under a common generic service discovery and provision framework suitable for addressing the autonomy needs of any mobile environment (see Figure 2).

To begin with the indirect reciprocity mechanisms as an autonomy facilitator, it is easy to understand that such mechanisms may be of great importance also in MANET environments. Especially if we take into account that service provision in MANETs is costly (nodes acting as providers or intermediaries expend valuable and scarce resources like energy and bandwidth) the network must provide a mechanism for enforcing fairness and ensuring cooperation between service providers and requestors. If this has to be done in a decentralized way without authorities, an indirect reciprocity mechanism, explained in previous paragraphs, is the perfect candidate. Also, in a centralized environment these mechanisms can be used in order for the system to implement autonomic self-protection by filtering out of the system malware services and nodes, based on reports/receipts published to the system by nodes that have used those services.

Finally, election mechanisms and tunable components as autonomy facilitators except for MANET environments can be proven useful also for centralized and decentralized environments for service provision. In the case of service provision, election mechanisms may be used to automatically decide the preferred provision policy and its rules. There is indeed much work yet to be done on realizing autonomic service

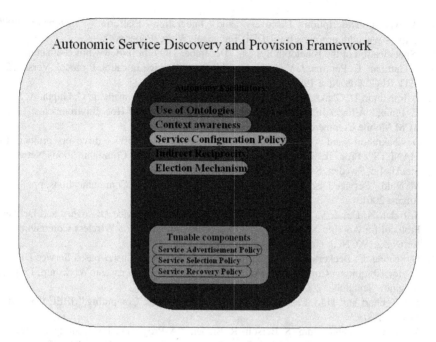

Fig. 2. Framework for Autonomic Service Discovery and Provision

discovery and provision for mobile computing, since until now research focused on the functional and optimization part of related architectures, instead of on their autonomic characteristics. We have identified the current lack of flexible-enough service discovery and provision protocols due to their monolithic nature and proposed, among others, breaking them into tunable components and implementing negotiation mechanisms among nodes (in order to select the preferred way of operation) as possible ways to increase their autonomy. This is certainly just the beginning of the investigation, but our proof of the existence of mechanisms for driving distributed systems with no central authorities and cheap identities towards evolutionary stability with near-zero configuration cost, is a step in the right direction. Towards this direction, we have tried to envision a generic framework by synthesizing autonomy facilitators, flexible enough for adding self-configuration, self-protection, self-healing and self-managing properties to any mobile computing environment dealing with service discovery and provision.

References

1. Sun Microsystems, "JINI Architecture Specification," November 1999.
2. Salutation Consortium, "Salutation Architecture Specification," 1999 (http://ww.salutation.org/specordr.htm).
3. Microsoft Corporation, "Universal Plug and Play: Background," 1999 (ww.upnp.org/resources/UPnPbkgnd.htm).

4. "Universal Description Discovery and Integration Platform," September 2000 (http://www.uddi.org/pubs/Iru_UDDI_Technical_White_Paper.pdf).
5. "Specification of the Bluetooth System," December 1999 (http://www.bluetooth.com).
6. E. Guttman, C. Perkins, J. Veizades, and M. Day, "Service Location Protocol, Version 2," IETF RFC 2608, June 1999.
7. O. Ratsimor, D. Chakraborty, S. Tolia, D. Kushraj, A. Kunjithapatham, G. Gupta, A. Joshi, T. Finin, "Allia: Alliance-based Service Discovery for Ad-Hoc Environments," Proc. ACM Mobile Commerce Workshop, September 2002.
8. D. Chakraborty and A. Joshi, "GSD: A novel group-based service discovery protocol for MANETS", Proc. IEEE Conference on Mobile and Wireless Communications Networks, Stockholm, Sweden, September 2002.
9. M. Nidd, "Service Discovery in DEAPspace," IEEE Personal Communications, pp. 39-45, August 2001.
10. S. Helal, N. Desai, V. Verma, and C. Lee, "Konark – A Service Discovery and Delivery Protocol for Ad-Hoc Networks," Proc. 3rd IEEE Conference on Wireless Communication Networks (WCNC), New Orleans, Louisiana, March 2003.
11. G. Schiele, C. Becker and K. Rothermel, "Energy-Efficient Cluster-based Service Discovery for Ubiquitous Computing," Proc. 11th ACM SIGOPS European Workshop, Louven, Belgium, September 2004.
12. J.O. Kephart and D.M. Chess, "The Vision of Autonomic Computing," IEEE Computer, January 2003.
13. Y. He, C.S. Raghavendra, S. Berson, R. Braden, "A Programmable Routing Framework for Autonomic Sensor Networks," Proc. Autonomic Computing Workshop, Fifth Annual International Workshop on Active Middleware Services (AMS'03), Seattle, WA, pp. 60-68, June 2003.
14. E. Valavanis, C. Ververidis, M. Vazirgiannis, G.C. Polyzos, and K. Nørvåg, "MobiShare: Sharing Context-Dependent Data and Services from Mobile Sources," Proc. 2003 IEEE/WIC International Conference on Web Intelligence (WI 2003), Halifax, Canada, October 2003.
15. E.C. Efstathiou and G.C. Polyzos, "Self-Organized Peering of Wireless LAN Hotspots," European Transactions on Telecommunications, 2005 (special issue on Self-Organization in Mobile Networking, in press).
16. M. Feldman, K. Lai, I. Stoica, and J. Chuang, "Robust Incentive Techniques for Peer-to-Peer Networks," Proc. 5th ACM Conference on Electronic Commerce, 2004.
17. J.R. Douceur, "The Sybil Attack," Proc. First International Workshop on Peer-to-Peer Systems (IPTPS'02), Cambridge, MA, 2002.

Context Dissemination for Autonomic Communication Systems

Nadeem Akhtar, Klaus Moessner, and Ralf Kernchen

Centre for Communication Systems Research,
University of Surrey, Guildford, UK
{n.akhtar, k.moessner, r.kernchen}@surrey.ac.uk

Abstract. Autonomic Communication is a new communication paradigm that has been proposed as a way to design new self-organizing, self-healing, self-optimizing, self-protecting and evolvable networks. The motivation comes from the problems created by the unstructured and haphazard growth of the Internet. Among the many guiding principles of Autonomic Communication is context-awareness. In this paper, we discuss architecture for context dissemination in Autonomic networks based on the Autonomous Decentralized Community Communication for information dissemination.

1 Introduction

The rapid and often chaotic growth that the Internet has seen over the past few years has resulted in an extremely complex network. Furthermore, the introduction of newer communication technologies, services and applications is leading to a growing patchwork of interconnected networks. Therefore, the management of the increasingly unwieldy Internet is becoming extremely difficult with each passing day. It is in this context that the Autonomic Communication paradigm has been proposed that aims to create self-organizing, self-managing and context-aware autonomous networks in order to meet the diverse demands and challenges confronting the Internet and to allow for a scaleable and manageable growth.

Context-awareness is an important property of Autonomic systems [1]. Research in this area is rather diverse and broad because the notion of context has many different connotations and the nature of context varies a lot depending on the application scenario. According to Moran and Dorish [2], context refers to physical and social situation in which computational devices are embedded. Anind Dey *et al* [3] define context as any information that can be used to characterize the situation of an entity where an entity is a person, place, or object that is relevant to the interaction between a user and an application. Most of the earlier work in the area of context awareness has focused on human-machine interactions. In contrast, Autonomic systems can also be deployed for business-business, human-human and machine-machine communications. Therefore, the notion of context acquires a much broader meaning.

For Autonomic systems, context needs to be considered as a dynamic process in which context is generated as a consequence of continuous interactions between

I. Stavrakakis and M. Smirnov (Eds.): WAC 2005, LNCS 3854, pp. 237–242, 2006.
© IFIP International Federation for Information Processing 2006

users, networks, network elements and the physical environment itself. This gives rise to the notion of *context state* which can be described as the instantaneous view of the context. The context state consists of a set of information elements corresponding to the parameter set that characterizes the system under consideration. Depending on the specific application that requires context awareness, whole context state may not be useful or relevant and only a subset of the state maybe needed. Therefore, the context sub states are constructed out of the overall 'global' state. The lowest level of granularity is the so-called feature context that corresponds to individual information elements. Fig. 1 illustrates how the context hierarchy is organized.

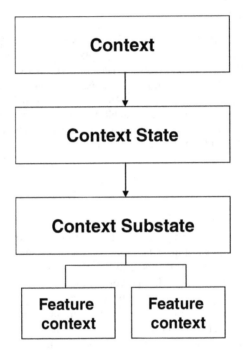

Fig. 1. Context has to be regarded as a function over time, hence it can be seen in different states and sub states which again can be described as context features forming a "context hierarchy"

Considering existing context aware systems and applying them to Autonomic Communications puts a set of requirements on the structure of such an approach. The extensible distribution of the network has to be taken into account. Further the characteristics of such a system have to be researched in general. We identified three main challenges in designing context aware systems in this area.

Some sort of context representation, often referred to context abstraction [4], is necessary in order to provide a common language for communication between entities of such a system. For Autonomic Communication Systems in particular this has to support and follow the distributed character of these networks. Apart from common characteristics for context-aware systems the distributed context representation has to consider certain network parameters and possible dependencies and relations between different entities as well as relevance of context items for certain levels of context based decision.

Another important factor for the successful decision process is the collection of the context itself on the very deepest information layer: the entity itself. This data has to be reprocessed in order to distinguish between network or entity-level relevant information. This needs to be done to know which information has to be sent to higher level decision units in the network or can be kept in the entity itself.

Context dissemination is the third part of the design problem. A distributed context dissemination mechanism is required for Autonomic systems. Furthermore, for dynamic network, function and service composition/decomposition, the mechanism should be fast and efficient. Security is also a very important component of the dissemination mechanism.

Each of these needs to be researched in detail with reference to Autonomic systems. In this paper, we focus on the context dissemination problem. In the following, we consider the Autonomic Decentralised Community Communication and investigate its suitability as an efficient and scalable context dissemination solution for Autonomic systems.

2 Context Dissemination

Context dissemination is an important component of the overall context awareness design. A distributed, efficient, secure and scalable architecture needs to be designed for this purpose. In the following, we present an overview of information dissemination architecture for community communication and then investigate its suitability to the requirements of context dissemination in an Autonomous Communication System.

2.1 Autonomous Decentralized Community Communication

The Autonomous Decentralized Community Communication (ADCC) system proposed in [5] is designed to provide an infrastructure for information dissemination in order to help end-user groups communicate and share information efficiently. This system is based on a decentralized architecture and application-level multicast is used to distribute information from a particular source of information to all the other members of the community. The ADCC system was originally developed to meet the growing demand for real-time content delivery in the Internet and aimed to provide a scalable dissemination infrastructure. We propose to use it for context dissemination between network elements of an Autonomic Communication system. In the following, we first provide an overview of ADCC and then describe how it can be used for dissemination of context in an AC system.

2.2 ADCC Overview

The basic idea of the ADCC system is to enable members of a 'user' community to exchange information that is of 'interest' to all the members while ensuring their autonomy at the same time. In ADCC, a community consists of members that may have individual objectives but share common interests and have similar information requirements. No distinction is made between senders and receivers of information and thus, any member can be a source as well as sink with respect to information flow. The community members organize themselves into a logical network. To form such a network, member nodes maintain information about immediate neighbors in a table and share this information with other nodes. Nodes autonomously decide to join or leave the community based on their own requirements and each member has to carry out the same set of network duties.

Since there is no central co-coordinating entity in ADCC, the responsibility of construction and maintenance of the community is shared by all the members. This self-organized community is created with three main objectives. Firstly, the resulting network must support efficient broadcast. Secondly, the traffic is distributed evenly to avoid hotspots. Finally, the network must provide redundancy so that the effect of node failures can be mitigated.

In the ADCC system, the nodes that constitute a community network are organized in a regular 2-dimensional graph: $G = \{V, E\}$, where V is the set of nodes and E is the set of edges. The graph G consists of a set of n Hamiltonian Cycles (HCs) that are all edge-disjoint with each node having $2n$ neighbors. Each such cycle connects all the nodes in the graph and each node is traversed only once. The advantage of HC lies in localized impact of a node joining or leaving of a node.

When a node, say X, wishes to join the community, it first has to discover at least one existing member. Node X sends a join request to the discovered node Y. The latter has to find out the $2n$ neighbors so that the former can join the n Hamiltonian Cycles. The join request is forwarded by node Y to all nodes that are within $O(log_{2d}M)$ distance, where M is the current number of nodes in the network. The nodes that receive this message then decide whether to accept the request or not. A repair mechanism is in place in case there is a node failure. Keep-alive messages are used to keep the neighbor information fresh and when a node failure is detected, the fault tolerance algorithm comes into play to repair the network.

Communication within the network is multilateral. When a node receives new information that needs to be shared with other members, it is sent to the neighbor nodes which then send it to their own neighbors while ensuring that it is not sent to the sender itself. This process is repeated until every member of the community has received the information. ADCC uses a hybrid pull-push model as well as a request-reply-all model. In the first case, when a member has new information to share, it forwards it to the neighbor nodes as described above. In the second case, when a member wants to find a specific piece of information, it sends a request to neighbor nodes. If a particular neighbor does not have the desired information, it forwards the request to its own neighbors but if the neighbor does have the information, it replies to the sender with its results. This is disseminated to all members of the community, thus updating the whole community.

The performance of the ADCC system has been compared with traditional techniques such as sequential unicast and peer-to-peer methods. It has been shown that the mean communication cost of the ADCC approach is much lower compared to the other two while its mean delay is significantly less than the unicast approach. Furthermore, the cost of community construction and maintenance increases logarithmically as the number of member nodes increases. Results also indicate that the load on the physical links connecting the nodes in the network is distributed more evenly across the network.

2.3 ADCC for Context Dissemination

The ADCC system described above has been designed for information dissemination between members of a community in a co-operative, efficient and scalable manner. In an Autonomic Communication system, network elements are inherently members of one or more groups on the basis of physical or logical proximity, similar functionality, need for same set of configuration data etc [1]. The notion of *entity-group* has been mentioned in the context of Autonomous Communication. It refers to a type of group communication where network elements can join and leave a group based on a set of *membership rules.* Within a group, the behavior of an entity is dictated by its *group-behavior* definition. This results in a programmable and controllable group. Thus, we see that there are similarities between the idea of a community in ADCC and the notion of group in AC. Furthermore, no central controller is required to create a community and nodes can join and leave the network as and when they wish. Furthermore, the network heals itself when a node leaves the network or when one or more nodes fail. Thus, the ADCC community network is self-organizing and self-healing thereby making it an attractive choice from the AC point of view.

In an AC network, context could mean many different things. It may include information about the networking environment, the physical environment itself and even rules and policies could be treated as context. Thus, it is very important that these different types of information are described in such a way that it is easy to distribute them to interested parties. In ADCC, a *content code* is used to describe the information that is being sent out. Also, a *characterized code* is used to indicate further details of the content. This approach is well-suited for the purpose of context dissemination in the AC network.

ADCC implements a multilateral communication between community members for information flow using the bilateral links between members. Once again, it does not require any central source of information and any member can send information and all members are receivers. In the AC system, many scenarios are possible. For example, there maybe a controller node that is responsible for information collection and dissemination. Alternatively, these responsibilities maybe distributed throughout the network. Finally, it may not be required to send all new information to all the group members. The ADCC system is flexible to cater to these different requirements.

Network elements may require regular context updates as well as instant updates based on the type of context. In some scenarios, a specific type of information maybe requested by a particular element. In the ADCC system, the two communication protocols used are *hybrid push-pull* and *request-reply-all.* They could be used for context dissemination as well. In the second model, when a node replies to an information request from another node, the reply with the desired data is sent to all members. This is done to avoid further requests for the same data. While this works very well when all the members have more or less the same information requirements but it may not be efficient when this is not the case. Thus, there is a trade-off which needs to be studied further for the communication dissemination scenario.

3 Conclusion

Context awareness is an important requirement of AC systems. Context dissemination is an important component of a context-aware system. In this paper, we have reviewed the Autonomous Decentralized Community Communication system and discussed its suitability as the basis of context dissemination architecture in AC network. The ADCC was primarily designed to provide an efficient and scalable infrastructure for data dissemination on the web. It supports cooperative information exchange in a loosely-connected network consisting of members with similar information requirements. We observe that ADCC is a suitable choice for context dissemination. The community network in ADCC is self-organizing and self-healing which is extremely important from the AC perspective. Furthermore, the way content is coded in ADCC can serve as the starting point for describing context and addressing it to the right destinations. The communication model in ADCC is flexible for request-based context delivery as well as periodic and instantaneous context updates. Further work is required for a more thorough and detailed investigation of the ADCC with respect to context dissemination, especially the analysis of communication models as well the coding of context information.

References

1. Smirnow, M.: Autonomic Communication: Research Agenda for a New Communication Paradigm. Fraunhofer FOKUS White Paper (2004)
2. Moran, T.P., Dorish, P.: Introduction to the Special Issue on Context-Aware Computing. Special Issue of Human-Computer Interaction, Volume 16 (2001) 87-95
3. Dey, A.K., et al: Understanding and Using Context. Personal and Ubiquitous Computing Journal, Volume 5 (1), (2001) 4-7
4. Dey, A.K., Abowd, G.D, and Salber, D.: A Conceptual Framework and a Toolkit for Supporting the Rapid Prototyping of Context-Aware Applications. Human-Computer Interaction, vol. 16, (2001) 97-166
5. Ragab, K., et al: Autonomous Decentralized Community Communication for Information Dissemination. IEEE Internet Computing, Vol.8, No.3, (2004) 29-36

On Natural Mobility Models

Vincent Borrel, Marcelo Dias de Amorim, and Serge Fdida

LIP6/CNRS – Université Pierre et Marie Curie,
8, rue du Capitaine Scott – 75015 – Paris – France
{borrel, amorim, sf}@rp.lip6.fr

Abstract. There is an increasing consensus that existing mobility models, such as the well-known random walk or random waypoint models, are insufficient to represent real node mobility. In this paper, we discuss the need for a better characterization of natural mobility. Our contributions rely on recent advances of real-life network analysis and modelling, and in particular on the observation that natural networks behave on a scale-free basis. We devise then a novel mobility modelling approach that focuses on the behavioral aspect of individuals and the interactions between them. This fulfils a gap between individual and group mobility models. Our first results show a strong relevance of the scale-free distribution in mobility modelling, and open further directions in modelling the costs associated to building a network structure in general.

1 Introduction

The increasing demand for mobile networking has raised a number of complex problems ever addressed by the network research community. Many of these problems do not have all the required elements for a complete solution.

In this context of autonomic communications, mobility management will play a major role. Indeed, given the rapid growth of the radio equipped population, future networks will face serious challenges in terms of node density and complexity of the communication environments. A key functionality to adapting to rapid changes in the environment is to tailor network configuration and routing algorithms according to the spatial characteristics of the real world they rely upon. Considering that a part of future networks are based on mobile terminals, a mobility model matching reality at its best is highly demanded [1, 2].

Existing mobility models are either too simplistic or do not represent the real characteristics of user mobility. The current model used to represent mobile scenarios is the Random Waypoint Model, despite its obvious flaws and the lack of similarity it has with real-life situations [3]. We acknowledge that it is difficult to define the *real* characteristics that a mobility model should capture. We adopt an extrapolative approach by inferring mobility from observations made in real-life networks. Starting from the simple parallel analysis between man-to-group interactions and dynamic principles of real-life network models, we devised a mobility model from the ground up, bridging the gap between individual and

I. Stavrakakis and M. Smirnov (Eds.): WAC 2005, LNCS 3854, pp. 243–253, 2006.

group mobility models. The objectives are to bring new enhancements and fine-grained population modelling by matching the main observation spanning many real-life domains: *scale-free* spatial distribution.

Such a model could be useful in many aspects. Being a behavioral model, it may give us clues on some statistics of the behaviors it implements. This may serve, for example, as a feedback to sociologists to validate hypothesis on dense and large populations, which might be difficult to measure in practice. In the context of autonomic communications, it would serve as a basis for innovative heuristics for routing, connectivity establishment, self-healing, and security. With this work, we also intend to strengthen the links among complementary disciplines. We are interested, in particular, in how biological or sociological observations could be integrated in our framework.

The remainder of this paper is organized as follows. In Section 2, we survey existing mobility models. In Section 3, we focus on the explanation of the most distinctive characteristics of real-life networks, and quickly overview where such characteristics have been observed. We summarize, in Section 4, the considerations that led us to reconsider mobility models, and describe the objectives, characteristics, and first results concerning our approach. Finally, Section 5 concludes this paper.

2   An Overview of Existing Mobility Models

A number of mobility models have been proposed in the literature [1]. Generally, two types of mobility have been addressed so far: individual mobility and group mobility. We define them in the following and present the most important approaches proposed in the literature.

2.1   Individual Mobility Models

Individual mobility deals with the movement at the node level, where each node is considered independently from the others. We present in the following the most important models proposed so far.

The Random Walk mobility model. First proposed by Albert Einstein in 1926 to characterize Brownian motion, is also called "Drunkard's Walk", and is the *de facto* mobility model used for mobile network analysis. In this model, a node travels by changing its direction and speed at random, at regular time or distance intervals. However, since its behavior is independent of past motion (memoryless), it generates very unrealistic displacements.

The Random Waypoint mobility model. In this model, nodes travel between randomly chosen locations. The speed of displacement and pause periods are also randomly determined. This model is also widely used in mobile network simulations; however, since the performance is obtained in a bounded space, the density of nodes at the center of the simulation area tends to grow indefinitely.

The Random Direction mobility model. This model has been conceived to overcome this drawback of the Random Waypoint model. Here, a node chooses a random direction and follows it until it reaches a border, pauses there for a random duration, then restarts by choosing its next direction. The problem is that nodes tend to stay at the borders of the simulation area, which is likely to generate network partitions and big hop counts in simulated networks.

The Boundless Simulation Area mobility model. It focuses on a boundless space, by wrapping a rectangular zone around its opposite borders, in a toroidal manner. Each node has a direction and a speed, respectively updated at random following a maximal angular change speed and a maximal acceleration. This model has an even spatial repartition, and accounts for a quite realistic user movement. However, it does not account for pauses, and the boundless situation is not considered to be the most representative one in simulations of mobile scenarios.

The Gauss-Markov mobility model. This model considers that nodes have an initial speed and direction, which they update at each time step, taking into account the previous speed/direction (s_{n-1}/v_{n-1}), the mean speed/direction $(\overline{s}/\overline{v})$, and a random value $(s_{x_{n-1}}/v_{x_{n-1}})$ having a Gaussian distribution. These parameters are considered with different weights, according to a *randomness parameter* α. The current speed and direction are then given by:

$$s_n = \alpha s_{n-1} + (1 - \alpha)\overline{s} + \sqrt{1 - \alpha^2} s_{x_{n-1}}, \tag{1}$$

$$d_n = \alpha d_{n-1} + (1 - \alpha)\overline{d} + \sqrt{1 - \alpha^2} d_{x_{n-1}}. \tag{2}$$

The City Section mobility model. This model uses a street network map, upon which nodes go from random place to random place choosing their shortest time path, with possible speed limitations and minimal distance between nodes, and pause upon arrival before restarting. It is very specific, being designed to render cars or pedestrians in constrained maps.

2.2   Group Mobility Models

In a group mobility model, the mobility of a node is computed relatively to the mobility of a reference point in the subset of nodes (group) it belongs to. A number of such models have been proposed in the literature, and some of them are described in the following.

The Reference Point group mobility model. It is the most generic group mobility model. It implements groups of nodes which follow more or less loosely a *reference point* whose motion can be dictated by various ways, such as using one of the previously discussed individual mobility models. Its specialization gives birth to mobility models presented in the following.

The Exponential Correlated mobility model. This model uses a motion function to compute the next movement vector $\overrightarrow{b}(t)$, in the complex space, given by:

$$b(t+1) = b(t)e^{-\frac{1}{\tau}} + \left(\sigma\sqrt{1 - e^{-\frac{2}{\tau}}}\right)r, \qquad (3)$$

where r is a random Gaussian variable of variance σ, and τ accounts for the rate of change (the smaller τ, the quicker the change). The main problem here is that it is difficult to set appropriate parameters to obtain a particular effect.

The Nomadic Community mobility model. In this model, groups move from point to point, inside which every member wanders in a Brownian motion locally. It corresponds in fact to a combination of the Random Waypoint that uses locally, in a smaller scale, a Random Walk.

The Pursue mobility model. This model defines a particular node which is the target, while the other ones are its prosecutors. The target follows its own mobility model while the other nodes have a motion accelerating toward it, plus a random vector.

3   Aspects of Real-Life Networks

Many recent studies have found, in various areas of real-life ranging from biology to computer networks, via sociology, scientific citation, literature, movie acting, ecosystems or economics, some fascinating common features of the networks, or graphs, modelling the many relationship that pervade them. These common features, not present in the traditional Erdös-Rényi Random graph model, lie on two major aspects: the scale-free property and the high clustering coefficient.

The scale-free property relates to a power-law distribution of the degrees of nodes in the network. This distribution is different from the usual Poisson distribution, also called exponential, defining node degrees in the random network model.

The power law distribution means that the probability of having a node of degree k is

$$P[k] \propto k^{-\lambda}, \qquad (4)$$

where λ, the exponent, can be seen in a log-log graph as the pent of the linear fit of the distribution. Fig. 1 shows an example of the resulting node distribution for both the random and scale-free cases.

The clustering coefficient defines the propensity of nodes to be gathered in small groups that are highly interconnected. Its mathematical formula is derived from the "fraction of transitive triples", defined by Wasserman and Faust in [4]. At the node level, it is the effective number E_i of links relating the node's k_i direct neighbors over the total possible number of links between them:

$$C_i = \frac{2E_i}{k_i(k_i - 1)}. \qquad (5)$$

Averaged over all its nodes, C_i becomes the clustering coefficient of a graph. In random graphs, as each edge exists with a probability p, the clustering coefficient is $C = p$. In many real-life networks, it has been found to be several orders of magnitude higher than in random graphs.

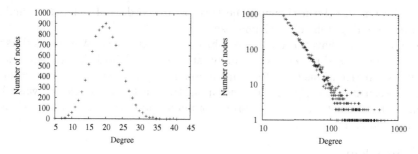

Fig. 1. A comparison of Random graphs versus Scale-Free graphs. On the left, the node degree distribution of an Erdös-Rényi random graph with 10000 nodes and a link probability of 0.02, in a linear scale. On the right, the node degree distribution of an Albert-Barabási scale-free graph grown to 10000 nodes with preferential attachment in a log-log scale. Both have the same number of nodes and links, but their structures are significantly different.

Various observations have been made of these two characteristics in many domains, as presented in the following.

3.1   Biology

Jeong *et al.* [5] studied the metabolism of 43 organisms in networks where the nodes are substrates (ATP, ADP, H_2O, ...) and edges account for the predominantly directed chemical reactions in which these substrates participate. The distribution of incoming and outgoing edges are both in power-law with respective degrees of 2.4 and 2. Wagner and Fell [6] concentrated on the metabolism of *Escherichia coli* bacterium. Here in addition to a power-law distribution, the undirected version of the substrate graph has a small average path length and a high clustering coefficient (around ten times that of an equivalent random graph).

Another important network is the one of interactions between proteins. Here the nodes are proteins, and the edges represent the fact that they bind together. Jeong, Mason *et al.* [7] studied such a network in the yeast (*S. cerev.*) and found a power-law with an exponent of 2.4.

Scala, Amaral and Barthélémy [8] studied the networks formed by the conformation of a two-dimensional lattice polymer. The clustering coefficient is much larger than the one of an equivalent random graph. However, here, the degree distribution is consistent with a Gaussian.

The network of gene expression (here a node represents the expression of a gene, and the impact that the expression of a gene has on the expression of another gene is an edge) also exhibits power-law distributions.

3.2   Computer Networks

In the Internet, partial graphs, both at the router and domain levels, have been shown to have the scale-free property, with a degree distribution of the nodes in power law, with an exponent between 2 and 3 [9, 10]. At the domain level, the clustering has also been shown to be two to three orders of magnitude higher than in random graphs of the same size.

Similar results have been observed in the context of the Web (hyperlinks). Although being a directed graph, many studies found both its in-degree and out-degree distributions to be following a power-law of the same exponent range, also between 2 and 3 [11, 12]. Furthermore, the clustering coefficient was also found [13] to be between 2 and 3 orders of magnitude higher than the ones of a similar random graph.

3.3   Sociology

This domain is one of the most prolific, and also the oldest. It relates directly to humans, and network structures depend on their behaviors.

The Internet Movie Database [14] contains all movies and their casts since the 1890s. Here the nodes are actors and two actors having participated in a movie have an edge between them. This graph can also be seen as a bipartite graph, consisting of a set of actors, a set of movies, and a set of participations, linking an actor to a movie. Watts and Strogatz, in their study of 1998 [15] found its clustering coefficient to be 0.79, around 3000 times more than the one of the equivalent random graph, 0.00027.

Citation networks in scientific publications are made of nodes which represent publications, and edges that go from a publication to the other one it cites. The first study of scale-free properties in citation networks of scientific publications were conducted in 1965 by Derek de Solla Price. He noted that these networks have power-law distributions, and subsequently built a model, close to the one of Albert and Barabási, but directed and adapted to publications networks that are directed. A more recent study by Redner [16] found that the in-degree of these networks follow a power-law distribution, with an exponent of 3.

In phone call networks, where nodes are phone numbers and edges telephonic conversations between them are a kind of social network. Abello *et al.* [17] and Lu [18] studied the graphs of long distance telephone calls made during a day and found that the distributions of incoming and outgoing calls follows a power-law with an exponent of 2.1.

3.4   The Albert-Barabási Model

As shown by Albert and Barabási in their eponym model [11, 19], scale-free distributions can be obtained mimicking the dynamics of groups of elements following two simple rules: growth and preferential attachment. Albert and Barabási start from an initial random network, and progressively add new nodes. This is the growth principle. In the preferential attachment principle, the probability that a new node be connected to an existing node i is proportional to i's connectivity degree:

$$\Pi(k_i) = \frac{k_i}{\sum_j k_j} \qquad (6)$$

This process gives rises to graphs with scale-free degree distributions, as illustrated in figure 1.

They also showed the generated graphs to be more resilient to random failures [20].

4   A Novel Approach for Mobility Modelling

In the models presented in Section 2, complex interactions are faintly represented, if not at all. For example, in group mobility models, groups are fixed and cannot evolve in time. Zhou *et al.* [21] address this issue by creating the Group and Swarm Mobility model, where groups follow virtual tracks and swarm at intersections. However, to be able to regroup at intersections, two subgroups must arrive at an intersection at the same time. This constraint leads to a plethora of small groups, and large groups cannot be well represented.

Furthermore, none of these models minds the spatial density dynamics of populations. In real-world situations, groups forms and dislocate, crowds evolve, people mimic others, the total population grows or diminishes. Another point is that the notion of group is too rigid for many situations. Nodes must belong to a precise group, and only one, or be independent. Group don't evolve, and moreover, the behavior space is discretized in few classes, if more than one, which doesn't allow for fine behavior materialization. We would like to fine-tune the tendency to behave more or less following one or more groups.

4.1   Why Scale-Free Characteristics in Mobility?

It is now a known fact that mobility strongly influences the results of ad-hoc protocols simulations. As there currently exist no precise large scale traces of users mobility, one is forced to resort to the mobility models for such work.

A mobility model must be as close as possible to the reality, and thinking about matching reality we must strongly consider its prevalent characteristics. As we have seen earlier, one of these prevalent characteristics has been extensively reported in many different areas. It is the scale-free distribution of graphs representing real-life situations.

Although this reason is by far not sufficient to justify by itself a Scale-Free behavior in human mobility, several other aspects clearly point to its usefulness.

In the first place, the preferential attachment behavior in human crowds is a known fact in sociology. From the 'rich get richer' comportment, at the base of the Albert-Barabàsi model itself, to the positive feedback of crowd sizes on their growth, passing by the propensity of people to admit the most common idea as the most valid. If then, as in many other aspects, preferential attachment in human motion decision behaviors can be found, the question of whether it drives to scale-free spatial density distributions becomes prominent.

This supposition becomes especially relevant in the light of recent studies by Chaintreau *et al.* [22], which observe scale-free inter-contact distributions in different crowds of humans equipped with Bluetooth devices. In fact, scale-free spatial distributions might greatly account in these surprising observations.

This is very important since mobile ad-hoc networks structures are highly dependant on the connectivity of the nodes, itself depending on the spatial density of them. If this density is scale-free, the network structure inherited will be far different from a random network, and knowing this property can greatly help design routing protocols, prior to help better test them.

The lacunas we have seen in existing mobility models led us to consider a new category of mobility model, where a collective behavior arises. Such a model must have, however, a finer-grained definition of the traditional concept of group, which is too rigid to efficiently represent the evolution of real populations. This new type of mobility is called *gathering mobility*, where nodes meet in space and evolve almost independently one from another (with some level of interaction).

We want to model displacement of crowds in free spaces, with centers of interests. Such a model finds its application in many scenarios, as for example in an exhibition hall, where different booths exist. This can also represent a school courtyard, where children go from occupation to occupation, from group to group. This can also be seen as a market or shopping center, where people aggregate around stores and stands. More generally, this can be seen as the way people go: from an interesting place to another interesting place. This model would have people tending to judge the interest of a place in function of the interest other people have in this place, following trends and mimicking others.

4.2   Characterization of the Model

We consider the network as a constrained rectangular space of configurable dimensions, where two different types of objects coexist: *individuals* and *attractors*.[1]

Attractors are landmarks toward which nodes move; they appear for a certain period of time, do not move, then disappear. They model centers of interest for individuals (*e.g.*, stands in a show). Each attractor is associated with a force, which influences its propensity to gather individuals.

Individuals are the main focus of our work. They behave in cycles. A cycle consists of a displacement, a pause period, and a decision of leaving or not the current position. Displacements are characterized by the origin and a target attractor. We will show that in the choice of the target lies the main characteristic of the model.

The preferential attachment principle is implemented in the attractor decision process as follows. The probability $\Pi(a_i)$ that an individual z_k chooses an attractor a_i among all possible ones is proportional to the portion of the total attractiveness it carries: $\Pi(a_i) = \frac{\mathcal{A}_{a_i, z_k}}{\sum_j \mathcal{A}_{a_j, z_k}}$. The attractiveness of an attractor is

---

[1] The term *attractors* used here bears similarities with another term, *Attraction points*, defined by Jardosh *et al.* in [23]. They both model centers of interest for nodes displacements. However, their function and the mobility they generate are different in essence. In the Jardosh paper, nodes move from point to point along enhanced Voronoi graph. During some time intervals, a random selected Voronoi point becomes and attraction point, the destination of a certain percentage of all the nodes. In our proposal, Nodes or *Individuals* move freely in space, and all destination points are *Attractors*. Our mechanism implements positive feedback, in the form of preferential attachment to select the desired attractor among others, weighted by inverse of distance.

relative to an individual, and is proportional to its popularity (number of other individuals who chose it), and inversely proportional to the distance separating it from the considered individual:

$$A_{(z_i, a_l)} = \frac{\left(1 + \sum_{z_j \in \mathbf{Z}, z_j \neq z_i} B(z_j, a_l)\right)}{\sqrt{(X_{a_l} - X_{z_i})^2 + (Y_{a_l} - Y_{z_i})^2}}. \tag{7}$$

where $B(z_j, a_l)$ is a Bernoulli variable, with $B = 1$ if the individual z_j is going toward or staying at attractor a_l and 0 otherwise, and X and Y are the coordinates of a node (individual or attractor). Observe that Eq. 7 includes the effect of other nodes' decision on z_i to represent the collective behavior of individuals.

4.3 Experiments

We have built a simulator to test the premises of our mobility model (see screenshots shown in Fig. 2). The scale-free nature of the Albert-Barabási model was

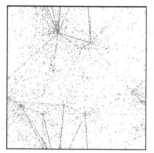

Fig. 2. A view of the simulated mobility showing the spatial distribution of nodes in both planar and toroidal representations. One can clearly see paths forming, and a much diversified spatial distribution of individuals.

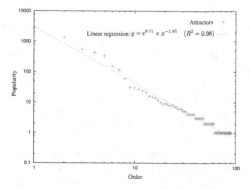

Fig. 3. First results of our model, in situation of population growth

observed in situation of population growth. Thus, we place our model in this same context, where attractors and individuals arrive and never leave.

In a simulation run of 10,000 individuals, we obtain the distribution shown in figure 3. A linear fit in log-log scale gave us a power-law distribution, with an exponent of −1.85 and a confidence of 0.98. These results exhibit the scale-free characteristic of the attractor population.

A corollary result is that, given an equiprobable spatial distribution of nodes, and in situation of population growth, the preferential attachment, even weighted by the inverse of distance, gives scale-free results.

5   Conclusion

Following observations in nature of a scale-free distribution, we devised a mobility model using preferential attachment, aimed at fulfilling lacunas seen in the domain of mobility modelling. This mobility model belongs to a new kind of mobility paradigm, we call gathering mobility. In this model, individuals evolve independently and do not explicitly belong to groups, although they exhibit strong collective behavior and are influenced by others. They gather around centers of interest of varying popularity levels.

Our first tests show that, in situation of growth, such as defined by recent scale-free models, our mobility model leads to scale-free spatial density, where preferential attachment is weighted by inverse of distance.

Further works will focus on the characterization of the parameter space of our model. We will investigate in particular other types of population dynamics: not only growth, but also steady and renewing populations, as well as decreasing ones. In these situations it is important to verify if the scale-free distribution of attractors is maintained.

References

1. T. Camp, J. Boleng, and V. Davies, "A survey of mobility models for ad hoc network research," *Wireless Communications and Mobile Computing (WCMC)*, vol. 2, no. 5, no. 5, pp. 483–502, 2002.
2. M. E. J. Newman, "The structure and function of complex networks," *SIAM Review*, vol. 45, no. 2, no. 2, pp. 167–256, 2003.
3. J. Yoon, M. Liu, and B. Noble, "Random waypoint considered harmful," 2003.
4. S. Wasserman and K. Faust, "Social network analysis : Methods and applications," *Cambridge university*, 1994.
5. H. Jeong, B. Tombor, R. Albert, Z. N. Oltvai, and A.-L. Barabási, "The large-scale organization of metabolic networks," *Nature*, vol. 407, pp. 651–654, 2000.
6. D. A. Fell and A. Wagner, "The small world of metabolism," *Nature Biotechnology*, vol. 18, pp. 1121–1122, 2000.
7. H. Jeong, S. Mason, A.-L. Barabási, and Z. N. Oltvai, "Lethality and centrality in protein networks," *Nature*, vol. 411, pp. 651–654, 2001.
8. A. Scala, L. A. N. Amaral, and M. Barthélémy, "Small-world networks and the conformation space of a lattice polymer chain," *Europhys. Letters*, vol. 55, pp. 594–600, 2001.

9. M. Faloutsos, P. Faloutsos, and C. Faloutsos, "On power-law relationships of the internet topology," pp. 251–262, 1999.

10. R. Govindan and H. Tangmunarunkit, "Heuristics for internet map discovery," *Proceedings of IEEE Infocom 2000*, vol. 3, p. 1371, 2000.

11. R. Albert, H.jeong, and A.-L. Barabási, "Diameter of the world-wide web," *Nature*, no. 401, pp. 130–131, Sept. 1999.

12. R. Kumar, P. Raghavan, S. Rajagopalan, D. Sivakumar, A. S. Tomkins, and E. Upfal, "Stochastic models for the web graph," in *Proceedings of the 42nd Annual IEEE Symposium on the Foundations of Computer Science*, (NY), pp. 57–65, IEEE, 2000.

13. B. A. Huberman and L. A. Adamic, "Growth dynamics of the world-wide web," *Nature*, no. 401, p. 131, Sept. 1999.

14. "Internet movie database." http://www.imdb.com.

15. D. J. Watts and S. H. Strogatz, "Collective dynamics of 'small worlds' networks," *Nature*, vol. 393, pp. 440–442, 1998.

16. S. Redner, "How popular is your paper ? an empirical study of the citation distribution," *Eur. Phys. J. B*, vol. 4, pp. 131–134, 1998.

17. W. Aiello, F. Chung, and L. Lu, "A random graph model for massive graphs," in *Proceedings of the 32nd Annual ACM Symposium on Theory of Computing* (A. of computing machinery, ed.), (New York), pp. 171–180, 2000.

18. L. Lu, "The diameter of random massive graphs," in *12th Annual Symposium on Discrete Algorithms (SODA)*, ACM-SIAM.

19. R. Albert and A.-L. Barabási, "Statistical mechanics of random networks," *Review of Modern Physics*, vol. 74, pp. 47–97, 2002.

20. R. Albert, H. Jeong, and A.-L. Barabási, "Attack and error tolerance of complex networks," *Nature*, vol. 406, pp. 378–382, 2000.

21. B. Zhou, K. Xu, and M. Gerla, "Group and swarm mobility models for ad hoc network scenarios using virtual tracks," (Monterey, California, USA), 2004.

22. A. Chaintreau, P. Hui, J. Crowcroft, C. Diot, R. Gass, and J. Scott, "Pocket switched networks: Real-world mobility and its consequences for opportunistic forwarding," tech. rep., Cambridge, UK, Feb. 2005.

23. A. P. Jardosh, E. Belding-Royer, K. C. Almeroth, and S. Suri, "Real-world environment models for mobile network evaluation," Mar. 2005.

Nomadic Wireless Sensor Networks for Autonomic Pervasive Environments

Iacopo Carreras[1], Antonio Francescon[1], and Enrico Gregori[2]

[1] CREATE-NET, via Solteri 38, 38100 – Trento, Italy
`name.surname@create-net.it`
[2] CNR-IIT, Via Moruzzi 1, 56124 – Pisa, Italy
`enrico.gregori@cnr.iit.it`

Abstract. Pervasive computing is one of the most promising research directions for the next future. More and more interest is devoted to the definition of protocols and paradigms for such challenging scenarios. It is envisioned that almost every object surrounding us will be accessible via some electronic device and will become, to some extent, a node of the communication super-structure. This, of course, will entail completely new problems to be addressed, since it will not be possible to manage a network composed by billions of nodes with traditional Internet protocols.

In order to overcome the aforementioned problems, we propose a novel communication paradigm that, despite its simplicity, provides a viable solution to the new all embracing pervasive environments, exploiting the implicit heterogeneity of the network nodes and the time/space dependence of the information circulating in the network. This article presents the approach and evaluates it through simulations in a real application scenario: a parking lot finding system.

1 Introduction

The term Pervasive Computing generally refers to an explosion of interconnected "smart devices" from watches to cars that can make our lives easier and more productive. According to this, in the future pervasive environments, we can expect the number of nodes to grow by multiple orders of magnitude as tags, sensors, PDAs, watches etc., get fully integrated into the communication super-structure [1, 2]. This will dramatically increase the amount of information to be managed, while reducing, at the same time, the processing and communication capabilities of the devices participating in the network.

The vast majority of the devices will be constituted by tiny small nodes that will be required both to sense and to communicate with other nodes in the near proximity. The limited capabilities and dimensions of these nodes, together with the limited energy available, pose severe constraints on their complexity and on the protocols they will be able to run.

As opposed to these small tiny nodes, there will be powerful users devices (e.g., PDAs, smart phones, laptops etc.), capable of intensive processing operations, of

I. Stavrakakis and M. Smirnov (Eds.): WAC 2005, LNCS 3854, pp. 254–265, 2006.

storing high volumes of data, and of performing high data rate communications. Today's cell phones will evolve into personal devices, which will be used not only for communicating, but also for supporting people in their daily life operations. Through the exploitation of the tiny devices information, users will be able to interact with a living environment, and their communication devices will represent the key driver for accessing such a digital ecosystem, and for starting to interact with it .

Hence, we envision a scenario where there will be a clear distinction in the role of the network nodes, and the network will be organized according to a hierarchical architecture. Nodes complexity and communication capabilities will scale with their role. The tiny nodes will act primarily as source of information, while the user devices as consumers of the generated information.

Moreover, the information circulating in the next generation networks is drastically changing in its significance, since it will be constantly localized in space and time, which means that, most of the time, information will be outdated and therefore useless with respect to the context where the user is moving in. It will be always possible to define a *local sphere* (both in time and space) within which the data represents useful information to the user.

Nomadic Wireless Sensor Network (NWSN) is a novel paradigm, firstly proposed in [3], for dealing with the described new pervasive environments. It exploits the implicit hierarchical structure of Next Generation Networks (NGNs) together with the physical mobility of users, in order to achieve an effective diffusion of the information in a totally distributed fashion. Sensor nodes will have the only role of broadcasting their information to mobile users in proximity, while all the complexity needed for transporting the gathered information, and for diffusing it, is shifted at the user nodes. Information is exchanged among users exclusively through single-hop broadcast communications. The applicability of such a network model is confined to a class of services requiring massive amount of data retrieved locally and with relaxed delay constraints.

The use of a hierarchical architecture and of mobility to improve network performance has already received some attention. In [4] a multi-tier network architectures is utilized to mitigate the scalability problems of creating a self-organizing network composed by thousands of heterogeneous nodes. In [5,6] a multi-tier architecture is introduced for collecting data in a sparse sensor network. By exploiting the mobility of some nodes of the network, sensor data is gathered from the environment and transferred to the final users.

In Delay Tolerant Networks (DTNs) [7] problems related to intermittent connectivity, variable delay and asymmetric links are faced by adopting a store-and-forward policy. Most of the work is related to the analysis of packet delays, buffer dimensioning and routing strategies of the storing nodes.

All the referred work focuses on ensuring the delivery of packets from a source to a destination, either in the case of a disconnected network, or in the case of a network where the high number of nodes is too prohibitive to be managed. On the contrary, NWSN aims at a the pure diffusion of the information, which has been generated from sensors, in the environment where the users are moving in.

The major contribution of this paper is the definition of the NWSN architecture and related protocols, and the analysis of a parking lot finding system running on top of the NWSN. The performances of the NWSN network are evaluated through simulations and compared with the case of a centralized system. It is shown how, with an adequate mobility and number of users, the NWSN performance is comparable with a centralized system.

The article is organized as follows: in Sec. 2 the nomadic approach is presented in terms of architecture and protocols. In Sec. 3 results of simulations are showed. Finally, in Sec 4 some conclusions and future research directions are presented.

2   Nomadic Wireless Sensor Network Architecture and Protocols

In the near future, it is reasonable to expect the surrounding ambient to be equipped with a halo of small tiny devices with sensing functionalities, and limited communication capabilities. These devices will be able to identify objects (RFIDs), or to measure physical phenomena surrounding us (sensors). As opposed to these embedded devices, there will be user devices, which will be constantly increasing in their communication, storage and processing capabilities.

The described scenario suggests a multi-tier network architecture. This direction was followed in [3], where it was shown that, by exploiting the users' physical mobility, it was possible to efficiently diffuse the information in an urban environment without the support of any backbone. Following a similar approach, we propose the Nomadic Wireless Sensor Network (NWSN) in order to maximally exploit the peculiarities of future pervasive environments and of the devices that will be composing them.

In the following the network architecture, and the related protocols, are detailed.

2.1   NWSN Network Architecture

NWSN try to fit its network architecture into the technological trend of extremely simple devices as opposed to particularly powerful ones. It is therefore assumed a hierarchical architecture with two kind of nodes:

– *sensor nodes*, which will be simple tiny nodes deployed in the environment, with limited functionalities of sensing and communication. We expect these nodes to be extremely low power and to run extremely simple communication protocols. The unique role of these devices will be the broadcasting of the sensed information to user nodes in the near proximity. As opposed to traditional Wireless Sensor Networks [8], we are freeing these nodes from the burden of running store-and-forward policies. Their network address might simply be their geographical location (e.g., GPS position), or identification number;

- *user nodes*, which correspond to users devices (e.g., PDAs, cell phones etc.). We assume these nodes to be capable of intensive processing operations, and of running complex information exchange protocols. These nodes will be moving in the environment as a consequence of the physical mobility of users, and will collect information from sensor nodes, when in their communication range, and store this information in their device's memory.
  User nodes will exchange the collected sensor information with other users encountered on-the-move. The exchange of the information will occur through single-hop broadcast.

The basic NWSN network architecture does not suppose any connection to the backbone, since it is expected that all the useful information, needed for running services on top of the NWSN, will be available from a nearby sensor or from user nodes encountered while moving.

Two user nodes, when in the communication range, will *opportunistically* exchange the information gathered from the environment.

2.2 NWSN Communication Protocols

As emerged from the NWSN architecture, there are two possible communications: user-to-user and sensor-to-user.

Information circulating in NWSN is expected to be always localized in time and space, meaning that the information, whenever is gathered from the environment, will be stored in the user devices together with the *age*, representing the time elapsed from reading of the sensor[1], and with a geographical position (e.g., GPS position)[2]. Hence, the smallest information unit exchanged from the user nodes will be a tuple $< value : age : location >$.

The sensor-to-user communication will be "one shot", where the sensor source broadcasts a single packet to mobile users in the communication range. The packet will consist of the tuple described above, where value is read in real-time through the sensing functionalities of the node. The user node stores the received information in the internal memory of the device in an ordered data structure, where the order may be time-based or location-based.

The user-to-user communication will contribute to the diffusion of the information gathered from the environment (e.g., the sensor sources), and physically transported by the user devices. This communication follows a simple handshake:

- user 1 sends a request for interest (RFI) packet to user 2. This packet contains some metric resembling the sensor data a user is transporting, i.e., the mean age and location of the information gathered from the sensor nodes;

---

[1] We are not assuming, in principle, all nodes to be synchronized to a common clock. Indeed, we assume that when a sensor reading is relayed from a node to a new one, the node increases the age field of the tuple with the time the reading was stored in his device's memory.

[2] We can safely assume the GPS position to be set in the device at installation time, or to correspond to the mobile user position.

- user 2 receives the packet and decides whether he is in interested in the information that the second user is transporting. If so, user 2 sends a request for data (RFD) packet to user 1. If not, the communication between the 2 users ends;
- user 1, if a RFD packet is received, sends a bundle of information, containing the sensor data gathered from the environment.

Clearly, this will entail an exponential growth of the data exchanged as the number of sensor grows, and, thus, a mechanism to drop out the outdated information is also needed. We call this mechanism *Information filtering* and represents the policy, according to which information is discarded from the mobile users. User 2, when receiving the bundle of information, will merge the received data with his own. This is done by means of *Information Filtering* policies, where the information locality is exploited in order to drop data, which is considered as useless to the user, and merge the received useful information with the information already present in user 1 device's memory. Filtering can be done on a "bundle basis", or on an information data unit basis. In both cases, the information filtering policy, which determines whether to drop or merge the received information, will be in the form:

$$F(Age, Distance) < ServiceThreshold \qquad (1)$$

where *Age* is the age of the sensor information, *Distance* is the user distance from the sensor source (the sensor node) and *Service Threshold* is a parameter that depends from the specific service constraints, and determines whether the information is useful or not to the user. In case filtering is done on a bundle basis, Time and Distance are the average age and distance of the sensor data units within the bundle.

It assumed that the services running on top of the NWSN will determine the Information Filtering policies, and that services will have a specific tolerance to delays of the sensor information exchanged. Hence, the service tolerance to delays will determine the specific values of the Service Threshold.

3   Simulation Environment

As a show case of the potential performance of the NWSN network architecture we choose a *parking lot finding system*, which is supposed to assist drivers in the search of a free parking spot in a city, suggesting the best destination according to its knowledge and, eventually, updating the destination of the users if better information is received on the way.

The aim of the simulation is to evaluate the performance of the parking lot finding service running on top of the NWSN network architecture, and to compare it when run on a centralized system. Hence, in this preliminary work, a simplistic model is assumed for the NWSN communication protocols and for the resources allocation, while the focus is on the number and speed of users needed for efficiently run such a service.

The model has been simulated in the freely available tool Omnet++ [9].

3.1   Parking Lot Finding System Application

We assumed each parking spot of the city to be equipped with a sensor, and the city to be uniformly divided in blocks. Users drive randomly around the city and, after a random driving time, decide to look for a free parking spot in a random block of the city. This would correspond to, let's say, "look for a parking spot near the train station" or "look for a parking spot near the theater".

The *parking lot finding system* assists drivers in the search of the free parking spot in the destination block, suggesting the destination that most likely will be free and, eventually, updating the suggested destination on the way, if more updated information is retrieved.

Due to simulation's scalability problems, we assumed two classes of users to be driving in the environment:

- *served users*, which correspond to users assisted by the parking lot finding service, thus benefiting from the system assistance in the search of a free parking spot;
- *unserved users*, which correspond to users not assisted by the service, thus transparently occupying parking spots for a random *ParkingTime* and leaving the parking unoccupied for a random *FreeTime*. The unserved users model is depicted in Figure 1.

Unserved users will keep occupying and freeing parking spots, and the less is the FreeTime, the less is the probability to find a free parking spot for the served users.

Fig. 1. Unserved users parking occupation model

The same service has been evaluated on the NWSN, on a centralized system, and compared also with the case of a random search, where no support is provided to the users in the search of a free parking spot.

In the following, the three simulated models are described. Please refer to [10] for more detailed description of the three simulated models.

Random Search Model. In the *random search model* it is assumed that mobile users nodes do not have any assistance in the search of a free parking spot, and they behave according to the following steps:

- move randomly in the playground size for a random driving time *Driving-Time*;
- decide to park in a random block k;
- move in block k, and, once entering it, start to move randomly as long as they do not find a free parking spot on their way. It is assumed that users are not aloud to leave the destination block before having parked;
- once parked, sleep for a random parking time *ParkingTime* and then starts from the first step again.

Centralized Network Model. In the *centralized network model*, we tried to imagine the way we would run the same parking lot finding system utilizing state-of-the-art technology. We assumed the network to be organized according to a 3-tier hierarchical architecture, as shown in Figure 2, with 4 kinds of nodes: sensor nodes, sink nodes, user nodes and a central control node.

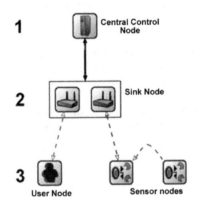

Fig. 2. Centralized model network architecture

Sensor nodes are deployed on every parking spot of the city. They sense the presence of a car, and transmit the change of their status to the nearest sink node in a multi-hop fashion. Routing is done according to the AODV [11] protocol. Clearly, there are several other routing protocols that are more efficient for a WSN, but the aim of this work is not to analyze the efficiency of the network, but rather to compare the performance of the service in the case of different communication paradigms.

Sink nodes are supposed to communicate with sensors for gathering the information on the parking lots status, with the central control for updating a centralized controller, and with the user nodes for answering to their service requests.

Central control node is the network node with a global knowledge of the parking spots status of the city. On the central control node resides the parking spot finding service, and mobile users, when looking for a free parking spot, send requests to this node, which answers with the best available destination. The available destination is simply the nearest free parking spot to the mobile user sending the request.

User nodes correspond to mobile nodes, randomly searching for a free parking spot. Hence, they will be able to communicate with the sink nodes, for sending requests to the Central control node, and for receiving answers from it.
Users behave according to the following steps:

- move randomly in the playground size for a random driving time *Driving-Time*;
- decide to park in a random block k, and, therefore, query the central control for the best parking, according to its knowledge;
- move towards the destination suggested from the central control, and update the destination on the basis of possible updates from the central control;
- once parked, sleeps for a random parking time *ParkingTime* and then starts again from the first step.

The central control node implements a virtual reservation mechanism. When a mobile user sends a request for a free parking spot in a block, the central control answers with the best destination available and virtually reserves this destination for other users searching in the same block. This is introduced in order to avoid the central control node sending several mobile users to the same destination.

Nomadic Wireless Sensor Network Model. The *NWSN model* is based on the Nomadic Wireless Sensor Network, as described in section 2.

Information gathered from the sensors is stored in an array of data, where each entry contains the reading and location of the sensor and the timestamp of the reading.

Without loosing in generality, in this first implementation we assumed the mobile users to be periodically sending a beacon message, for detecting other mobile users in the communication range. When a beacon message is received, the total information carried by mobile users is broadcasted.

The Information Filtering process consists of a simple time-based merge of the information received with the information carried: older information is dropped, while fresher information is kept. This is the simplest policy we can think at, and it unrealistically assumes infinite resources in the user device, i.e. it is possible to store one entry for every sensor node. Nonetheless, in this work we wanted to study if a totally distributed approach, such as the one NWSN, can yield to a system performance comparable to a centralized system, which has been considered as the optimal. Current work is dealing with a more accurate characterization of the Information Filtering and of the allocated resources.

Each user behaves according to the following steps:

- moves randomly in the playground size for a random driving time *Driving-Time*;

– decides to park in a random block k, and, on the basis of the knowledge stored in his device, selects the best destination , which is the nearest free parking place to the user. The mobile then starts moving towards it;
– eventually updates the destination when exchanging information with other users;
– once parked, sleeps for a random parking time *ParkingTime* and then starts from the first step again.

3.2   Simulation Details

The simulation scenario consists of a 4000 m. x 4000 m. playground size, which represents the simulated "city". It is adopted a Manhattan network, constituted by 13x13 streets, starting from 5 m. and ending to 3995 m.. Each street is 2 meters width and has 2 lanes, with two opposite directions. The city environment is subdivided into 16 blocks. Each block is 1000 m. width and 1000 m. height.

Sensors are uniformly distributed over the grid, with a distance of 50 m. among 2 of them, and a communication range of 50 m.. Totally there are 2028 sensor nodes.

Mobile users are moving over the manhattan network at a constant speed and according to a random waypoint mobility model [12], if they have a destination, or a random walk, if they do not have a destination.

Mobile Users implement an IEEE 802.11b-compliant PHY and MAC layer protocols [13, 14], with a communication range of 150 m..

According to the simulation scenario, mobile users will communicate only when meeting along the streets. The introduced parameters are the same for the three simulated models. Clearly, not all of them are completely realistic, but are consistent with the aim of this work.

Time to Park. The metric adopted for evaluating the system's performance is the *Time to Park*, which represents the time, measured in seconds, needed for a user to find a free parking spot starting from the instant he enters the destination block (in each one of the three models we assume that a user is looking for a parking place in a specific block of the city). The Time to Park represents a metric that is independent from the position of the user when he decides to start looking for a parking spot.

3.3   Simulation Results

Simulations have been run varying the speed and the number of mobile users, and the FreeTime of the unserved users. Each one of these parameters has a different impact on the performance of the systems, even though they are strictly related.

In Figure 3 the three analyzed models are presented with 2 and 4 minutes FreeTime of the unserved users. Both the NWSN as well as the centralized system perform better than the random search. How it's intuitively clear, the less is the sensor FreeTime, the more is the network supposed to react fast for updating the users with possible alternative destinations. This is shown in

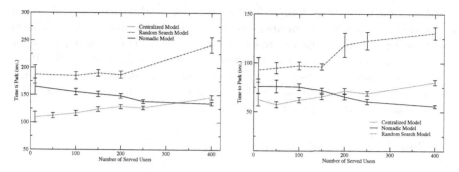

Fig. 3. Time to park in the case of a variable number of served users moving at a speed of 14 m/s speed, and with 2 min. (left) and 4 min. (right) freeTime of the unserved users

Figure 3, where, with 4 minutes of FreeTime all three models perform better than the case of 2 minutes FreeTime.

When comparing the three models, it is possible to see how with a number of served users high enough, i.e., around 300 with a 2 minutes $FreeTime$, and 400 with 4 minutes $FreeTime$, the NWSN system performs better then the centralized one. This is due to the effect of the virtual reservation mechanism implemented in the centralized model, which badly influence the assignment of free parking spots when the competition for the a free parking is extremely high.

While scaling the number of users, the random search and the centralized system decrease their performance due to a higher number of served users competing for the same free parking spots. Differently, the increased number of served users leads to more efficient diffusion of the information, and, thus, to a more stable performance of the service.

In Figure 4 the effect of the users speed (left graph) and the unserved users FreeTime (right graph) is analyzed in the case of 400 served users. As expected, the performance of the system increases with higher speeds. Nonetheless, above

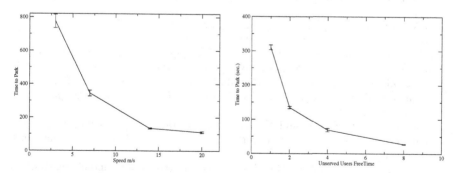

Fig. 4. Time to park in the case of 400 served users and a variable speed of the unserved users (left), and a different FreeTime of the unserved users (rigth) running the NWSN model

a certain threshold (e.g., 15 m/s) further increases in the users speed does not correspond to a similar improvement in the system's performance. A lower Free-Time corresponds to a sensor source changing faster in time, thus requiring the system to spread extremely fast the information in the network for having a good performance of the parking finding system. It is possible to observe that for a sensor source changing slower then 3 minutes does not seem to be a correspondent increase in the performance of the system.

4   Conclusions and Future Work

This paper presents the Nomadic Wireless Sensor Network, which is a communication paradigm specifically tailored to future pervasive environments.

The new challenges deriving from future ubiquitous environments are first introduced and described. The NWSN network architecture, and related protocols, are then introduced and evaluated in a specific case study: a parking lot finding system. The NWSN communication model has been evaluated and compared with the case of a centralized system. Simulations show how the NWSN, despite its simplicity, can perform as well as a centralized system, if an adequate number and mobility of the users are present.

Future work will be devoted to the evaluation of specific communication protocols. This will be reflected in a more fair and realistic utilization of the system resources. An analytical framework for the analysis of the Information Filtering will also be developed.

Finally, we are also working in the implementation of an experimental set-up of the system, in order to have an on-the-field assessment of the expected performances.

References

1. J. M. Kahn, R. H. Katz, and K. S. J. Pister, "Next century challenges: Mobile networking for "smart dust"," in *Proc. of ACM MobiCom*, Seattle, 1999, pp. 271–278.
2. M. Weiser, "The computer for the 21st century," *SIGMOBILE Mob. Comput. Commun. Rev.*, vol. 3, no. 3, pp. 3–11, 1999.
3. I. Carreras, I. Chlamtac, H. Woesner, and H. Zhang, "Nomadic sensor networks," in *Proc. of EWSN*. Istanbul, Turkey: Springer-Verlag, Jan. 2005, pp. 166–176.
4. I. S. S. Zhao, K. Tepe and D. Raychaudhuri, "Routing protocols for self-organizing hierarchical ad-hoc wireless networks," in *Proc. of IEEE Sarnoff 2003 Symposium*, March 2003, pp. 65–68.
5. S. J. R.C. Shah, S. Roy and W. Brunette, "Data mules: modeling a three-tier architecture for sparse sensor networks," in *Proc. of the IEEE SNPA*, May 2003, pp. 30–41.
6. L. Tong, Q. Zhao, and S. Adireddy, "Sensor networks with mobile agents," in *MILCOM 2003 - IEEE Military Communications Conference*, October 2003, pp. 688–693.
7. K. Fall, "A delay-tolerant network architecture for challenged internets," in *Proc. of ACM SIGCOMM 2003*, August 2003.

8. I. F. Akyildiz, W. Su, Y. Sankarasubramaniam, and E. Cayirci, "Wireless sensor networks: a survey," *Computer Networks*, vol. 38, pp. 393–422, 2002.
9. OMNeT++ discrete event simulation system. [Online]. Available: http://www.omnetpp.org
10. I. Carreras, "Nomadic wireless sensor network: a parking lot finding system example," CREATE-NET, Tech. Rep. 3, 2005.
11. C. E. Perkins and E. M. Royer, "Ad-hoc on-demand distance vector routing," in *WMCSA '99: Proceedings of the Second IEEE Workshop on Mobile Computer Systems and Applications*. Washington, DC, USA: IEEE Computer Society, 1999, p. 90.
12. J. Yoon, M. Liu, and B. Noble, "Sound mobility models," in *Proc. of ACM Mobi-Com*, San Diego, CA, 2003.
13. *IEEE Standard for Wireless LAN Medium Access Control (MAC) and Physical Layer (PHY) Specifications*, IEEE Std., Aug 1999.
14. *Supplement to 802.11-1999, Wireless LAN MAC and PHY specifications: Higher Speed Physical Layer (PHY) extension in the 2.4 GHz band*, IEEE Std., Sep 1999.

Adaptive Scheduling in Wireless Sensor Networks

A.G. Ruzzelli[2], M.J. O'Grady[1], G.M.P. O'Hare[2], and R. Tynan[2]

[1] Adaptive Information Cluster (AIC), Department of Computer Science,
University College Dublin (UCD), Belfield, Dublin 4, Ireland
`michael.j.ogrady@ucd.ie`
[2] Practice & Research in Intelligent Systems & Media (PRISM) Laboratory,
Department of Computer Science, University College Dublin (UCD),
Belfield, Dublin 4, Ireland
{`ruzzelli, gregory.OHare, richard.tynan`}`@ucd.ie`

Abstract. As the number of Wireless Sensor Networks (WSNs) applications is anticipated to grow substantially in coming years, new and radical strategies for effectively managing such networks will be needed. One possibility involves endowing the network with an autonomic capability to dynamically adapt itself to the prevailing network operating conditions, even while communications sessions are active. This may involve the network adapting itself either partially or completely. The approach suggested in this paper proposes that a suite of intelligent agents autonomously monitor the various network nodes and, depending on the status of certain parameters, actively intervene to alter the scheduling mechanism used, thus ensuring continuous operation and stability of the network together with an an improved performance yield.

1 Introduction

Wireless Sensor Networks (WSNs) technology consists of a large number of small electro-mechanic devices (in the order of hundreds or thousands) called sensor nodes and a low number of gateways or sinks (in the order of units or tens). Sensor nodes are of low-cost and of low-memory capacity; their task is to collect data from the surrounding and to relay them to the gateway in multi-hop fashion. Gateways are usually considered more powerful and they act as medium between the network and the user. A sensor network is that it should remain active, functional and unattended for long period of time. Hence the need for a strategy to minimize the energy consumption of the nodes. MAC and Routing protocols have a big impact on the overall performance of the system in terms of energy saving. As suggested from the ISO/OSI architecture, network protocols are normally treated separately. Such an approach helps to reduce the descriptive and organizational complexity and is very useful in terms of solving problems independently and autonomously. However, a difficulty with this approach is the increased computational overhead of managing the system that can lead to an increase both in latency of messages from source to

I. Stavrakakis and M. Smirnov (Eds.): WAC 2005, LNCS 3854, pp. 266–276, 2006.

destination, as well as overload on the low memory of the sensor node processor. Ruzzelli et al. [1], proposed the lightweight MERLIN protocol to integrate MAC and Routing into one simple architecture. MERLIN has been designed to reduce the energy consumption of nodes by trading off its intrinsic low-latency properties. In this paper, the ability of the network to adapt in response to node failures or application requirements is studied. In particular, the solution proposed encompasses the use of mobile intelligent agents as a basis for realizing dynamic adaptivity in situations of unexpected or erratic network behavior. In such cases, an appropriate opportunistic scheduling mechanism will be adopted to reduce the energy consumption according to the prevailing circumstances. The paper is structured as follows: Section 2 provides a further reflection on the need for adaptivity and includes a review of other work in this area. In Section 3, a description of MERLIN, an energy efficient MAC and routing integrated protocol for WSNs is described. The use of intelligent agents for realizing dynamic adaptivity is discussed in Section 4.

2 Motivation

As the deployment of WSNs increases, the number of applications dependent on their reliable operation likewise increases. Thus the need for a strategy for handling the irregularities and anomalies that will undoubtedly arise. As an example, consider the case of a network of sensors that notify a variation of temperature above a certain threshold. In case of fire, all nodes in the vicinity will continuously send messages, leading to possible network overload as well as increasing the energy consumption. Furthermore, another application may be required to effect a dynamic change in the sampling rate, depending on the frequency of incoming measurements. To address such scenarios and ensure that the WSN continues to operate in an optimum manner, it is proposed that the network be initially subdivided into virtual sectors for node localization. Using a lookup table, intelligent mobile agents can migrate to the relevant sector, or indeed migrate to as many nodes as necessary to ascertain the status of the network, and then proceed to identify and take the necessary corrective action. Although the literature offers many relative energy-efficient MAC protocols [2, 3, 4, 5], and routing protocols [6, 7], only few protocols are focused on the integration of different layers and functionalities as in [8]. Moreover, TDMA protocols like TRAMA [3] or EMACS [9] suffer from a very high latency of messages that makes them very slow to adapt to changes in the network.

Traditionally, there has been a strong research focus on the use of mobile agents in the telecommunications area. As well as network management [10], the areas of resource allocation and management have successfully demonstrated the potential of agent technologies for these tasks [11, 12]. The use of agents in WSNs is a logical continuation of this research.

3 The MERLIN Protocol

3.1 Assumption

Sensor networks communicate in a multi-hop fashion hence packets are relayed from one node to another until a gateway is reached. Their number is usually very low compared with the total number of nodes in the network. We assume gateways to be synchronized (i.e to know the "perfect time"), to be responsible for data collection from the network and to be the medium, through which the user can both access the network and make changes to the network parameters if necessary.

In order to integrate the MAC, routing protocols and to include a possible localization procedure as an upper layer, MERLIN is designed to optimize the following data traffic patterns:

To gateway transmission by node. Packets are sent to nodes located in the neighbouring zone closer to the gateway (i.e. lower zone).

Subnet flooding by gateway. Gateway packets are forwarded to all nodes in the subnet;

Local broadcast by node. Nodes send packets to all of the direct neighbors. No forwarding is performed;

Sector flooding. Gateway packets are sent to all the nodes located in the sector of the network specified. The latter data traffic pattern is the result of a new feature of MERLIN and will be described in detail in section 3.4.

3.2 Overview

The protocol MERLIN [1] integrates characteristics of MAC and routing in a simple, single architecture. Fundamental to MERLIN is the natural division of the network obtained when gateways start flooding the network simultaneously with an initialization message, for example init-msg, containing the transmitting time and the gateway ID number. At the beginning of the session, nodes are in receiving mode waiting for any message. Nodes receiving the initialization message are categorized as being in the first time-zone. The nodes will proceed to update the init-msg with their own node ID and forward it to further nodes. Once network flooding is complete, the network may be regarded as being subdivided into subnets each containing one gateway. Usually the gateway of reference is the closest one. Every subnet will then be organized into time zones. All nodes will belong to one time-zone only. Hence they can join the network by using the scheduling table provided. During this procedure, collisions can occur and some nodes may be notified later through an alternative path; this leads to an initial imprecision of the time-zone that will be corrected when the later messages are received. One of the primary benefits obtained from such an approach to flooding the network is the resultant time and space divisions of the nodes. This allows the potential reuse of the medium through effective scheduling tables as described in section 3.5.

3.3   MAC Features of MERLIN

In this section, we describe characteristic of Medium Access Control (MAC) of MERLIN, the composition of slots and mechanisms for collided packet recovery. The basic technique of collision avoidance is the carrier sense multiple access CSMA, between nodes in the same zone. In fact, such nodes follow a channel contention procedure before they start transmitting the packet. Every slot starts with a *contention period CP* of about 30 times shorter than the total transmitting period. Any node, willing to transit a packet, firstly follows its scheduling table before picking up a *random time Tr* during the contention period at the beginning of its scheduled slot. The node will monitor the channel until Tr and then start transmitting if the channel is free. If the channel is busy then the packet is rescheduled for the next assigned slot. At the end of this procedure, the node switches to sleeping mode. The slot composition must also make provision for a *collision report period CR*, located between the end of the slot and the next CP. At such a time, the transmitting node switches to receiving mode so as to obtain a possible collision report, which is in the form of a short burst message. The collision report period has a double function:

- A collision can be notified back in order that the transmitter may reschedule the packet;
- Receiving nodes, which will apply CSMA to forward the packet, will implicitly acknowledge back to the transmitter their successful reception.

Finally, as required by some applications like in [13], MERLIN has a buffer where messages are stored while awaiting dispatch.

3.4   Routing Properties of MERLIN

The division of the network into time-zones has the advantage of generating an implicit routing to the nearest gateway by means of the data-traffic field contained in each packet. Initially, nodes store their number of hop-counts to the nearest gateway. The hop-count number identifies the node's time-zone. Packets can only be forwarded in two directions: towards the gateway, or away from it. Furthermore, a local broadcast data type is also possible. Gateways are considered to be located in zone 0 with respect to nodes in their subnets. Nodes receiving a packet will forward it to nodes in a zone either one level higher or one level lower according to both the packet type and the scheduling tables.

Virtual Sectors in MERLIN. In this section, the concept of *Virtual Sectorization* of the network as a novel feature of MERLIN is introduced. Virtual sectors are generated by simply using the CSMA approach already in use by nodes in the same zone. During initialization, nodes in zone 2, after receiving the init-msg from zone 1 nodes, will generate a *sectorID* based on the parent node ID in zone 1 and then proceed to flood the network with their *sectorIDmsg*. The sectorID is a unique number related to the zone 1 node ID-number. Because of the Contention Period (CP), only some nodes that are far apart can win the

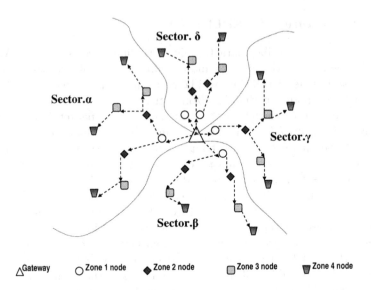

Fig. 1. The division of the network into sectors generated by zone 2 nodes by using the parent node ID

channel simultaneously and then transmit. Receiving nodes in zone 1 will set their sectorIDmsg, then switch to sleeping mode. The rest of receiving nodes will now set both their time-zone as 3, generate their sectorID and then forward the message to further zones. The new init-procedure incorporates a *notify-msg* back from every node containing: node ID, timeZone and sectorID. For further information see Fig. 1. The notify-msg will enable gateways to address the correct location when looking for a specific node or a group of nodes.

The Table of Neighbours. The protocol MERLIN support a relative low mobility of nodes. As a result, every node must maintains a table of neighbours. The table is updated any time a node receive a packet from a new neighbouring node. Moreover the protocol includes periodical broadcast for the table of neighbours to be refreshed. For each neighbour, the table maintains the following information:

- Nodes ID;
- Time-zone;
- SectorID;
- Energy level.

As described in section 4.1, the table of neighbours is an important resource which facilitates network adaptation.

3.5   X and V Scheduling

MERLIN has been designed to support several scheduling tables that can be opportunistically switched when network conditions change. As described in [14],

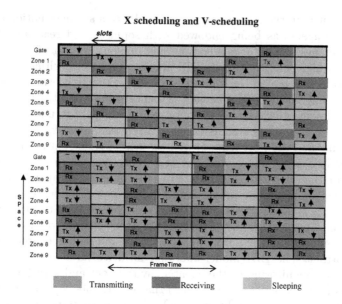

Fig. 2. Scheduling tables of MERLIN called X and V scheduling implemented and tested on the OmNet++ simulator

the X and V schedule tables in figure 2 have been studied and tested. Results derived from the implementation of the scheduling showed that the performance of the X-scheduling than twice that of the V-scheduling in terms of latency of messages and throughput. On the contrary, the V scheduling performs better in terms of network lifetime. Simulations reported a maximum setup time of 5 and 10 seconds for the X and V scheduling respectively. In case of a user request or a sudden application alteration, it is obvious that a facility for the autonomous and dynamic switching of scheduling is desirable. How this can be achieved is described in the next section.

4   Enabling Adaptive Scheduling Within MERLIN

Realizing adaptive scheduling calls for a solution that is flexible, efficient and responsive. Potentially, a number of approaches exist that exhibit these characteristics. However, the intelligent agent paradigm was identified as one that encompassed the necessary criteria. Intelligent agents by their very nature imply a number of inherent characteristics. These include amongst others:

Autonomy: Agents act autonomously without a need for explicit interaction;
Reactivity: Agents can listen for and react to external events and changes in operating conditions;
Proactivity: Agents seek to be proactive in fulfilling their tasks;
Mobility: Agents have a capability to migrate to different nodes in the network while fulfilling their objectives.

Some researchers in the AI community subscribe to a stronger notion of agency and envisage agents as being endowed with sophisticated reasoning facilities. One popular and computationally tractable implementation of such agents is that of one which conforms to the BDI architecture [15]. Such agents maintain a mental state that may include information about themselves and their operating environment. In the BDI scheme, the mental state takes the form of beliefs. Naturally all agents have a number of objectives or tasks to undertake. In BDI parlance, these are represented as desires. In practice, an agent can only realize its desires under certain predefined circumstances. When a situation arises that an agent is in a position to fulfill one of it objectives (or desires), it proceeds to do so. In the BDI case, the objective is formulated as an intention which the BDI agent immediately proceeds to fulfill. When realizing a software solution around agents, it is the software designer's prerogative to judge as to the necessity and appropriateness of the various agent characteristics. In the case of adaptive scheduling, the autonomous, reactive and mobile characteristics are most important. The BDI model offers an intuitive mechanism for modeling the continuous monitoring of the agent's environment and specifying the various criteria (usually in the form of rules) that determines certain agent actions, for example, under what conditions should the agent migrate to a certain node and under what circumstances should it commence a procedure to switch to a different scheduling model.

How a Change in Scheduling is Effected. The X and V scheduling tables have the same slot length, same frame time and same number of zone as depicted in figure 2. As a result,they can be interchangeable under certain conditions and timing. Such an expedient makes the agent able to order a change of scheduling after the related parameters are identified, whenever it is considered appropriate. The agent can order a change of scheduling for the entire sector or a portion of it, for example from the zone N to the zone N+M in a sector. Such a situation will imply for nodes in the border zones (N and M) an adoption of both the old and the new scheduling to keep the continuity of message flow. In order to have a simultaneous scheduling adoption for the entire group of nodes involved, the migrating agent should firstly identify the number of zones that will join the change, secondly calculate the overall time necessary so that the packet can be forwarded to all the nodes interested, through the graph provided in [14]. Finally, the agent should generate two messages for:

1. *AdoptSector(V-SCHEDULING)* or *AdoptSector(X-SCHEDULING)*, the *Time* at which the event occurs and *timeZones* involved.
2. *AdoptSector(V-SCHEDULING)* & *AdoptSector(X-SCHEDULING)*, the *Time* at which the event occurs and *timeZones* of nodes in the border.

4.1   Agents in MERLIN

In MERLIN, agents reside at the gateway and continuously monitor the status of the network. Information monitored includes:

- Total number of messages per minute received on average: *totalReceptRate* that exceeds a certain threshold *NetThreshold*
- Number of messages per minute received from an individual sector: *sectorReceptRate*
- Number of messages per minute received from an individual node: *nodeReceptRate*
- Percentage of data with the same sensed value received from a node, a sector, or a zone;
- Percentage of message with a value over a certain threshold received from a node, a sector or a zone: *NodThreshold*
- Changes to the sampling rate used in a node, sector or zone.

Though dependent on the purpose of the WSN, usually a number of scenarios could potentially arise that either require an alteration to the scheduling currently employed by the network as a whole, or more likely, in a sector or a zone. In the latter case, an agent would migrate to a particular node in order to investigate further. As an illustration of this, consider the following scenarios (note that the use of Agent Factory notation [16, 17] is used to express the various commitment rules):

1. The total number of messages received at the gateway from all the nodes in the subnet increases significantly in a short space of time. Such a situation implies that nodes are increasing their energy consumption considerably. Thus the agent needs to review the operation of the network and, if possible, identify a strategy for optimizing the longevity of the network. As previously mentioned, V scheduling is more suitable in situations where ensuring longer network lifetime is a priority. Therefore, the agent, based on parameters received, can make a unilateral decision to flood the entire network with an order to adopt V scheduling. In such circumstances, it does not need to migrate to any nodes in the network and can take the decision while remaining at the gateway. A commitment rule that it might adopt in such circumstances could be as follows:

$$BELIEF(totalReceptRate(?val))\&BELIEF(NetThreshold(?trigger)$$
$$\&BELIEF(currentScheduling(X\_SCHEDULING))$$
$$\Longrightarrow$$
$$COMMIT(Self, Now, Belief(True), AdoptGlobal(V\_SCHEDULING))$$

2. The agent can from the gateway order a change in one more sectors that present an anomaly (e.g an extraordinary increase of the sector reception rate), then the commitment rule would be:

$$BELIEF(sectorReceptRate(?val))$$
$$\&BELIEF(SectorThreshold(?trigger)$$
$$\&BELIEF(currentScheduling(X_S CHEDULING))$$
$$\Longrightarrow$$
$$COMMIT(Self, Now, Belief(True), AdoptSector(V\_SCHEDULING, ?T)$$

3. The number of messages received from a particular node increases beyond a predefined threshold. Clearly, this warrants further investigation. The agent can then decide to migrate by using the *nodeSector*, *nodeTimeZone* and *nodeID* provided at the gateway. The triggered commitment role is:

$$BELIEF(nodeReceptRate(?val, ?nodeID))$$
$$\&BELIEF(NodeThreshold(?trigger$$
$$\&BELIEF(NodThreshExceed(TRUE)$$
$$\Longrightarrow$$
$$COMMIT(Self, Now, Belief(True),$$
$$SEQ(Migrate(?nodeSector), Migrate(?nodeTimeZone))$$

4. On arriving at the relevant node, the agent proceeds to examine a number of parameters including:

   – The number of messages in the buffer;
   – The values of the messages encountered whether exceed a percentage of similarity *similPercent*.

By using the node's table of neighbors, an agent can determine what nodes are nearby and migrate to each of these nodes to acquire further information concerning their status. For instance, the commitment rule to migrate to the neighbouring node is:

$$BELIEF(nodereceptRate(?val, ?nodeID))$$
$$\&BELIEF(NodeThreshold(?trigger$$
$$\&BELIEF(NodThreshExceed(TRUE)$$
$$\Longrightarrow$$
$$COMMIT(Self, Now, Belief(True), Migrate(?nodeID))$$

Should the agent require, (in case it has enough information to believe that one or more sectors present an anomaly), it can order a change of node scheduling at time T or an adoption of the two scheduling tables for nodes in the border zones as described in section 4. Consequently, the agent can commit to further migrating to a node in a further time zone or sector; In such a case the agent determines the *timeZone*, *sectorID* and prevailing scheduling mechanism in use before proceeding to migrate. The related sequence of commitment rules is:

$$BELIEF(BufferSize(?val))$$
$$\&BELIEF(similPercent(?threshold))$$
$$\Longrightarrow$$
$$COMMIT(Self, Now, Belief(True),$$
$$SEQ(AdoptZone(?ZoneID, V\_SCHEDULING, ?T),$$
$$Migrate(?timeZone + 1, ?sectorID, ?nodeID)))$$

5 Conclusion

This paper has investigated the use of intelligent agents in the delivery of adaptivity at the networking layers. This is achieved by using two preexisting technology sets developed in part by the authors; these are: the energy-efficient integrated MERLIN protocol and Agent Factory a rapid prototyping environment for agent deployment. Effective and efficient use of limited energy resources is of paramount importance within WSNs. This research has described a method of optimizing energy resources in times when unexpected or heavy network activity occurs. Three instruments facilitate this: the provision of two efficient and interchangeable scheduling tables; the ability to generate virtual network sectors; the adoption of autonomous mobile agents. Such agents offer the deductive apparatus by which WSN adaptivity is delivered. Agents monitor network activity and determine which of the two scheduling regimes would be most appropriate at the network either level or alternatively through the creation of virtual network sectors at the sector level. Autonomous agents can deliberate and dynamically apply the respective schedules at either network or sector level. Determination of the network context in order to inform such decisions will often necessitate agent migration. This approach maximizes network performance while minimizing energy usage. Agent characteristics of autonomy, social ability and mobility suggest that they represent an intuitive choice for this task. Ongoing research is investigating utility functions, which underpin network adaptivity based on incomplete, localized, conflicting, or partial network information.

Acknowledgments

Gregory O'Hare, Antonio G. Ruzzelli and Richard Tynan gratefully acknowledges the support of Science Foundation Ireland under Grant No. 03z/IN.3/1361. Michael O'Grady gratefully acknowledges the support of the Irish Research Council for Science, Engineering & Technology (IRCSET) though the Embark Initiative postdoctoral fellowship programme.

References

1. Ruzzelli, A., Evers, L., Dulman, S., Hoesel, L.V., Havinga, P.: On the design of an energy-efficient low-latency integrated protocol for distributed mobile sensor networks. IWWAN International Workshop on Wireless Ad hoc Networks (2004)
2. Ye, W., Heidemann, J., Estrin, D.: Medium access control with coordinated adaptive sleeping for wireless sensor networks. Twenty-First AnnualJoint conference of the IEEE Computer and Communication Societies (INFOCOM) **3** (2002) 1567–1576
3. Rajendran, Obrazka, Garcia-Luna-Aceves: Energy-efficient, collision-free medium access control for wireless sensor netwoks. Conference on Embedded Networked Sensor System (2003) 181–192
4. Dam, T.V., Langendoen, K.: An adaptive energy efficient mac protocol for wireless sensor networks. ACM Sensys (2003)

5. Hoiydi, A.E., Decotignie, J.: Wisemac: An ultra low power mac protocol for multi-hop wireless sensor networks. In: Proceedings of the First International Workshop on Algorithmic Aspects of Wireless Sensor Networks (ALGOSENSORS 2004), Lecture Notes in Computer Science, LNCS 3121. (2004) 18–31

6. Johnson, D.B., Maltz., D.A.: Dinamic source routing in ad hoc wireless networks. Mobile computing **353** (1996) Kluwer Academic publishers.

7. Akan, O.B., Akyildiz, I.F., Sankarasubramaniam, Y.: Event-to-sink reliable transport in wireless sensor networks. IEEE-ACM Transactions on Networking (2004)

8. Lu, G., Krishnamachari, B., Cauligi, Raghavendra, S.: An adaptive energy-efficient and low-latency mac for data gathering in sensor networks. International workshop on Alghoritms for Wireless, Mobile, ad Hoc Sensor Networks (WMAN 04) (2004)

9. Hoesel, V., Chatterjea, Havinga: An energy efficient medium access protocol for wireless sensor networks. proRISC 2003 (2003)

10. Bieszczad, A., P.B., T., W.: Mobile agents for network management. IEEE Communications Surveys **1** (1998)

11. Haque, N., J.N.R.M.L.: Resource allocation in communication networks using market-based agents. International Journal of Knowledge Based Systems (2005)

12. Wang, Y., C.L.B.J.: Intelligent radio resource management for ieee 802.11 wlan,. IEEE Wireless Communications and Networking Conference (WCNC), Atlanta, Georgia USA (2004)

13. Kevin Mayer, K.T., Ellis, K.: Cattle health monitoring using wireless sensor networks. The 2nd IASTED International Conference on Communication and Computer Networks, Cambridge Massachusetts (2004) 8–10

14. Ruzzelli, A., Tynan, R., G.M.P.O'Hare: A low-latency routing protocol for wireless sensor networks. To appear on SENET'05 Advanced Industrial Conference on Wireless Technologies. Montreal (2005)

15. Rao, A.S., G.M.: Modelling rational agents within a bdi architecture,. Principles of Knowledge Representation. and Reasoning, San Mateo, CA. (1991)

16. G.M.P., O.: Agent Factory: An Environment for the Fabrication of Multi-Agent Systems, in Foundations of Distributed Artificial Intelligence. John Wiley and Sons (1996)

17. Collier, R.W., O.G.L.T.R.C.: Beyond prototyping in the valley of the agents, in multi-agent systems and applications iii. Proceedings of the 3rd Central and Eastern European Conference on Multi-Agent Systems (CEEMAS'03), Prague, Czech Republic **3** (2003) Lecture Notes in Computer Science (LNCS 2691), Springer-Verlag.

Keynote Talk Summary: Algorithmic Aspects of Sensor Networks

Paul Spirakis

Research Academic Computer Technology Institute,
University of Patras, Greece
spirakis@cti.gr

In this talk, we discuss some abstract models of sensor networks and also some basic algorithmic problems. We first go through the basic architecture and communication capabilities of a single, ultra small sized, sensor. We distinguish broadcast capability (at a certain radius r around the sensor) and directional broadcasting at an angle (which can be reduced to almost zero in case of optical communication). Based on such features, we first go through the (well known) model of Random Geometric Graphs and its threshold properties for connectivity, average degree, chromatic number etc. Then, we define the model of Random Sector Graphs, in order to capture unidirectional sending at an angle. We provide thresholds for connectivity, clique and chromatic number there also and point out its differences from the Random Geometric Graphs model. We also discuss other models related to ways of configuring sensor nets, such as "for each point (sensor) connect to its k nearest neighbours" or "for each sensor, connect to k points within its communication range r". We also refer to the Combinatorial model of Random Intersection Graphs (which abstracts the geometry out). It is interesting to see that some properties (e.g., sensors density necessary for connectivity) are "invariant" in all these models. We then go through some fundamental algorithmic issues and their existing solutions. We first discuss the problem of finding efficient local protocols aiming in propagating a local event E to a sink. We discuss several efficiency measures like hop count, energy spent, short paths. Randomised local protocols are then sketched, useful for robust and energy efficient local event reporting in case of a sensor net with failures at some nodes (or with energy depleted nodes). In this framework we indicate some energy-time tradeoffs for propagation of local information and notice their similarity with tradeoffs between area and time in VLSI circuits. The problem of propagation of local information in areas with obstacles (lakes) is also discussed. We then examine efficient routing policies and especially refer to greedy techniques in the case of existence of virtual coordinates. The topology control problem is the next topic. We present several local protocols for configuring the net and establishing global properties like connectivity or short paths. Finally, we discuss problems related to scenarios where a sensor net is "embedded" into a larger fixed-connectivity network. There, we provide the abstraction of a "sensors cloud" and discuss the possible enhancement of properties like maximum flow or connectivity or diameter of the resulting hybrid network. Our talk aims in showing the emergence of a new algorithmic subfield, useful to the pragmatic considerations in the actual design and control of such networks.

I. Stavrakakis and M. Smirnov (Eds.): WAC 2005, LNCS 3854, p. 277, 2006.
© IFIP International Federation for Information Processing 2006

Invited Talk I Summary: Opportunistic Spectrum Access for Wireless Ad Hoc Networks: Research Challenges*

Cesar A. Santivanez

Internetwork Research Department, BBN Technology, Cambridge, MA 02138, USA
csantiva@bbn.com

Traditionally, the frequency spectrum has been rigidly allocated to users/services. This rigid allocation has led to inefficient utilization and an apparent scarcity [1].

More recently, technological advances in a number of areas (software defined radios, wideband sensing, DSP receivers and waveforms agility) have enabled the development of a new communication paradigm, namely Opportunistic Spectrum Access (OSA) that promises to eliminate the apparent scarcity problem.

In OSA, wireless nodes' spectrum usage is not pre-determined (wired in hardware) with a fixed frequency/modulation assignment, but instead radios become aware of their environment, in particular of the presence of "primary" or "protected" spectrum users, and based on this decide on a spectrum usage that is compatible with the regulatory policy in effect at the current place and time.

OSA promises a significant improvement on spectrum utilization. However, while conceptually simple, OSA turns out to be a very complicated concept to realize, especially under a dynamic mobile ad hoc network where the decisions need to be taken on a distributed and autonomous manner. We revise current efforts underway to realize the OSA vision. In particular, we cover work on two enabling blocks for OSA in a distributed ad hoc network: policy-driven operation, and algorithms for coordinated spectrum allocation.

1 Policy Driven Operation

A radio operation is subject to rules or policies. Such rules are typically issued by government regulators, and are intended to avoid or reduce interference among users. For instance in the USA, radio equipment is tested and certified to fulfill FCC emission regulations (policies) before they are put into operation.

Now consider an opportunistic radio, able to transmit and receive in various forms on a number of frequencies. How do we ensure that the radios behavior is always consistent with the policies in effect on a particular place and time? How can we assure regulators and primaries that a radio will behave in accordance with established policy without building a custom radio for every situation?

One approach to solve this problem [2] is to divide the policy-driven operation into two modules: a simple YES/NO validator, namely Policy Conformance

---

* This talk was supported in part by the IST FET Coordination Action ACCA (IST-6475).

I. Stavrakakis and M. Smirnov (Eds.): WAC 2005, LNCS 3854, pp. 278–280, 2006.

Reasoner (PCR) and a more sophisticated policy Strategy Reasoner(SR). The PCR determines whether a given emission profile (power, frequency, etc.) is valid given the policies currently in place and the environment conditions (e.g. primaries present/absent). The PCR will have to be consulted each time a packet needs to be transmitted. The SR, on the other hand, reads and *reasons* about policy and based on this and its mission goals/constraints searches the opportunity space to determine opportunities to explore. The separation:

- Significantly reduces the accreditation burden. If we have m regulatory policy sets, and n radio parameters/methods (e.g. "sense in-band received power", or "set transmit power") only $n + m + 1$ accreditation steps are needed (1 to accredit the PCR, m to accreditate the policy sets, and n steps to accredit the radio methods/sensors). Without the clear PCR/SR separation, all the nxm combinations would have to had been tested/accredited.
- Provides a well defined interface for accessing radio state.
- Policies do not tell the radio what to do, they only define what is a valid usage of spectrum.
- PCR is light-weight, radio-independent and reusable. It can handle the load in a per-packet basis.
- Provides support for current and future implementations decoupling accreditation from innovation. Technology can be developed in advance to policy. Policy interactions can be worked out in advance to deployment.

From the above, it is clear that a language to express policies is needed. Such policy language must not only be able to handle the complexity of current spectrum policy (that evolved as a patchwork written for human interpretation) but also be extensible to future ones. This language must support a logical framework for validation of completeness and consistency of policies, and verification of policy-conformant usage. To this end, BBN developed a language [3] based on DAML/OWL [4]. This declarative language – following knowledge representation and rule-based approaches – enables deductive inference, allows reification, inheritance, and extension. It has inference and theorem proving support.

The main challenge left in this area corresponds to the design of an SR module that performs cognitive optimization of device operation by efficient search and prune of combinatorial decision space. Indeed, finding an algorithm that produces an optimal solution for any possible policy set is extremely hard. However, fast system-dependent optimizations are possible by reducing the search space to a smaller set of good candidates (based on either *a priori* knowledge or radio capabilities/shortcomings or on pre-defined semantical properties of opportunities – forfeiting exploiting opportunities that do not conform to them).

2   Algorithms for Coordinated Spectrum Allocation

Different nodes located at different locations will encounter different sets of transmission opportunities. For example, nodes closer to a primary node will not be able to transmit at that primary assigned frequency while nodes further away may transmit at a low power. Overall, nodes will need to exchange their transmission opportunity information and jointly decide on which (common) subset of them to use to communicate.

One way to solve the bootstrapping problem associated with disseminating opportunity information over links built based on this information, is to have a small common channel dedicated to coordination. Such a channel will be small, so special care will have to be taken to prevent overloading it. Among the techniques used to alleviate the "coordination channel" load are : (1) limiting the scope/granularity (i.e. resolution) of the opportunity information dissemination and (2) increasing the MAC achievable throughput by exploiting the periodicity of most of the control packets generated to determine loose-schedules (rendezvous times) that limit/prevent collisions.

BBN designed a complete OSA-based system employing the above mentioned-techniques[2] and conducted a set of experiments to explore the fundamental trade offs in such a system. Among the main results are that:

- Topology control has a higher impact in performance on a OSA-based system than it has in conventional networks.
- A small (5 dB) increase in interference tolerance by primaries unleashes a large increase in total capacity for OSA users.
- Even under full deployment of primaries - provided that they have long range links - *underlaying* (i.e. transmitting simultaneously with the primaries but at a much smaller power, i.e. equivalent to "whispering") allows to achieve the similar (high) capacity gains as under partial deployment of primaries.
- That for the class of carrier-sensing MACs, a small margin in the maximum transmit power – tied to the carrier sensing threshold – is enough to prevent the combined interference from a group of OSA users from exceeding the tolerable interference at the primaries. Therefore, policies can be written from a single-user perspective, as long as the proper margin is included.

Lastly, here is an area in need of much research. One of the most challenging open problems is that of developing a Common Control Channel Acquisition Protocol (CCAP) that is able to adaptively build a common channel to use for coordination between OSA users, without relaying on a dedicated one. The other extremely challenging problem is that of performing optimal frequency allocation to satisfy traffic or mission requirements. This is a cross layer problem that implies joint power/rate/frequency/time scheduling and routing.

References

1. Federal Communications Commission, "Spectrum Policy Task Force: Report," ET-Docket 02-135, November 2002, http://www.fcc.gov/sptf
2. R. Ramanathan and C. Partridge, Final Report: XG Architecture and Protocol Development, 30 December 2004.
3. http://www.ir.bbn.com/projects/xmac/pollang.html
4. OWL Web Ontology Language Guide, W3C. http://www.w3.org/TR/2004/REC-owl-guide-20040210/

Invited Talk II Summary: Incentive Schemes in Memory-Less P2P Systems*

Costas Courcoubetis

Department of Computer Science,
Athens University of Economics and Business,
76 Patision str., Athens, GR 10434
courcou@aueb.gr

The asymptotic analysis of certain public good models for p2p file sharing systems focusing on content availability, suggests that when the aim is to maximize social welfare, a fixed contribution scheme in terms of the number of files shared per unity of time can be asymptotically optimal as the number of participants n grows to infinity (see [1] and references therein). Such an incentive scheme is very simple and attractive, and is also suitable for other p2p applications with similar public good characteristics such as WLAN peering. However, its enforcement is not straightforward in cases where no trusted software or central entity accounting for peers' transactions can be assumed and peers are free to change their identity with no cost. That is, when no sort of user memory is available to be able to identify and punish the potential free riders.

A 'memory-less' p2p system should rely only on the time peers are consuming resources to ensure that they contribute adequately. BitTorrent is an example of a successful real world application focusing on bandwidth provisioning for content distribution, which implements a reciprocative incentive scheme without relying on past transactions of peers but on a direct exchange of resources (i.e. upload bandwidth). BitTorrent, however, does not tackle the objective of improving content availability.

Recent articles in the popular press discuss the importance of the 'long tail' of content; that large part of the set of content in which individual files are not popular, but which together constitute the majority of the total requests. The provision of this part of the content in a p2p system requires different types of incentives than the ones usually discussed in the p2p economics literature, which look at uploading cost rather than the cost for contributing to the overall content availability.

We therefore choose to consider a system where the probability of a certain file being requested is low but the overall value of satisfying such requests is much greater than for popular items (since it is more difficult to find unpopular or rare items, in many cases even if one wishes to pay for them). Thus we will attempt to give incentives for providing any content item regardless of request rate, considering uploading cost to be of limited importance, especially since we require that this cost is incurred only while peers are consuming resources. But

---

* Supported in part by the IST FET Coordination Action ACCA (IST-6475).

I. Stavrakakis and M. Smirnov (Eds.): WAC 2005, LNCS 3854, pp. 281–282, 2006.

we also need to give incentives to remain in the system and providing content, not simply to have provided at some time in the past.

We thus propose the following "contribute while consuming" incentive mechanism enforced by uploading peers, which

1. check that the downloaders share a predefined number of valid files
2. use a certain (not too low) upload throughtput in order to ensure that these files are made available for a significant amount of time.

Of course, there are some important implementation and incentive issues that arise in this context (e.g. ensuring the validity of files shared, the need for super peers in order to avoid the requirement of cycles of requests to be formed, and more), which are discussed in detail in [2]. In any case, we believe that time spent in p2p systems will become a critical parameter of the contribution of the participating peers, and especially in the case of file sharing as access speeds increase and people store more content in their PCs for their own use.

So, in [2] we made a first effort to formulate a corresponding economic model in order to provide a sound theoretical framework for the study of the qualitative characteristics of our memory-less incentive mechanism and provide insights for the appropriate tuning of its basic parameters (the number of files shared and the upload throughput used by all peers in the system). Our results show that the resulting efficiency is comparable to the one achieved using the theoretically optimal schemes, which is very encouraging taking into account the very limited implementation requirements of this mechanism.

Our on-going work includes the in-depth analysis of the proposed economic model (using both simulations and analytical tools). We also wish to explore to what extend our public good model and/or memory-less enforcement mechanism are applicable in other p2p systems with public good characteristics.

References

1. Courcoubetis, C., Weber, R.R.: Incentives for large p2p systems. Accepted for publication in IEEE Journal on Selected Areas in Telecommunications, available at http://nes.aueb.gr/p2p.html (2005)
2. Antoniadis, P., Courcoubetis, C., Strulo, B.: Incentives for Content Availability in Memory-less Peer-to-Peer File Sharing Systems. ACM SIGecom Exchanges 5(4) (2005) 11–20

Invited Talk III Summary: Coordination and Resilience in Wireless Ad Hoc and Sensor Networks

Leandros Tassiulas

Computer and Communications Eng.,
University of Thessaly, Volos, Greece
leandros@inf.uth.gr

In wireless adhoc and sensor networks a close synergy and coordination is required among entities at different layers of the network architecture to achieve the robust behavior that is expected from these systems in the potentially harsh environments where they may operate. The volatile wireless channel, the unpredictability of traffic due to unknown traffic generation scenarios as well as variability of the network topology itself due to mobility and node failures set a challenging stage for the network designer. A mathematical network model that captures the interaction of mechanisms at the different layers, from physical to transport as well as the intricacies of the time varying network topology was considered in [1, 2, 3] and refined and generalized later in several other papers. A brief description of that model is as follows. All the physical and access layer parameters including power selection, channel allocation, coding rate etc are collectively represented through a vector $I(t)$. The relevant parameters of the environment that affect the communications as well as the topology of the network itself are represented collectively by the topology state variable $S(t)$. The topology state might not be fully available to the access controller, who may observe only a sufficient statistic of that. The collection of bit rates of all the communicating pairs of nodes at each time, i.e. the communication topology, is represented by a function $C(I(t), S(t))$ where $I(t)$ is selected by the physical/access layer controller. Over the virtual communication topology the traffic flows from the origin to the destination according to the network and transport layer protocols. Packets may be generated at any network node having as final destination any other network node potentially several hops away. The network control mechanism determines the access control vector and the traffic forwarding decisions in order to accomplish certain objectives. An important performance attribute is the capacity region of the network defined as the set of all end-to-end traffic load matrices that can be supported under the appropriate selection of the network control policy. That region is characterized in two stages. First the ensemble of all feasible long term average communication topologies is characterized. The capacity region includes all traffic load matrices such that there is a communication topology from the ensemble for which there is a flow that can carry the traffic load and be feasible for the particular communication topology. An approach to characterize the performance of a control policy for the network is by the policy capacity region, i.e. the collection of traffic load

I. Stavrakakis and M. Smirnov (Eds.): WAC 2005, LNCS 3854, pp. 283–285, 2006.
© IFIP International Federation for Information Processing 2006

matrices that are sustainable by the specific policy. The larger the capacity region is the better the performance will be since the network will be stable for a wide range of traffic loads and therefore more robust to traffic fluctuations. Such a performance criterion makes even more sense in the context of wireless ad-hoc and sensor networks where both the traffic load as well as the network capacity may vary unpredictably; in that case robustness is a valuable attribute. That perspective to the control of the network was introduced in [1]. A control policy was proposed there that achieves the objective in an optimal manner optimally since it has a capacity region that coincides with the capacity region of the network and is therefore a superset of the capacity region of every other policy. The selection of the various control parameters from the physical to transport layer is done in two stages in that policy. In one stage all the parameters that affect the transmission rates of the various wireless links are selected while on the other the assignment of the traffic classes to the different connections is done. Its description though is facilitated by starting with the traffic forwarding part first. Each traffic class is routed such that the backlog of the class is balanced across the network at each time. The traffic of class k backlogged at node i is forwarded to downstream nodes with smaller backlogs, towards equalizing the load while the flow is throttled towards downstream nodes with higher backlogs for the same reason. The link capacity is allocated to the different traffic classes waiting for transmission through the link to the benefit of the traffic class with most unevenly distributed backlog. More specifically, through the link from node i to node j the traffic class with larger difference between the backlogs at i minus that of j is given priority for transmission. Based on the above considerations a weight w_{ij} is determined for link (i, j), indicating how much the backlog distribution will be uniformized by the transmission through the link (i, j). If the link capacity is fixed and independent of allocation decisions in neighboring links, as is the case in wireline networks, the above resource allocation rules are adequate for traffic control to stability. That is not the case though for a wireless network where the link capacity is determined by the access control vector selection $I(t)$. Effectively a bandwidth allocation decision to the different links is done that way and the bit rates $C(I(t), S(t))$ are specified at t. This is done such that the links with higher backlog weight w_{ij} are favored in their neighborhood and they are given a higher rate C_{ij} through the selection of the physical and access layer parameters. More specifically $I(t)$ is selected such that the sum of the resulting bit rates C_{ij} weighted by the corresponding weights w_{ij} is maximized. Since the bit rate of the link usually depends on the access parameters in a complicated way while the links interact due to interference, the optimization for different links may need to be performed jointly. As a result the access optimization problem might be both computationally hard and it may require centralized coordination. Several subsequent works focused on dealing with the challenges posed by implementable distributed versions of the policy. In various occasions a wireless network might be operating in overload conditions, i.e. outside of its stability region as defined above. A smooth and balanced system response in those stressful situations is essential for effective crisis management in the network. That problem is studied in [4]. A network consisting of an arbitrary spatial arrangement of nodes is considered where information may be generated at any node in the network and needs to be forwarded to a collection of hub (sink) nodes. When the traffic load lies outside the feasibility region of the system, there is no feasible flow to transfer the information to the sinks, given the capacity of the system. In that case traffic backlogs will occur in the nodes. The distribution of the backlog build-up is an indication of the behavior of the system. A fluid model is considered in [4] where the information flow induced by the routing policy is represented by superflows. A superflow is a generalized notion of flow, where the aggregate incoming flow in a node may exceed the outgoing. The difference of incoming minus the outgoing flow from a node is the backlog buildup rate at the node. That difference is called "node overload". The vector of node overloads under a certain routing policy is the quantitative performance objective that represents the overload response of the network to the routing policy. It is shown in [4] that in the space of node overload vectors there is one that is lexicographically minimal and is characterized. The overload corresponding to this vector also maximizes the information rate that reaches the sinks. Furthermore it is shown that this vector is the unique solution for a wide class of optimization problems where the optimization objective function is the sum of any non-decreasing convex function of node overloads. That vector is called "most balanced" overload vector and any superflow that induces the most balanced overload vector, "most balanced" superflow. A distributed adaptive superflow reallocation policy converging to a most balanced superflow is presented finally. That initial work sets the framework for studying the overload behavior of other wireless adhoc network architectures as well, towards more resilient wireless networks.

References

1. L. Tassiulas and A. Ephremides, "Stability properties of constrained queueing systems and scheduling policies for maximum throughput in multihop radio networks,"IEEE Transactions on Automatic Control, vol. 37, no. 12, pp. 1936–1949, December 1992.
2. L. Tassiulas and A. Ephremides, "Dynamic server allocation to parallel queues with randomly varying connectivity," IEEE Transactions on Information Theory, vol. 39, no. 2, pp. 466–478, 1993.
3. L. Tassiulas, "Scheduling and performance limits of networks with constantly changing topology,"IEEE Transactions on Information Theory, vol. 43, no. 3, pp.1067–1073, 1997.
4. L. Georgiadis and L. Tassiulas, "Most balanced overload response in sensor networks," in IEEE International Symposium on Information Theory, Adelaide, Australia, September 2005.

Panel 1 Report: Autonomicity Versus Complexity*

Ioannis Stavrakakis and Antonis Panagakis

National & Kapodistrian University of Athens,
Dept. of Informatics & Telecommunications,
Panepistimiopolis, Ilissia, 15784 Athens, Greece
{ioannis, apan}@di.uoa.gr

Abstract. The first panel in WAC2005 focused on the relation between autonomicity and complexity. It is widely believed that autonomicity is a principle that can reduce complexity, but there is also concern that autonomicity itself is complexity-producing. Autonomicity promotes all "self-*" attributes of a system and naturally distributes responsibilities and costs, but it can also bring the system close to a state of "anarchy" (modern Greek interpretation of "autonomous") if not properly handled. It appears that the overall system complexity may increase, but it is distributed and shared (hence, it is potentially easier to manage), in a similar way in which Integrated Circuits encapsulate the increased complexity and hide it from the bigger system. In addition to reducing complexity in the above sense, autonomicity can also help design truly adaptable, self-tuning and "all-weather" near-optimal systems, something not possible under traditional system design that are difficult to cope with the combined fine-tuning of a very large number of parameters.

The panel was composed by the following researchers from Academia, Research Organizations and the Industry: Paul Spirakis of University of Patras - Research Academic Computer Technology Institute in Greece (coordinator), Radu Popescu-Zeletin and Mikhail Smirnov of Fraunhofer FOKUS in Germany, David Lewis of Trinity College Dublin in Ireland, Tom Pfeifer of Waterford IT in Ireland, Stefan Schmid of NEC Europe in Germany and Cesar Santivanez of BBN Technologies in USA.

According to [1], complexity may be understood in a number of different ways (e.g., computational complexity in computer science, emerging complexity in a physical system or process); the term is used to characterize a system that is hard to control (complicated nets, myriads of interactions) and might have a dynamic character (fast changes in huge structures, failures, or even updates that may "move" slower than the rate of changes).

Autonomicity is a word of Greek origin. It literally translates to "self-lawed" and in Modern Greek almost to "anarchy". For people in the networking field it means all the "self-*" properties, e.g. self-managed, self-configured, self-healing,

---

* This report was supported in part by the IST FET Coordination Action ACCA (IST-6475).

I. Stavrakakis and M. Smirnov (Eds.): WAC 2005, LNCS 3854, pp. 286–292, 2006.

self-organised, self-improving; this also includes "selfishness" and thus antagonism. Autonomicity is perceived to presume a local "intelligence" of some degree
and can be studied from a very low components level, up to the highest system
level.

Comparing the terms, complexity is both a problem and a property. It is
easy to "see" and hard to understand. Autonomicity, on the other side, is both
a method and a property. It might provide an answer to complexity or it may
create worse problems (chaos, anarchy ...).

Autonomic systems design is motivated by the fact that large systems disallow
global control and therefore, central management becomes impossible. Examples
of large systems that work nice are the market, the society and animal groups.
These examples indicate that an autonomic system may start from simple principles and that evolution helps it.

Nevertheless, convincing arguments are needed to provide answers to questions such as: Do we attempt to hide some problems via autonomicity "magic"?
For example, the following issues should be addressed in the near future:

- How far does "self-*" become implementable?
- How can we verify the correctness of a "self-*" implementation of a property?
- Can we convince that autonomic protocols are "stable" (think, e.g, about
  BGP)?
- What are the measures of quality of service in autonomic systems?
- Can we really design/derive self-improving code (and get rid of software
  designers)?

An additional issue is the exact difference between "autonomicity" and other
research areas. For example, the foundations of distributed computing have many
resemblances with autonomicity goals (e.g. Dijkstra's self-stabilizing code); thus,
the question whether modern distributed computing is the same as autonomicity
in communications, but just renamed, arises. In distributed computing local
protocols and communication are utilized in order to achieve global goals (e.g.,
leader election protocols, byzantine agreement). Also, many impossibility results
(a la FLP) indicate that not everything is possible.

Modern approaches to study complex systems include mathematics of local interactions from Physics, emerging nets/structures/behaviour theories and
evolutionary processes. Evolutionary game theory is mathematically very precise
and can be used to study evolution under antagonism. Under the framework of
evolutionary game theory, "dynamics" and structure are connected in a beauti-
ful way and individuals "learn" or even better "copy" behaviours from others.
Motivating locals for "better" global behaviour is a new way to control complex
systems; at the same time, it is also old, if one thinks of traffic lights, taxes,
or advertisement. Modeling and controling the time-varying aspects of complex
systems is a challenging task (dynamic control theory is obsolete).

The presentation in [2] focuses on the tradeoff of complexity vs. autonomicity.
Complexity and autonomicity can not be viewed as two separate notions, but as
the two sides of the same coin. On one hand, complexity calls for autonomicity
especially in large dynamic systems. Since such systems are very complex by

nature, they require autonomic support in order for them to be managed in an economic manner; otherwise, it would seem impossible to manage them. On the other side of the coin, achieving full autonomicity in large systems (like the Internet) is very complex in terms of network engineering. Since the problem is too complex to tackle as a whole, a possible approach would be to divide the problem into many sub-problems, solve the sub-problems individually, and then merge the sub-solutions into a solution to the entire problem. However, while trying to "divide" a large, complex system in order to "conquer" it, there is always the danger to "divide" the problem in the wrong way.

One possible way to proceed would be to start with a "bottom-up" approach. That is, to build simple autonomic components that solve certain aspects of the overall problem space and then try to put them together to facilitate a good overall solution. Then, one should try to address issues like the possible interference between autonomic components that are put together, the extent to which the combination of two autonomic components form an autonomic component, the functionality that might be still missing and the extent to which the composite might be optimal. Finally, the above approach would require the application of an iterative/evolutionary methodology.

In order to manage the interoperability issues among different autonomic entities, the point was made that standards would be needed. The interaction among these autonomic entities should happen at various levels of the system hierarchy, depending on what is made autonomic each time and hopefully at not that many different levels. In the initial phase of the development of an autonomic system the required additional complexity might lead to an increase in the required capital expenses, but it is expected to lead to reduced operating expenses since it reduces management cost in the long-run.

The presentation in [3] addressed the panel question by contributing a comparison with the development of complexity in a well-known mass-market consumer device - the TV set. A short overview of the television history was presented as an example for how electronics industry dealt with the increasing complexity of electronic circuits.

Here we find early devices from the 30's with a very limited number of active elements (vacuum tubes), where functional overload was commonplace, and the signal to be received was designed in a shape that could be decomposed with such simple circuits. Over time, with the active elements getting cheaper (transistors), they increase in number, providing supportive and stabilising functions to the core functional blocks. However, without the technological need anymore, traditional functional overload remains.

With the appearance of ICs in the 80's, functional blocks get treated as black boxes, only specified by their interface parameters. While the internal complexity increases to hundreds of active elements, the task for the system designer becomes simpler, due to encapsulation. Reliability improves.

Regarding the functional overload, it is eventually the appearance of a new requirement, e.g. multi-sync computer screens, that leads to the separation of these functions into separate building blocks. Nowadays, with one-chip VLSI

TV-sets the complexity in this area has skyrocketed but the system design is easier and the reliability better than ever.

These landmarks in the history of television – representing iterative transitions that are characterized by an increase in complexity which is, however, encapsulated making the task of system design easier while at the same time increasing reliability – indicate that adding complexity to a complex system might be the answer to many of the problems that the complexity of the system creates.

The key question here is the following: can this successful, from the hardware perspective, approach be transferred to software? In order to answer this question several issues have to be addressed, such as the difference between hardware and software reliability and the definition of software quality and appropriate standards. Examples of self-healing hardware range from simple (e.g., capacitors where foil vaporizes at shortcut), to complex (hard disks reassign defective sectors), to redundant logic arrays (FPGAs) that are in an experimental phase.

As far as biological complexity and the potential impact of nature on autonomic research is concerned, it is argued that nature is complex (from quantum to the universe) and that there are several natural phenomena that could inspire autonomic research. Replication of information in every living cell poses the question as to whether abundance of stored information is needed. Ants are often quoted as examples for simple components forming a complex system; their behavior could be used as a paradigm for building pervasive systems providing redundancy and abundance. A plastic foil is simple but vulnerable, while the human skin is self-healing but far more complex. But are there any significant differences between human-made and natural complexity? If yes, one should keep that in mind when trying to apply the natural processes of autonomicity to the human-made world of engineering.

Finally, it is concluded that adding complexity to complexity (in the design) to achieve simplicity (for the end-user) might not be wrong, if it is well treated, with structure and encapsulation, with well-defined interfaces (APIs) hiding complexity.

As pointed in [4], network oriented R&D these days is largely driven by commercial interests, by expectations of new services that would natively support various types of mobility, user-centred ubiquity, personalisation and context awareness on top of increasing network heterogeneity. The IST FET proactive initiative on Situated and Autonomic Communication (S&AC) is a rare exception. The Commission provides this funding for a long-term basic research and the research community should take the opportunity for making the right strategic choices in research framework and road mapping. It is argued that a radical increase in complexity of network infrastructure is unavoidable, and that network autonomicity is the right solution within the new, service-oriented architecture.

In service-oriented computing, autonomous platform-independent computational entities are dynamically assembled into massively distributed evolvable systems. The enabler, the Service Oriented Architecture (SOA) is recognized as a mainstream trend in the design of software intensive systems. Are we prepared for research and development towards network SOA, in which network-level services (features) will operate not only media and media signaling objects but business objects properly defined at each architecture level with proper cross-layer business relations?

As of now, application services are traded by ISPs to end-users on the retail interface with almost all needed trade sophistication in place; however, the required trade sophistication is missing at the wholesale interface between ISPs. Obviously, the end-to-end services are broken without network SOA that promises to turn complexity into in-network, self-organized trade sophistication. Within S&AC, the "end-to-end argument" should postulate that no functionality and/or intelligence that cannot self-recover should be placed inside the network; self-organization and self-recovery being the advantages of autonomicity.

To fulfil the promise, network SOA must support a variety of multi-tier dependencies between in-network state data and policies spanning generic Internet infrastructure, service- and application- specific infrastructures and propagating down the protocol stack to network functions and function surrounds that are translated at datagram level to media and media signalling processing workflows ultimately controlled by workflow access controls.

Finally, network SOA requires research and development towards two new abstraction layers – requirements abstraction and language abstraction – that must enable true end-to-end seamless inter-working of yet unknown advanced Internet services.

The presentation in [5] notes that ultimately Autonomic Systems aim to provide benefits by dramatically reducing operating costs of complicated systems. This is achieved primarily by off-loading the monitor-analyse-plan-execute operational cycle from human to system intelligence. Human operators then deal with managing high level policies, thus reducing the cognitive load on operators, allowing them to be more productive and inducing fewer mis-configuration errors.

However, autonomic systems themselves involve additional capital cost and are a source of additional complexity in system operation; thus, they are a potential source of further operational cost. In developing autonomic systems we, therefore, need a means to perform the cost-benefit evaluation on a given autonomic system in order to assess whether the additional capital and operational cost its deployment incurs is justified. In this respect, the autonomic communications research has yielded, to date, little in terms of guidance into suitable metrics and benchmarking techniques.

However, some evaluation criteria for autonomic computing systems [6] have been introduced, involving the metrics such as:

- the quality of service in achieving the primary goal, with emphasis on achieving user satisfaction;
- both the capital cost of acquisition and deployment, as well as the cost of operation over time;
- granularity and flexibility;
- the ability to avoid the negative cost and QoS impacts of operational failures

- the degree of autonomy in terms of the level of decision making that can be undertaken by the system rather than its human operators;
- adaptivity to changes in operational context and the latency in reacting to such changes;
- sensitivity to changes in operational context and the ability to attain operational stability after such operational perturbations.

The autonomic communications therefore needs to develop a comprehensive and holistic set of benchmarks that address the total cost of ownership. This may build on existing frameworks, such as TL9000, but must focus on the critical assessment of the introduction of any autonomic feature on the overall cost of ownership. This must consider interaction between all the "self-*" attributes; e.g., does a cost saving through a self-configuration feature render the system more vulnerable to attacks, thus making self-protection more problematic?

This is a challenging proposition as such operational cost - benefits analyses need to be performed over ever-changing network technologies, service portfolios and multi-provider value chains. More fundamentally, with many of the technologies currently being addressed, the solution seems to move from handing complicated systems to exploiting complexity, such that emerging behaviour in multi-agent systems is exploited. This presents a major cultural shift for network operators, which currently strive for full understanding of complicated systems, to one where they rely on the statistical behaviour of complex ones and thus a level of constrained non-determinism. Effective benchmarking also requires that we greatly improve our understanding of the lifecycle engineering costs for self-organising, adaptive systems, in terms of re-use, re-tasking and ameanability to innovation.

Finally, the presentation in [7] argues that complexity is not introduced by autonomicity but it is inherent in complex structures associated with today's networks. As an example, the case of mobile ad hoc networks is presented. In such networks there are numerous sources of complexity, such as setting the numerous parameters at the several layers. It is typically very difficult to determine the optimal setting for a given environment and this task becomes almost impossible when dealing with changing environments. Here is where autonomicity can have an important role by adding the "control stability" complexity in exchange for simplifying the parameter tuning.

"Control stability" complexity manifests itself in several ways. For instance, for the mobile ad hoc networking environment the feedback loop has to deal with many "conflicting" concerns such as forwarding, reliable delivery, resource sharing, channel access and utilization, security and trust management, etc. The various control knobs interact with each other at possibly fairly diverse time scales. The question that naturally arises is as to why one should go through all this "control stability" complexity. There is a very good reason for this: adapting to the environment can result in a great performance improvement!

Since complexity is unavoidable the key question is how to handle it. The general principle should be to keep it as simple as possible (KISS: Keep It Simple Stupid). A good approach would be to try to decouple the system's "intelligence" from "interaction monitoring". The boundaries introduced by such decoupling would also prevent the appearance of control loops and instabilities. Such a decoupling can be observed in the human nervous system: one part reasons and a different one monitors sensory information and reacts.

To reduce complexity, different levels of "intelligence" in the nodes could also be considered, allowing for simple instantiations first, that will be open to extensions, as well as be enhanced with more sophistication as nodes (and designers) evolve and learn over time. It should be noted that dumb individuals (or low "intelligence" nodes) may result in smart group behavior, as is the case, e.g., with ants. Simple users can still adapt/mutate and be excellent for a particular goal. There is an analogy here with FPGAs – as fast as specialized DSP but with the versatility of "multipurpose" microprocessors.

References

1. P. Spirakis, presentation at the panel "Autonomicity vs. Complexity", 2nd IFIP Workshop on Autonomic Communication (WAC2005), Oct. 2005, Athens, Greece, availabe at `http://www.di.uoa.gr/~istavrak/PDF_presentations_WAC/Panel1_wac_Spirakis.pdf`.
2. S. Schmidt, presentation at the panel "Autonomicity vs. Complexity", 2nd IFIP Workshop on Autonomic Communication (WAC2005), Oct. 2005, Athens, Greece, available at `http://www.di.uoa.gr/~istavrak/PDF_presentations_WAC/Panel1_WAC2005_StefanSchmid.pdf`.
3. T. Pfeifer, presentation at the panel "Autonomicity vs. Complexity", 2nd IFIP Workshop on Autonomic Communication (WAC2005), Oct. 2005, Athens, Greece, available at `http://www.di.uoa.gr/~istavrak/PDF_presentations_WAC/Panel1_Pfeifer_WAC2005.pdf`.
4. M. Smirnov, R. Popescu-Zeletin, presentation at the panel "Autonomicity vs. Complexity", 2nd IFIP Workshop on Autonomic Communication (WAC2005), Oct. 2005, Athens, Greece, available at `http:// www.di.uoa.gr/~istavrak/PDF_presentations_WAC/Panel1_WAC2005_Smirnov.pdf`.
5. D. Lewis, presentation at the panel "Autonomicity vs. Complexity", 2nd IFIP Workshop on Autonomic Communication (WAC2005), Oct. 2005, Athens, Greece, available at `http://www.di.uoa.gr/~istavrak/ PDF_presentations_WAC/Panel1_wac05-lewis.pdf`.
6. McCann, J, Huebscher, Evaluation Issues in Autonomic Computing, GCC 2004, LNCS 3252.
7. C. Santivanez, presentation at the panel "Autonomicity vs. Complexity", 2nd IFIP Workshop on Autonomic Communication (WAC2005), Oct. 2005, Athens, Greece, available at `http://www.di.uoa.gr/~istavrak/PDF_presentations_WAC/Panel1_WAC2005_Santivanez.pdf`.

Panel 2 Report: Autonomic Communication Roadmap

Mikhail I. Smirnov

Fraunhofer FOKUS,
Kaiserin-Augusta-Allee 31, 10589 Berlin, Germany
Mikhail.Smirnov@fokus.fraunhofer.de

Abstract. Situated and Autonomic Communication (AC) research roadmap needs to be addressed from a mixture of viewpoints. What are the market drivers for AC? Can we really automate SLA? Does it help AC to radically depart from TCP/IP? Do we know all new requirements for networking software, auto- and re-configuration? How autonomics shall transform network management? What's the role of governance in autonomic control hierarchy, and do we know how hierarchy should emerge? These and similar questions were used to set the scene for the panel discussion on AC roadmap.

1 Introduction

Project ACCA[1] coordinates the creation of a harmonised R&D programme to be implemented by the proactive initiative Situated and Autonomic Communication (S&AC), a part of EU IST framework programme six and beyond. The goal of the programme is long-term foundational research in computer communications with the focus on studies in the area of network infrastructure self-organisation (self-management, self-healing, self-awareness, etc.). One of the major target applications is the design of a network element's autonomic behaviour exposed by innovative (cross-layer optimised, context-aware, and securely programmable) protocol stack in its interaction with numerous often-dynamic network communities. This is seen as the major vehicle for a new generation of ICT services to meet the requirements of information society.

The panellists were the following project participants: Mikhail Smirnov of Fraunhofer FOKUS in Germany (chair), Lidia Yamamoto of University of Basel in Switzerland, Spyros Denazis of University of Patras in Greece and Hitachi SAL in France, Simon Dobson of University College Dublin in Ireland, Ioannis Stavrakakis of NKUA in Greece, James Scott of Intel Corporation (UK) Ltd., David Lewis of Trinity College Dublin in Ireland, Jaouhar Ayadi of CSEM in Switzerland, and Serge Fdida of UPMC in France. The two invited speakers were Fabrizio Sestini of European Commission Future and Emerging Technologies,

---

[1] ACCA – Autonomic Communication Coordination Action, IST-6475 http://www. autonomic-communication.org/projects/acca/

I. Stavrakakis and M. Smirnov (Eds.): WAC 2005, LNCS 3854, pp. 293–302, 2006.

and Nancy Alonistioti of University of Athens in Greece, also representing the IST integrated project E2R.

Prior to the panel the panellists have agreed that autonomic networks share a particular common characteristics. First, they are distributed and self-organized. Second, they must be elastic with regard to new services, goals and border conditions (Tschudin). Third, they must address the ongoing transfer of operational knowledge and decision-making authority from human operators to the system (Lewis).

The object of studies in S&AC is the autonomic network element as it is affected by and affects other elements and the often numerous groups to which it belongs as well as network in general. A design methodology is needed that shall empower autonomic network elements with the abilities to understand how desired element's behaviours are learned, influenced or changed, and how, in turn, these affect other elements, groups and network.

2   State of the Art in S&AC

After the initial white paper on AC [14] a number of conferences and journals have been considering S&AC as a topic, on which submissions were solicited. The majority of publications though did not address the difference between IBM's autonomic computing and AC. The overview [13] has an excellent motivation for autonomic computing; it can be largely borrowed to motivate the AC research as well. However, certain care should be taken when borrowing the IBM's self-management paradigm that is usually explained as "Monitor - Analyze - Plan - Execute" sequence. Unpredictable traffic and network load behaviour in packet switched networks place autonomic decision making under the condition of deep uncertainty, where some governance needs to be provided either from a Knowledge Plane (KP) [16] in a form of behaviour rules or from dynamic network communities in a form of community context or fitness. In this case the "Analyze - Plan" could be seen as being outsourced to KP, while autonomic network element shall instantly act following the "Sense - Assess Risk - Behave" sequence, while concurrently at more relaxed time scale being also in communication with the KP and/or communities [1]. On a road to AC the research community has to build a new *science of interaction* (aka, science of interfaces [17]) addressing the above, and perhaps other approaches, finding the way to create, learn and influence behaviours, to detect and to assess risks, to understand how to apply policy- based management in these settings, etc. - all these to assure end-to-end services guarantees.

The state of the art in AC research is being shaped not only from the academic interest but also from projects funded by the industry and by the EU Commission. The Future and Emerging Technologies (FET) part of IST programme that has a record of proactive initiatives has started to prepare the S&AC one in July 2003 with the four selected projects to start early 2006 [2]. FET views the S&AC research as the answer to many networking challenges identified, such as increasingly high complexity of management, emergence of multi-technology paradigms (e.g. embracing ad hoc and sensor networks), pervasiveness and ubiquity of computing and communication in support of ambient intelligence, etc. with a strong emphasis on multi-disciplinarily research. The four S&AC integrated projects that are set up to investigate the goal of task- and knowledge-driven, scalable, trustworthy, resilient, evolvable and society-friendly networking are BIONETS, ANA, Haggle, and CASCADAS.

The BIONETS (BIO-inspired NExt generation Services) project shall investigate a bio-inspired approach to localized communication services that should be able to evolve spontaneously, without centralised control. The project targets a communication system supporting millions of localized services in an environment consisting of billions of heterogeneous nodes, intermittently connected and extremely low-cost. The two types of nodes (static and mobile) are envisaged to form the project's peer-to-peer communication architecture, in which high-level services will adapt by evolution following the rules of genetics.

The ANA (Autonomic Network Architecture) project is addressing the architectural stress of the Internet. It will develop a novel network architecture that enables for flexible, dynamic and secure autonomic formation and adaptation of network elements and networks. Following the principles of atomisation, diffusion and sedimentation the project shall depart from the statically and globally layered protocol stacks aiming instead at dynamic flexible *functional composition* for wired and wireless networks. The project goal is to demonstrate the feasibility of situated and autonomic networking by 2010.

The project Haggle wants to support transmission when end- to-end contemporaneous connectivity is not available, taking advantage of local and global connectivity; it will build on the model of search engines such as Google, but with no centralised services and no prerequisite of network connectivity; it has no ambition to become an alternative to global services.

The goal of the project CASCADAS (Component-ware for Autonomic Situation- aware Communications, and Dynamically Adaptable Services) is to define the underlying technology for a new generation of composite, highly distributed pervasive services that addresses the configuration and complexity problem at the level of resources and services. The project is therefore driven by the ambition of identifying a fundamental, uniform abstraction for situated and autonomic communication entities, at all levels of granularity, and across stack layers. This abstraction will be the cornerstone of CASCADAS's component model, in which four driving scientific principles will properly converge: situation awareness, semantic self-organisation, self-similarity, and autonomic componentware.

Not only FET funded projects but also mainstream IST projects are addressing very relevant technical goals. The integrated project E2R (End-to-End Reconfigurability) is a part of wireless world initiative; it considers reconfigurability as the enabler of seamless experience in all-IP infrastructure [3]. In a tough international competition the E2R is expanding the principles of software defined radio beyond cognitive radio and brings autonomous cognitive and proactive end-to-end reconfigurability to the telecom sector. The project sets itself to design a reconfiguration management plane based on an abstract view of a network or network element in the standard-friendly manner using the 3GPP integration reference point specification stages. Practically, the project has created a UML profile for reconfiguration that together with other results contributes to standards bodies like OMG and TMF.

Considering the above state of the art research plans and results as the starting point of S&AC roadmap, the rest of this text will be structured in two parts - deliberations and S&AC research challenges. Deliberations part first, addresses the risk and potential reasons of S&AC failure [4], followed by reasoning originated within the eternal motivation for research - curiosity and knowledge enhancements [5]. Then, the myth of reduced complexity will be attacked [6]. The research challenges part examines selected facets of S&AC and provides per facet a partial roadmap. This part starts with the governance dynamic [7], followed by scenario-based design [8], trends and promises in microelectronics area [9], stressing the need of applied S&AC research [10], classical networking issues to be addressed in AC [11], and concludes with a phasing attempt of S&AC roadmap.

3   Deliberations

Despite the visible success of S&AC as a research direction it is important to alert the community on the risk factors that eventually might downgrade the initiative to yet another hype; to have early understanding of the required depth of the research, and to unveil any myth that might break the research.

The three failure risk factors were identified in [4]: ignorance, fear, and self-star [16]. The ignorance is bad because nobody knows what autonomics really is; at the same time, since nobody knows what autonomics is not, it might be good as well, for example more researchers will be attracted. The fear is bad because humans fear to loose control, and in fear of selfish behaviour of autonomic network will demand many control knobs, effectively preventing the idea of autonomics. From the positive viewpoint the fear will help to find fundamental limitations of the new technology. Self-star is risky since it might happen that its even very valuable solutions will open doors for new and serious problems that will require yet more effort to solve. The concern has been expressed that self-star could require even more human intervention than before thus perverting the idea of autonomics. Ironically, the "good news" about self-star is the risk that it becomes its own self-justification in an emergent way.

The panel presentation in [4] also proposed the cure for all the risks. To cure ignorance, the community needs to find a razor-sharp definition of AC; it's better to be too restrictive initially than to be too open or too late. To cure the fear, one needs to demonstrate that autonomic solutions are more robust than those relying on human intervention, which in turn requires research on novel ways of expressing "What we want?" instead of "How?". Finally, two things have been proposed as the cure for the risk of a self-justifying fate of self-star: a measure of autonomicity and the complexity handling. The former is to be inversely proportional to the amount of human intervention, the latter includes proper encapsulation.

Correct treatment of complexity that requires early understanding of what is needed but is yet unknown and defines the necessary depth of research was addressed in [5]. Complexity in any system typically arises from two different sources: in intrinsic complexity of individual components, and the interaction complexity between components. In a software context these might be paraphrased as application complexity and system complexity respectively. While we have a reasonable understanding of applications, and their focus on addressing possibly complex but bounded and well-specified problem, we have significantly less understanding of the systems aspects especially in the presence of adaptation.

These interactions should be an object of study in their own right. We do not have a good understanding of component composition in software, as interactions between components are often surprising. We have even less understanding of the composition of adaptive components, whose adaptations may typically be expected to be antagonistic rather than synergistic with one component negatively impacting the adaptations of another. These aspects are typically addressed (in other domains) through control theory, but even here there is only a limited understanding of complex interactions that are sensitive to on-going conditions, and it is not clear that such models provide a good basis for software.

It is not enough for an adaptive system to adapt: it must exhibit the correct adaptation for the circumstances, retaining (and possibly optimising) some properties. This in turn implies an external semantic frame of reference within which issues such as optimality and trade-offs may be expressed and studied, in terms of the process which the system is involved in supporting. This goes beyond simple static descriptions of component interfaces.

A foundational science of composition will allow us to state and study both individual adaptations and their composition in a way that supports open adaptation while maintaining core properties.

In a myth unveiling fashion the [6] did negatively answer the question "Will autonomicity reduce management complexity?". The rational of autonomic computing adopted by S&AC is to spread the cost of management to several entities; this does not imply that the overall complexity is reduced, From the history of consumer electronics we learn that management complexity can be reduced by integration. On contrary, S&AC attempts to disintegrate a system, it increases the number of interfaces we need to manage, it adds anarchy-inducing autonomicity. This may lead to spreading the cost between multiple entities however the overall management complexity increases. Contrary to automation that reduces management complexity, increases performance, capacity and efficiency, the autonomicity, as a disintegration mechanism increases the complexity. From a positive side [6] observes that autonomicity is not an invention of the S&AC community, it is emerging as the network naturally disintegrates following the process of shifting of the ownership of resources to autonomic entities (contributors). From this viewpoint the AC research has the two major challenges. First, behaviour management since the behaviour is the single most defining characteristic of an autonomous element that follows its own laws or lack of those. Second, interaction management (aka, science of interfaces [17]) that can be grounded by borrowings from the ecology by the notions of individualism (organisms), behaviours (rational, irrational, changing, random, unpredictable, etc.), interactions, equilibrium (slow changes), and evolution. In this new science complexity will be managed without central authorities and global rules.

4   Research Challenges

Stating that autonomic system must be governed rather than managed [7] argues that this will require two things. First, operational goals and constraints have to be expressed as policies, and second the process of definition of these policies by humans will need on-going human understanding of adaptive space and its governance potential. The latter can be formalised as a *governance space* - an operationally accessible portion of an adaptive space, while *adaptive space* in turn can be represented by contextual space with possible adaptive behaviours. There are certain restrictions within governance space, for example grouping must reflect organisational and social policy-making, and take into account such non-functional aspects of the latter as stability, responsiveness and potential for conflict.

Governance is a dynamic two-way process within a system hierarchy: delegation of decision making authority propagates governance policies downstream; escalation of decision making due to governance space violations propagates detected policy conflicts upstream. Note that violations of adaptive space lead to a semantic mismatch. The research vision here is to find ways of applying the governance process recursively between communities of agents within a system. This vision translates into the following set of research objectives. First, since the governance process needs to be managed, one needs to build a (e.g. community based) policy management mechanism with a requirement for fast stabilisation of policy set for a given governance space. Second, since the governance space is volatile due to changing contexts, service offerings, and value chains a handling mechanism for a governance space is needed supported by semantic mappings to convey adaptive space.

Roadmap-wise [7] suggests as the first step to gain understanding of human decision-making dynamics in a small cross-disciplinary research project. This can be followed by a larger project to establish benchmarks for assessing governance effectiveness and explaining benefits. As the third step a number of projects could develop forms of adaptive and governance spaces, build and evaluate solutions for the governance dynamic in different domains, such as communications, pervasive computing, electronic markets, and collaborative spaces.

From a traditional management perspective [8] argues that autonomicity is a property of *evolution in the making* that revolves around an intelligent self-centric control loop "Collect - Decide - Enforce" and should be addressed in conjunction with application field. In network management *Collect* translates into monitoring and building a network picture, helps to achieve self-awareness; *Decide* translates into inference, i.e. problem diagnosis process and into planning - a process of selecting a solution; *Enforce* translates into deployment i.e. adding functionality and configuration, i.e. changing the behaviours.

Autonomicity without tangible and universal scenarios is meaningless; the above control loop is lifeless without *semantic languages* that enable exchange between loop entities that operate at different levels and in different contexts. Collectively, these semantic languages describe purposeful behaviours of network and services composed of components. For example, with emerging modular router architectures where control and forwarding are separated the support for components is provided.

Roadmap-wise [8] suggests in the first phase to select a characteristic application field with a set of representative scenarios, and to select existing functionality to be transformed into an autonomic one. This can be done by defining abstractions for the intelligent control loop tailored for this functionality and suitable for low-level programmability, assuming modular router as a target platform. This can be followed by large-scale trials and testbeds.

Modular architectures are enablers for autonomicity, however they are in turn enabled by the progress in microelectronics, manufacturing and power supply. The [9] provides an outlook into 10-15 years from now with the goal to outline the wireless and mobile communication landscape in that future. Humans are in the centre of this future of communication, surrounded by ambient intelligence and autonomic networking. It is envisaged that by 2020 microelectronics industry will reach the level of 1 million MIPS[2] per 10 Watt, which is expected to be sufficient for autonomic communication demands. The basic element - an autonomic component - will be a universal building block for autonomic applications and systems. It will be a modular unit with specified interfaces, clear context dependencies and self-management mechanisms that shall manage behaviour within actual constraints based on policies and rules. Policy will govern not only self-management of individual components but also interactions between components, including agreements establishment towards consistency, robustness and system self-management.

Contributing to the *interface science* discussion [9] argues that today's physical interfaces to network and equipment might disappear at the time of autonomic communication being replaced by ubiquitous short-range radio communication. In this environment precise localisation, sensing, sounding, etc. will foster the ability to adapt to changes in traffic load, service, even functionality required. With ultra-low power consumption self-sustained operation over years is predictable, combined with ultra small size and low cost it brings autonomic communication to the size we can swallow.

However we should not need to wait until 2020 to start working on situated and autonomic communication; [10] calls for applied S&AC to solve real-life problems, which does not preclude long-term research. Even more, perhaps theory and practice must work more closely to constantly verify theoretic results in experiments. Without practically building new technology we can't understand the constraints; building actually means transfer of results between groups of developers. Technology evaluation requires more sets of measurement data available publicly; this will facilitate development of realistic models and trace-based

---

[2] Mega Instructions Per Second.

evaluation largely in place of simulation. Since applied S&AC calls for higher involvement of community members it is necessary to create proper incentives including funding and programme committees.

As observed by [11] foundation of the Internet technology can be compared with the foundation of S&AC; this view perhaps is helpful in the understanding of S&AC roadmap. There are always similar forces (applications, regulation, economy, management) that are shaping the R&D learning curve: early solutions tend to have higher complexity, while solutions in use converge over time to the level of complexity that is conformant to the level of understanding of the problem area, though only until the next innovation cycle starts. The state of the art understanding in AC is characterised in [11] as the one that lacks a single solution though has a clear networking focus proposing a very ambitious future for communication. While focussing mainly on self-star properties the AC will impact many networking issues: naming and addressing for autonomic entities, data gathering and knowledge management including interactions with the environment, soft-layering leading to time-dependant architectures, interoperability in multiple contexts, composition and behaviour modelling, service management and adaptation.

5   Conclusions

Any roadmap is a tool for strategy development, S&AC roadmap needs to show a path to full AC, and perhaps even beyond starting with the tailored description of "now". As it is outlined in [12] the current situation is characterised as proliferation of brittle systems that often require babysitting; the Internet appears to end systems as a black box, its end systems are vulnerable to blue screens and viruses. The full S&AC promises fully autonomic operation at many levels, so that underlying systems conspire to do what we want automatically, perhaps based on self-star. We do not know yet how to characterise the life beyond full S&AC, the [12] mentions intelligence, autonomy and creation and suggests four stages of the roadmap.

The zero phase is the one that is happening now, its main goal is to agree on a definition, to continue the dissemination activities and to work closely with funding bodies on improving the awareness. This will soon be followed by the first phase, in which the four funded projects mentioned above should investigate their domains. Interleaved with this, the second phase should embrace the most mature S&AC topics as part of the mainstream IST research, in EU as framework programme seven (2007-2013), and worldwide as activities coordinated by ACF, IEEE, etc; while long-term outstanding S&AC topics would still remain in FET. The third phase should continue the S&AC research but should also expand to the society to start using mature solutions.

The strongest consensus during the zero phase is the S&AC definition need; we must agree on what is S&AC and what is not. Tentatively, [12] suggest to agree that S&AC systems are those that do what we want without direct human supervision using for this context information and knowledge, policies and conflict resolution, embedded performance evaluation and feedback; these systems do things by themselves, meaning that their components self-organise and use emergence of e.g. control hierarchies as a part of self-organisation. The S&AC systems are self-managed; this property is seen as an integral character-istic comprised of abilities to be self-protected, including security and immunity, self-diagnosed including detection of failures and conflicts, self-healed, includ-ing correction, repair and recover actions, self-configured including updates of functionality at several levels, self-optimised including self-adaptation and evolv-ability, and self-deployed including self-deprecation.

What is not S&AC then? Perhaps we should agree on a fuzzy metric of au-tonomicity that will be computed as a weighted sum of different criteria in the definition and will be inversely proportional to the amount of human interven-tion. However the risk is to have too wide scope for S&AC and to be hardly distinguishable from pervasive and ubiquitous computing, from autonomic com-puting, and from networking. The phase one projects might provide answers in the following ascending sequence along the imaginable axis comprised of *risk, time and futuristic* orientations. The clear start is to be found within wireless and broadly defined opportunistic networking (project Haggle); these and other advances are likely to be abstracted to the S&AC networking architecture by project ANA; the next mark on the imaginable axis is to be provided by project CASCADAS that shall address pervasive S&AC, bring it based on its compo-nent model in line with context and knowledge. Finally, the project BIONETS is targeting evolutionary protocols and services reflecting socio-economic models. In general, the first phase outcomes are expected to be both well established theoretical foundations for S&AC and practical case studies covering handheld solutions, meetings and conferences support, and home platforms.

The society in general is expected to use, to enjoy and to benefit from S&AC by radical increase of productivity levels in engineering (adaptation and evolution), software synthesis (eternal software, autocatalytic systems), and even in research itself (e.g. in complex systems and emergence). However for this to happen S&AC has to receive much more attention from funding bodies, namely to become a part of mainstream ICT. Immediate ICT benefits will be technology- and society-wise. Technology-wise examples are improved naming and addressing as opposed to cur-rent DNS and DHCP, and new communication services which are not limited by the end/to-end obligation. Society-wise, higher usability levels of ICT solutions, higher acceptance rates and better addressed socio-economic issues.

References

1. Mikhail Smirnov, Autonomic Communication Roadmap, WAC 2005 Panel overview,   http://cgi.di.uoa.gr/~istavrak/PDF_presentations_WAC/Panel2_ WAC2005_Smirnov.pdf
2. Fabrizio Sestini, Situated and Autonomic Communications in FET, WAC 2005 Panel   presentation,   http://cgi.di.uoa.gr/~istavrak/PDF_presentations_WAC/ Panel2_Sestini_WAC2005.pdf

3. Nancy Alonistioti, Management of Reconfigurability, WAC 2005 Panel presentation http://cgi.di.uoa.gr/~istavrak/PDF_presentations_WAC/Panel2_Alonistioti_WAC2005.pdf

4. Christian Tschudin, Lidia Yamamoto, Reasons why AC may fail, and how to prevent it, WAC 2005 Panel position statement, http://cgi.di.uoa.gr/~istavrak/PDF_presentations_WAC/Panel2_ubasel_WAC2005_red.pdf

5. Simon Dobson, Everything interesting is composition, WAC 2005 Panel position statement, http://cgi.di.uoa.gr/~istavrak/PDF_presentations_WAC/PAnel2_WAC2005_Dobson.pdf

6. Ioannis Stavrakakis, Will Autonomicity reduce management complexity? No! WAC 2005 Panel position statement, http://cgi.di.uoa.gr/~istavrak/PDF_presentations_WAC/Panel2_WAC2005_Stavrakakis.pdf

7. David Lewis, An Autonomic Governance Dynamic, WAC 2005 Panel position statement, http://cgi.di.uoa.gr/~istavrak/PDF_presentations_WAC/Panel2_wac05-Lewis.pdf

8. Spyros Denazis, Autonomicity is a property, WAC 2005 Panel position statement, http:// cgi.di.uoa.gr/~istavrak/PDF_presentations_WAC/Panel2_WAC2005_Hitachi.pdf

9. John Farserothu, Jaouhar Ayadi, Communication 2020 and beyond - on the road to Autonomic Communication, WAC 2005 Panel position statement, http://cgi.di.uoa.gr/~istavrak/PDF_presentations_WAC/Panel2_WAC2005_CSEM.pdf

10. James Scott, Applied Situated and Autonomic Communications, WAC 2005 Panel position statement, http://cgi.di.uoa.gr/~istavrak/PDF_presentations_WAC/Panel2_jscott- wac2005.pdf

11. Serge Fdida, ACCA, A Networking View, WAC 2005 Panel position statement, http://cgi.di.uoa.gr/~istavrak/PDF_presentations_WAC/Panel2_WAC2005_Fdida.pdf

12. Lidia Yamamoto, Christian Tschudin, Autonomic Communication Roadmap Proposal, WAC 2005 Panel position statement, http://cgi.di.uoa.gr/~istavrak/PDF_presentations_WAC/Panel2_ubasel_WAC2005_green.pdf

13. Autonomic Computing Overview, IBM, on line at http://www.research.ibm.com/autonomic/overview/elements.html

14. Autonomic Communication: Research Agenda for a New Communication Paradigm, white paper, Fraunhofer FOKUS 2003, on-line at http://www.autonomic-communication.org/publications/doc/WP_v02.pdf

15. David D. Clark, Craig Partridge, J. Christopher Ramming, and John T. Wroclawski, A Knowledge Plane for the Internet, In Proceedings of SIGCOMM '03, 2003.

16. Ozalp Babaoglu, Ma'rk Jelasity, Alberto Montresor, Christof Fetzer, Stefano Leonardi, Aad van Moorsel, and Maarten van Steen, editors., Self-Star Properties in Complex Information Systems, volume 3460 of Lecture Notes in Computer Science, Hot Topics. Springer-Verlag, May 2005.

17. Paul Spirakis, Complexity vs Autonomicity, WAC 2005 Panel position statement,http:// cgi.di.uoa.gr/~istavrak/PDF_presentations_WAC/Panel1_wac_Spirakis.pdf

Author Index

Lecture Notes in Computer Science

For information about Vols. 1–3805

please contact your bookseller or Springer